Computational Fluid Mechanics

Computational Fluid Mechanics

Edited by Fay McGuire

Clanrye International
www.clanryeinternational.com

Clanrye International,
750 Third Avenue, 9th Floor,
New York, NY 10017, USA

ISBN: 978-1-63240-645-3

Cataloging-in-Publication Data

Computational fluid mechanics / edited by Fay McGuire.
p. cm.
Includes bibliographical references and index.
ISBN 978-1-63240-645-3
1. Computational fluid dynamics. 2. Fluid mechanics--Data processing. I. McGuire, Fay.
TA357.5.D37 C66 2018
620.106 4--dc23

For information on all Clanrye International publications
visit our website at www.clanryeinternational.com

CLANRYE
INTERNATIONAL

Contents

Preface

Every book is initially just a concept; it takes months of research and hard work to give it the final shape in which the readers receive it. In its early stages, this book also went through rigorous reviewing. The notable contributions made by experts from across the globe were first molded into patterned chapters and then arranged in a sensibly sequential manner to bring out the best results.

Computational Fluid Mechanics studies the properties of fluids by using numerical and computational approaches. This book on computational fluid mechanics discusses the complexities involved in analyzing data structures in a precise manner. Computer software can provide physical models and fluid flow capabilities to simulate various environments. Analyses are done on materials such as non-Newtonian liquids, real gases, water vapor and compressible gases. The book studies, analyses and upholds the pillars of computational fluid mechanics and its utmost significance in modern times. It will be of great help to students and researchers in the fields of automotive engineering, biomedical sciences and aerospace engineering.

It has been my immense pleasure to be a part of this project and to contribute my years of learning in such a meaningful form. I would like to take this opportunity to thank all the people who have been associated with the completion of this book at any step.

Editor

A time stepping method in analysis of nonlinear structural dynamics

A. A. Gholampour[a,*], M. Ghassemieh[a], H. Razavi[a]

[a]*Department of Civil Engineering, University of Tehran, Tehran, Iran*

Abstract

In this paper a new method is proposed for the direct time integration method for structural dynamics problems. The proposed method assumes second order variations of the acceleration at each time step. Therefore more terms in the Taylor series expansion were used compared to other methods. Because of the increase in order of variations of acceleration, this method has higher accuracy than classical methods. The displacement function is a polynomial with five constants and they are calculated using: two equations for initial conditions (from the end of previous time step), two equations for satisfying the equilibrium at both ends of the time step, and one equation for the weighted residual integration. Proposed method has higher stability and order of accuracy than the other methods.

Keywords: direct time integration, weighted residual, nonlinear structural dynamics, higher accuracy, second order acceleration

1. Introduction

There are two main methods for the analysis of structural dynamics problem; modal superposition and direct time integration. While for the analysis of linear structures both methods are applicable, for nonlinear analysis, the latter method is the only option.

In the structural dynamics problems, governing equation is a second order differential equation [4, 12]. For solving differential equations of nonlinear systems, the numerical procedure can be used in the incremental step [4]. Among different methods, those related to Newmark's method are the most common methods in structural dynamics. The direct time integration of the equations provides the response of the system as discrete intervals of time which are usually equally spaced. Determination of the response involves the computation of three structural responses; displacement, velocity, and acceleration at each time step.

In nonlinear analysis, stiffness is calculated at the beginning of each time step and then response is calculated at the end of this time step with assuming that stiffness is constant through out the step. Therefore nonlinearity is considered with calculating stiffness again at the beginning of next time step. Calculated responses will be considered at the end of each time step as the initial conditions for next time step. Therefore system nonlinearity behavior is replaced with a series of consecutive approximate linear differential equations [4,5,9,12].

In the explicit methods, in each time step, equation of motion is written at the beginning of the time step and the unknown values at the end of time step are calculated explicitly, but in the implicit methods, unknown values at the end of time step are calculated by writing the equation of motion at those points [1,4–6,8,9,11,12,14]. Because implicit methods require

*Corresponding author. e-mail: razavihadi@yahoo.com.

more calculation in each time step with a smaller number of time steps, in the past it has been shown that the implicit methods are more accurate than the explicit ones [7,14].

Because of the approximation in the formulation and calculation of these methods, it is expected to have some error compared to exact solution that the error is usually a function of time step length, frequency content of the load and also degree of nonlinearity.

In conditionally stable methods, the instability occurs when time step size is more than a specific value (critical time step). While in unconditionally stable methods, instability never happens, regardless of the time step size [8,9,12,14].

Because central difference method is very simple for implementation in the nonlinear systems, among explicit time integration methods, it is one of the most widely used methods [2,4, 9]. Other known method for analysis of nonlinear structural dynamics is a family of Newmark's method that these methods assume a constant or linear behavior for the variation of acceleration at each time step [4,8,9,12].

In this paper, a time integration method is proposed that is both implicit and explicit and it assumes a second order variation of the acceleration within each time step. The proposed method is shown to have higher accuracy compared to conventional methods.

2. Proposed Method

The differential equation describing a nonlinear system can have the general form:

$$\ddot{x} + f(\dot{x}, x, t) = 0. \tag{1}$$

Therefore the equation of motion for a nonlinear system (with nonlinear stiffness) is:

$$M\ddot{x} + C\dot{x} + K_i x = P, \tag{2}$$

where M and C are the mass and damping matrix; K_i is the stiffness matrix in the i-th time step; P is the vector of applied forces; x, $\dot{x}$ and $\ddot{x}$ are the displacement, velocity and acceleration vectors, respectively. The initial conditions are $x(0) = x_0$, $\dot{x}(0) = \dot{x}_0$ where x_0 and $\dot{x}_0$ are the initial displacement and velocity vectors, respectively.

The acceleration in each time step is assumed to be a second order function which results in the displacement to be a fourth order complete polynomial in each time step. Therefore the displacement function contains five constants. Those constants are obtained from; two initial conditions from the end of previous time step, satisfying the equation of motion at both ends of the time step, and setting the weighted residual of the method in the step equal to zero.

If the objective is to find the displacement in the time t_i, first time interval $[0, t]$ is divided to the i smaller sub-interval. In the beginning of the calculation, the displacement is determined in the time step Δt; and then in the second time step $2\Delta t$ to $i\Delta t$. In the i-th time step, displacement function in the $\delta \in [0, \Delta t]$ interval can be shown as:

$$x(\delta) = a_i\delta^4 + b_i\delta^3 + c_i\delta^2 + d_i\delta + e_i, \tag{3}$$

where a_i to e_i are the unknown coefficients that should be determined. Therefore, the velocity function is defined as:

$$\dot{x}(\delta) = 4a_i\delta^3 + 3b_i\delta^2 + 2c_i\delta + d_i \tag{4}$$

and the acceleration function is:

$$\ddot{x}(\delta) = 12a_i\delta^2 + 6b_i\delta + 2c_i. \tag{5}$$

That c_i, d_i and e_i are calculated using the above equations:

$$x(\delta = 0) = x_{i-1} \quad \rightarrow \quad e_i = x_{i-1}, \tag{6}$$
$$\dot{x}(\delta = 0) = \dot{x}_{i-1} \quad \rightarrow \quad d_i = \dot{x}_{i-1}. \tag{7}$$

By placing Eqs. (6) and (7) into the equation of motion at the beginning of this time step, we have:

$$M(2c_i) + C(d_i) + K_i(e_i) = P_{i-1}, \tag{8}$$

therefore c_i is:

$$c_i = (2M)^{-1} \cdot (P_{i-1} - C\dot{x}_{i-1} - K_i x_{i-1}). \tag{9}$$

Now by satisfying equation of motion at the end of the present time step, we have:

$$M\ddot{x}_i + C\dot{x}_i + K_i x_i = P_i, \tag{10}$$

which results in:

$$\begin{aligned} M(12a_i\Delta t^2 + 6b_i\Delta t + 2c_i) + C(4a_i\Delta t^3 + 3b_i\Delta t^2 + 2c_i\Delta t + d_i) + \\ K_i(a_i\Delta t^4 + b_i\Delta t^3 + c_i\Delta t^2 + d_i\Delta t + e_i) \quad = \quad P_i. \end{aligned} \tag{11}$$

The final equation is obtained from weighted residual integral. Because this method is approximate, it does not satisfy the equilibrium equation of motion in domain of $[0, \Delta t]$ interval. The residual of the method in satisfying the equation of motion is defined as:

$$R = M\ddot{x} + C\dot{x} + K_i x - P. \tag{12}$$

Then the residual is forced to be zero over the domain and using a unit weight function, we obtain:

$$\int_0^{\Delta t} 1 \times R\,\mathrm{d}t = 0. \tag{13}$$

Finally by solving Eqs. (11) and (13), the values of a_i and b_i can be determined. Therefore by calculating these five unknowns in i-th time step, displacement, velocity and acceleration vectors at the end of the i-th time step is calculated as follow:

$$x_i = a_i\Delta t^4 + b_i\Delta t^3 + c_i\Delta t^2 + d_i\Delta t + e_i, \tag{14}$$
$$\dot{x}_i = 4a_i\Delta t^3 + 3b_i\Delta t^2 + 2c_i\Delta t + d_i, \tag{15}$$
$$\ddot{x}_i = 12a_i\Delta t^2 + 6b_i\Delta t + 2c_i. \tag{16}$$

3. Stability, order of accuracy, and overshooting effect

For evaluation of stability of the present method, single degree of freedom system is considered and the magnification matrix is derived for calculating the eigenvalues of the matrix [3,10,13]. Absolute eigenvalues of the matrix must be smaller than or equal to one. For undamped systems, in softening conditions such as $K_i/K_0 = 0$ (stiffness at the end of time step to the stiffness at the beginning of time step i), proposed method has not any instability, but for $K_i/K_0 = 0.5$ has a small local instability at $\Delta t/T_0 = 0.72 - 0.76$ (T_0 is period at the beginning of first time step) which it can be resolved for damping ratio as $\xi = 3.4$ %. For $K_i/K_0 = 1$ has a small local instability too at $\Delta t/T_0 = 0.52 - 0.54$ which it can be resolved for damping ratio as

$\xi = 4.6\ \%$. The central difference and linear acceleration methods have smaller limitation of stability compared to proposed method.

By replacing the differential equation by the difference equation, the local truncation error is created in each time step which local truncation error is relative to the order of accuracy of one method. The order of the accuracy of the proposed method is three which is higher than the other methods.

The tendency to overshoot from exact solution is significantly important factor which should be considered in an evaluation of numerical solutions. Proposed method from the displacement responses has a tendency to overshoot linearly in the displacement term and from the velocity responses, has a tendency to overshoot quadratically in the displacement term and linearly in velocity term.

4. Examples

In order to see the results of the proposed method and to see its advantages over the other existing methods, two examples are considered which the results obtained from the proposed method are compared with the central difference and linear acceleration (Newmark's) methods.

Example 1 [12]: Consider a single degree of freedom with the frame as shown in Fig. 1. This system has an elastoplastic behavior as shown in Fig. 2.
Exciting force is applied on the spring damping system as shown in Fig. 3.

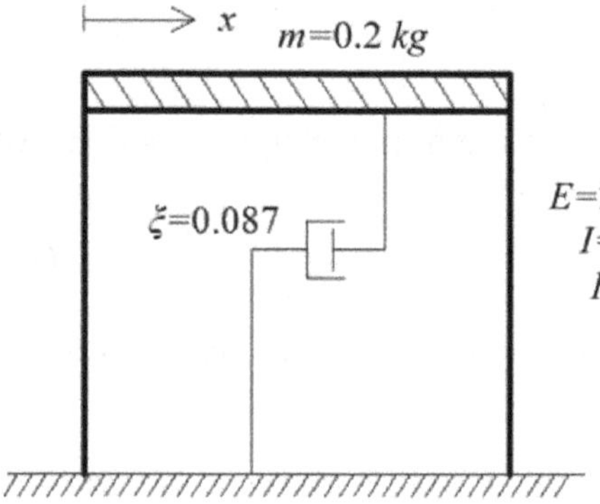

Fig. 1. Frame of structure

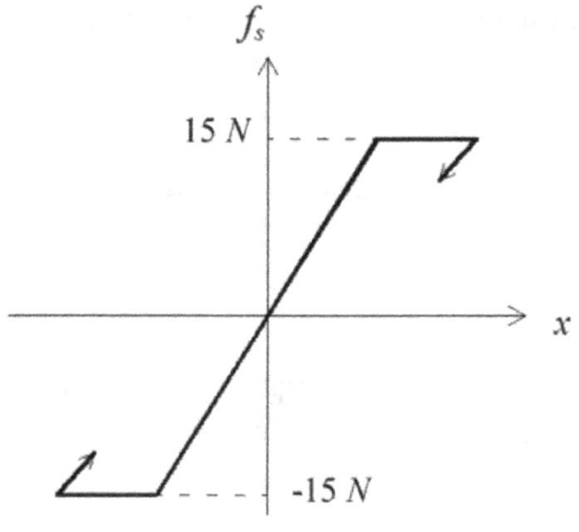

Fig. 2. Force-displacement relationship

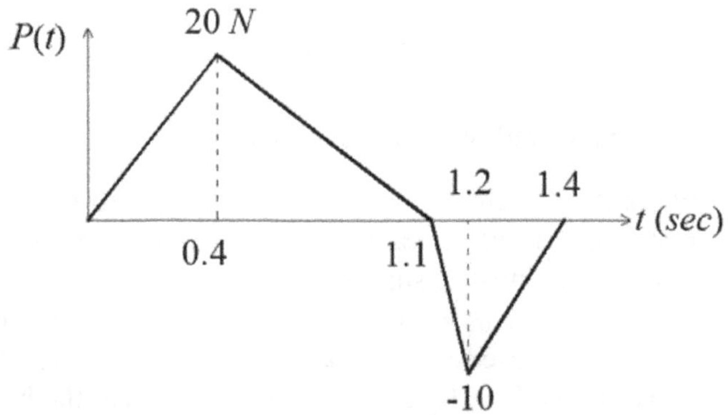

Fig. 3. Exciting force

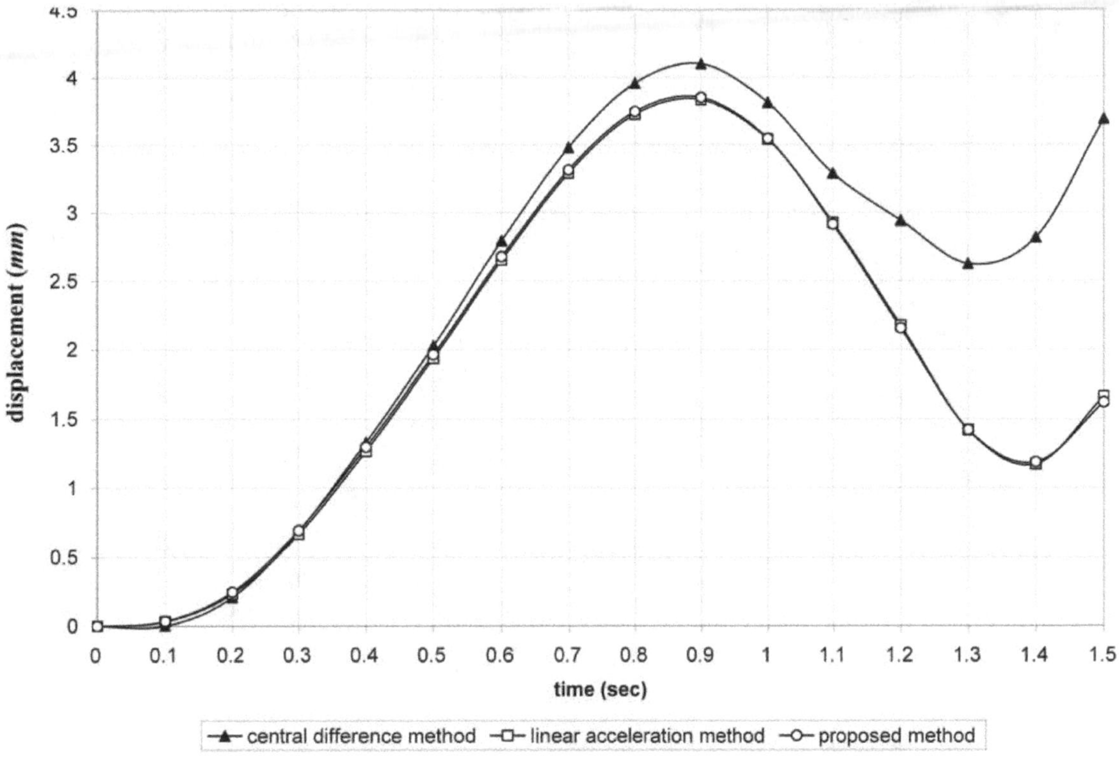

Fig. 4. Displacement responses versus time diagram for example 1

In Figs. 1, 2, and 3 the results of displacement, velocity, and acceleration of this system due to the applied loading $P(t)$ are shown for time step duration 0.1 sec. The results compare from central difference method, linear acceleration method, and the proposed methods.

The results obviously show that proposed method has a better responses in comparison with the other methods.

Example 2 [3]: Consider a two story shear building with initial conditions as; $x_0 = \dot{x} = [0, 0]^T$ in which has flexurally rigid floor beams and slabs. Nonlinear story stiffness for each story is defined as $k = k_0[1 + \lambda(\Delta x)^2]$, where Δx and k_0 are story drift and initial stiffness, respectively. Bottom and top stories have $k_0 = 10^7$ N/m and $\lambda = -100$, and $k_0 = 10^4$ N/m and $\lambda = -0.001$, respectively. Lumped masses are considered to be $1\,000$ kg and this system has been excited by a ground acceleration of $50 \cdot \sin(\omega \cdot t)$, $\omega = 1$ rad/sec, at the base of building. Natural frequencies of system are found to be 3.16 and 100.05 rad/sec, respectively. Displacement responses obtained from linear acceleration method (Newmark's method) by a time step duration of 0.001 sec resulted in exact solution. Figs. 7 and 8 show the comparison of displacement responses with $\Delta t = 0.02$ sec to exact solution for bottom and top stories, respectively.

According to the Fig. 7, central difference and linear acceleration methods have small jumps from the exact solution but the proposed method is on the exact solution line. On the other hand, according to the Fig. 8, all methods have similar response respect to the exact solution.

In this example, we presented only displacement responses, whereas the velocity and acceleration responses calculated using the proposed method are also more accurate than the other methods.

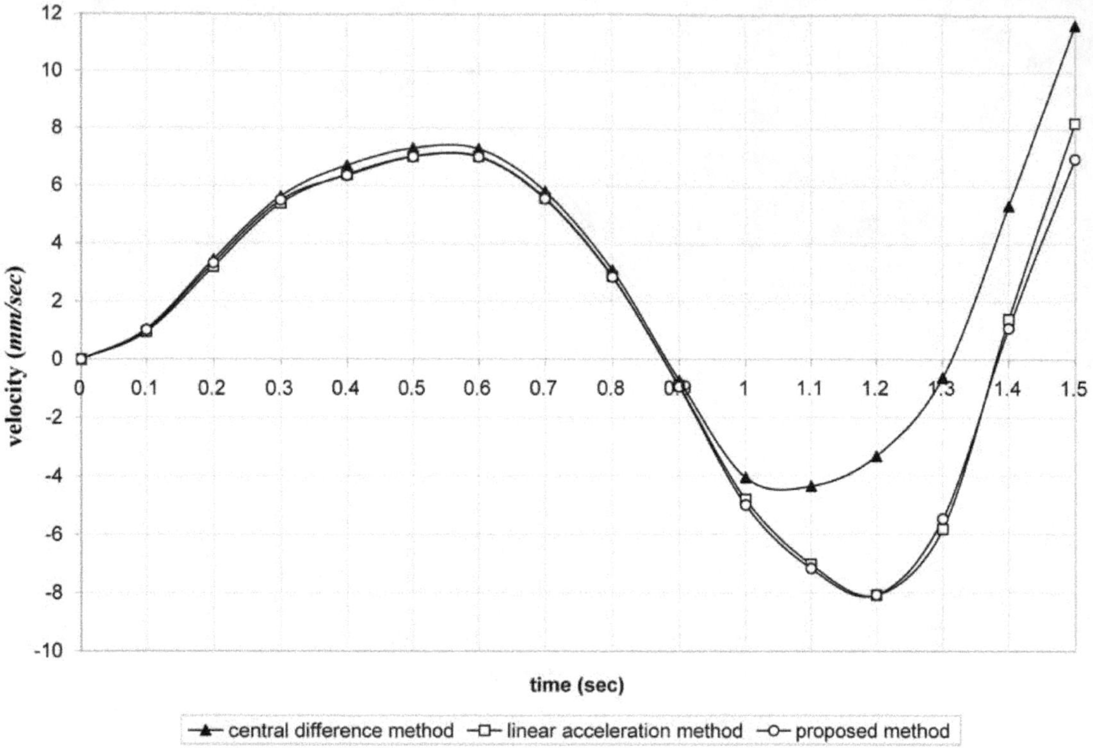

Fig. 5. Velocity responses versus time diagram for example 1

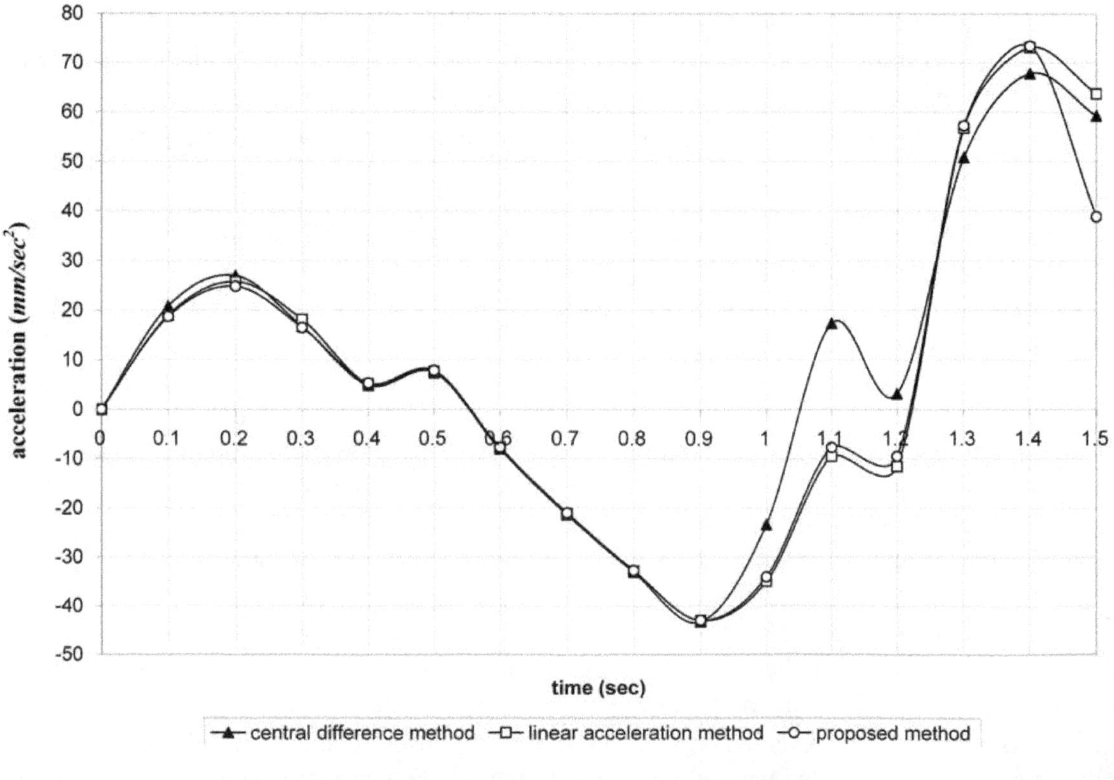

Fig. 6. Acceleration responses versus time diagram for example 1

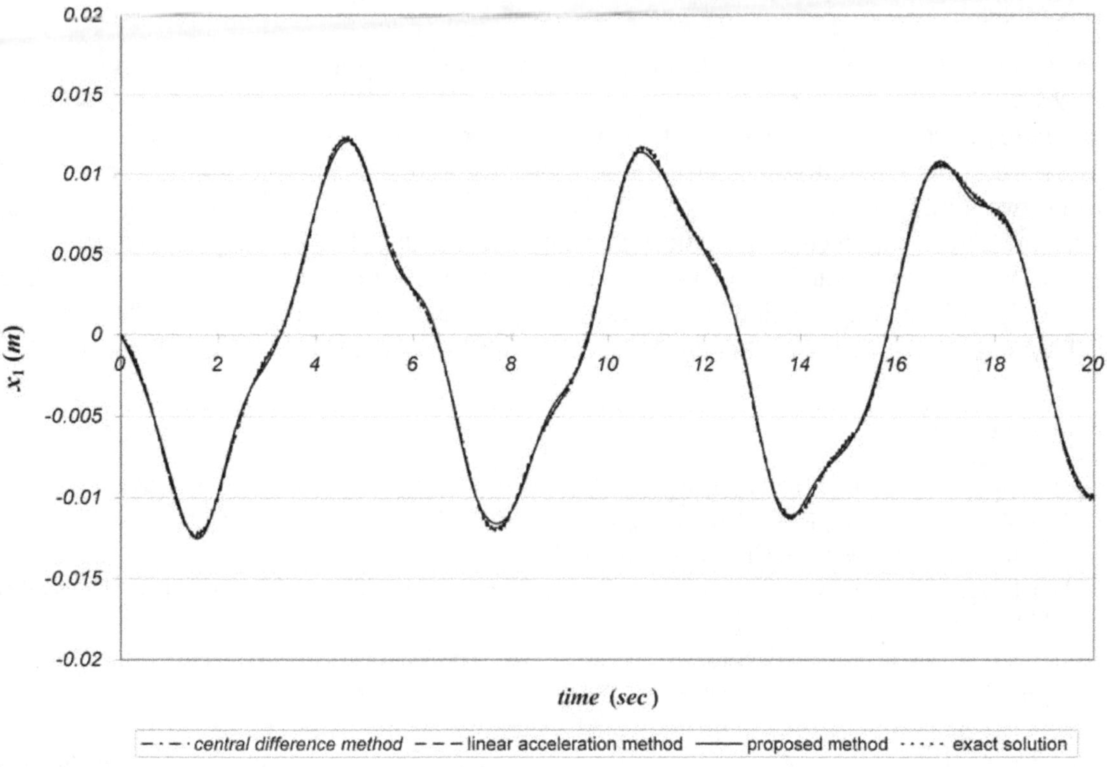

Fig. 7. Displacement responses versus time diagram for first story for example 2

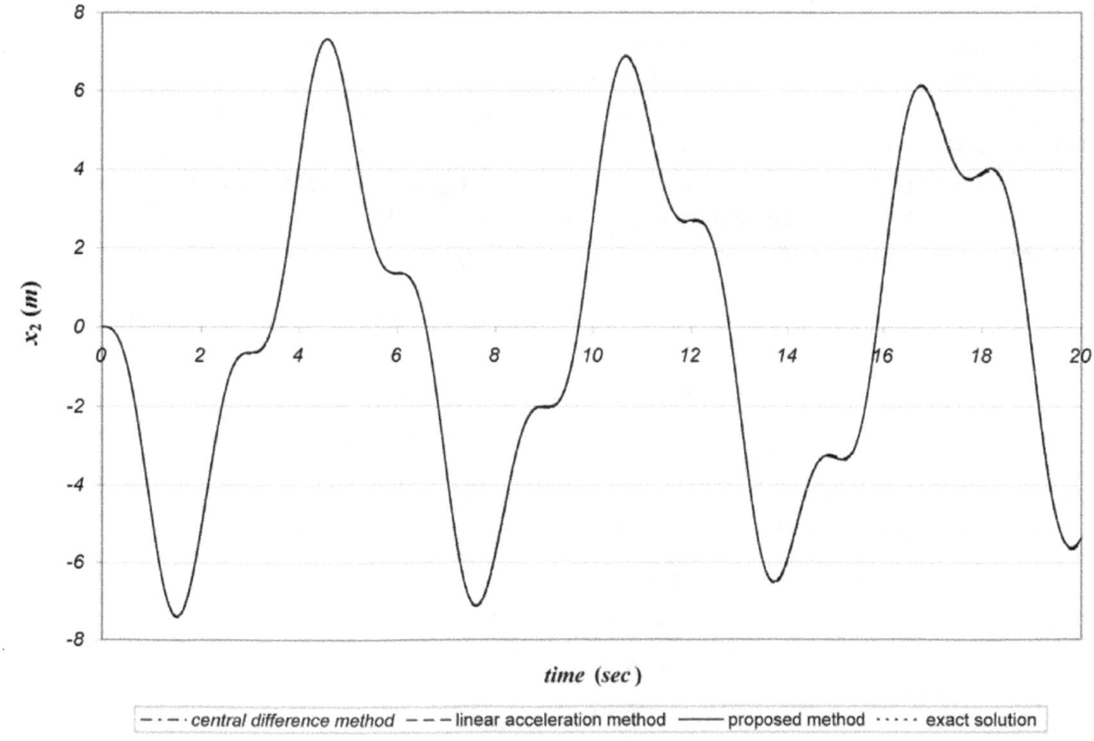

Fig. 8. Displacement responses versus time diagram for second story for example 2

5. Conclusion

A new method of time integration technique for problems in nonlinear structural dynamics was illustrated. To show the accuracy and response of the method, two examples were presented. A quadratic polynomial as a function of time was used in order to approximate the variation of acceleration in each time step. The displacement function had five constants that were calculated using: two initial conditions, from the end of previous time step, two equations from satisfying the equilibrium at both ends of the time step, and one equation for the weighted residual integration where the weight function is assumed to be unit function. The proposed method had a small local instability which it can be resolved by increasing the damping ratio that however had higher stability than the other methods. Also order of accuracy of the method was three.

References

[1] Bathe, K. J., Finite Element Procedures, Prentice-Hall, Englewood Cliffs, New Jersey, 1996.

[2] Belytschko, T., Liu, W. K., Moran, B., Nonlinear Finite Elements for Continua and Structures, 3rd ed., John Wiley & Sons, Chichester, UK, 2000.

[3] Chang, S. Y., Improved explicit method for structural dynamics, ASCE, Journal of Engineering Mechanics, 133 (7) (2007) 748–760.

[4] Chopra, A., Dynamics of Structures: Theory and Applications to Earthquake Engineering, 3rd ed., Prentice-Hall, Upper Saddle River, New Jersey, 2007.

[5] Clough, R. W., Penzien, J., Dynamics of Structures, McGraw Hill, 1983.

[6] Crisfield, M. A., Non-Linear Finite Element Analysis of Solids and Structures, John Wiley & Sons, Vol. 2., 1997.

[7] Dokainish, M. A., Subbaraj, K., A survey of direct time integration methods in computational structural dynamics. I. Explicit methods, Computers & Structures, 32 (6) (1989) 1 371–1 386.

[8] Hughes, T. J. R., Belytschko, T., A precis of developments in computational methods for transient analysis, Journal of Applied Mechanics, 50 (1983) 1 033–1 041.

[9] Humar, J. L., Dynamics of Structures, Prentice-Hall, Englewood Cliffs, New Jersey, 1990.

[10] Kavetski, D., Binning, P., Sloan, S. W., Truncation error and stability analysis of iterative and non-iterative Thomas-Gladwell methods for first-order non-linear differential equations, International Journal for Numerical Methods in Engineering, 60 (12) (2004) 2 031–2 043.

[11] Park, K. C., Practical aspects of numerical time integration, Computers & Structures, 7 (1977) 343–353.

[12] Paz, M., Structural Dynamics: Theory and Computation, 4th ed., Chapman & Hall, New York, 1997.

[13] Razavi, S. H., Abolmaali, A., Ghassemieh, M., A weighted residual parabolic acceleration time integration method for problems in structural dynamics, Computational Methods in Applied Mathematics, 7 (3) (2007) 227–238.

[14] Subbaraj, K., Dokainish, M. A., A survey of direct time integration methods in computational structural dynamics. II. Implicit methods, Computers & Structures, 32 (6) (1989) 1 387–1 401.

2

Blade couple with dry friction connection

L. Půst[a,*], L. Pešek[a], A. Radolfová[a]

[a] *Institute of Thermomechanics, AS CR, v.v.i., Dolejškova 5, 182 00 Prague, Czech Republic*

Abstract

Vibration of a blade couple damped by a dry friction contact in the shroud is investigated by means of hysteresis loops and response curves analysis. The studied system is excited by one harmonic external force in a frequency range near to the lowest eigenfrequency of real blades. Blades are connected by means of a damping element consisting of dry friction part linked in series with linear spring. This "stick-slip" damping element is supposed to be either weightless or of a very small mass which models the mass of elastically deformed parts of contacting bodies near the friction surface. Two approximate mathematical models of "stick-slip" dry friction elements are suggested and analysed. The response curves of blade couple connected by stick-slip damping element are presented for different values of slip friction forces and two values of mass of elastically deformed parts.

Keywords: stick-slip dry friction, 3V friction characteristic, tangential contact stiffness, hysteresis loop, response curves

1. Introduction

Dry friction connections are very often used in technical applications for quenching of dangerous resonance or self-excited vibrations [1–4,7,11]. Many theoretical, numerical and experimental investigations of dynamic properties of turbine blades have been done with the aim to develop means for reduction of dangerous resonance amplitudes of blades. The dynamic systems investigated, for example, in papers [5,6,8] contain dry friction elements modelled by relatively simple 2V (two variables) "force-velocity" characteristics. Application of these mathematical models enables easy calculation for the majority of engineering problems where the vibrating bodies can be assumed to be stiff and when these bodies in contact surfaces only slip against each other without any elastic deformation. Such models correctly describe the properties of dry friction process for sufficiently large amplitude of relative motion in contact surfaces.

However, in dry friction elements used for vibration damping, these relative amplitudes of friction couples are usually very small. Also the friction surfaces are sometimes placed on relative compliant parts of moving bodies. Therefore in such cases, it is necessary to use more sophisticated 3V computational "stick-slip" model with "force-velocity-displacement" characteristic for computational analyses. A detailed analysis of the influence of rough surfaces on friction properties is given in [12].

A large part of research activities in world literature is oriented on the investigation of dynamics of turbine disk, with blades connected by friction elements in shroud. For the purpose of a detail analysis of friction processes and their influence on blades vibrations, the dynamic tests of a separated blade couple were performed together with parallel theoretical investigations.

*Corresponding author. e-mail: pust@it.cas.cz.

Papers [5,6] describe the influence of various mathematical models of dry friction forces on the response curves of harmonically excited blade couple.

The dynamic systems investigated in both these papers contain a friction connection, described by the simplest types of 2V Coulomb dry friction mathematical model and its modified version without any elastic deformation of contacting bodies.

Presented work describes mathematical models of two types of 3V "stick-slip" damping elements with force-velocity-displacement characteristics and provides an analysis of their properties. These non-linear friction elements are used as damping connections between blades. The effects of tangential micro-deformations in contact surfaces as well as dry friction forces and excitation amplitudes on response curves will be shown and discussed.

2. Stick-slip contact with elastic micro-deformation

Dry friction characteristics described by 3V "force-velocity-displacement" are necessary to be used in the cases, when the friction surface is placed on some relative compliant parts of moving bodies and when the contacting bodies vibrate with small relative amplitudes. Such situation is shown in Fig. 1. The motion of bodies is usually defined in analytical or numerical solutions by motion $x_1(t)$ of the centre of gravity, which can be far from the position of friction contact defined as $x(t)$ in Fig. 1.

The mathematical model of such damping element consists of Coulomb dry friction part consecutively connected to a spring with characteristics

$$\begin{aligned} F_t &= k_t(x - x_1) \qquad && \text{if } \dot{x} = 0, \qquad && F_t \in (-F_{to}, F_{to}), \\ F_t &= F_{to}\,\mathrm{sgn}\,(\dot{x}) \qquad && \text{if } |\dot{x}| > 0, \qquad && F_t = -F_{to} \text{ or } F_{to}, \end{aligned} \tag{1}$$

where F_{to} is dry friction force at motion. The graphical presentation of this "stick-slip" damping element is given in Fig. 1a–b.

The point A of the spring-damper connection can be supposed either weightless (case shown in Fig. 1a) or with a very small mass m_1 modelling the mass of elastically deformed parts of contacting bodies near the friction surface (case shown in Fig. 1b). In this case, the excitation force F_t^* is slightly different from F_t in Fig. 1a.

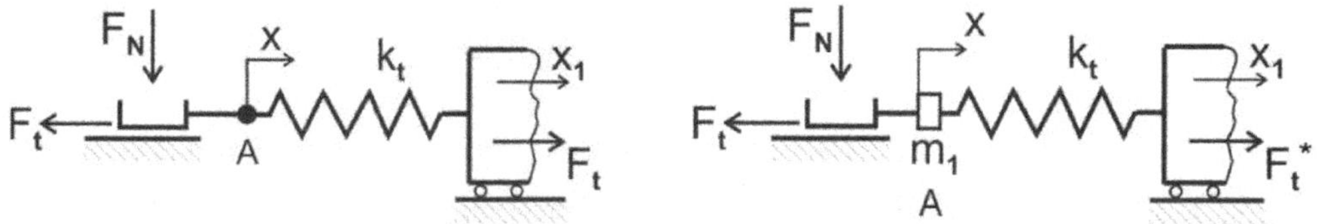

Fig. 1. Two types of stick-slip dry friction models: a) weightless model, b) with small mass m_1

3. Hysteresis loops of stick-slip dry friction elements

The hysteresis loop of the weightless "stick-slip" damping element (Fig. 1a) for the simple cosine excitation motion $x_1(t) = a\cos(\omega t)$ has a rhomboid-form and it contains four break points in one cycle [3,7]. This exact form is shown in Fig. 2, which is constructed, as well as Figs. 3–4, for dimensions of forces F in [N], displacements x and amplitudes a in [mm] and the stiffness k_t in [kg s^{-2}].

However, if the relative motion in the contact surface is not so simple (contains higher harmonic components) then the computation is much complicated and it is very inconvenient for

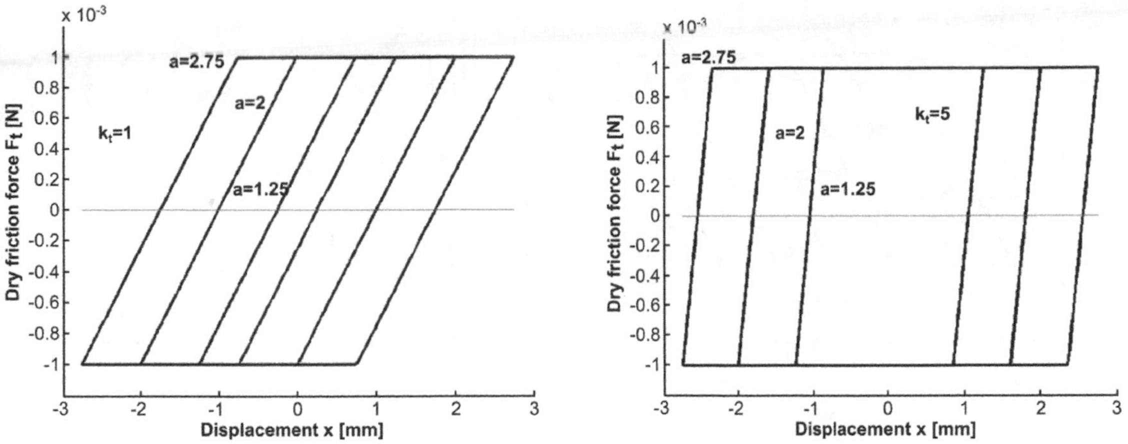

Fig. 2. Exact rhomboid forms

dynamic problems solution. Direct numerical solution (e.g. ODE solvers in Matlab) of ordinary differential equation based on the transformation into a set of first order equations becomes very complex due to the altering function *sgn* in friction characteristic. This complication can be removed by expressing the Coulomb law by means of "arc-tangent" function, which is continuous in the whole range of velocity $\dot{x}$:

$$F_t = F_{to} \tfrac{2}{\pi} \,\mathrm{arctg}(\alpha\, \dot{x}). \tag{2}$$

The parameter α [$\mathrm{s\,mm^{-1}}$] multiplying the velocity $\dot{x}$ controls the slope of smooth transition from positive to negative values of friction force in the points of the reversals.

Examples of hysteresis loops with rhomboid-like shape are plotted for three amplitudes $a = 1.25;\ 2;\ 2.75$ mm, for two spring stiffnesses $k_t = 1$ or 5 $\mathrm{kg\,s^{-2}}$ and for dry friction element with slip friction force $F_{to} = 10^{-3}$ N in following figures. The equations governing the motions x of the systems shown in Fig. 1 are

$$k_t\,(x - a\cos(\omega t)) + F_{to}\,\mathrm{sgn}\,(\dot{x}) = 0,$$

$$m_1\ddot{x} + k_t\,(x - a\cos(\omega t)) + F_{to}\,\mathrm{sgn}\,(\dot{x}) = 0,$$

where friction forces $F_{to}\,\mathrm{sgn}\,(\dot{x})$ can be replaced by equation (2).

The loops of the mathematical mass-less model of the damping element consisting of a spring connected with Coulomb-dry-friction-damper with the "arc-tangent" characteristic (Fig. 1a) are plotted for $\alpha = 25\ \mathrm{s\,mm^{-1}}$ (see relation 2) in Figs. 3a–b.

The calculation of hysteresis loops for the same parameters but for the mathematical model shown in Fig. 1b, i.e., the model of dry friction stick-slip damping element with a small mass $m_1 = 0.002$ kg of an elastic deformable part of contacting bodies, gives loops plotted in Figs. 4a–b.

Comparing the results presented in Fig. 2 with those in Fig. 3 and Fig. 4, it is evident that both mathematical models graphically presented in Fig. 1 give very similar properties in the investigated range of parameters both from the point of view of hysteresis loop's form and area proportional to the friction loss energy. They can be successfully applied for the solution of complicated cases (e.g. at general multi-harmonic excitations) due to the compact definition forms. They are also more advantageous than the stepwise piece-wise solution [3] when one period is divided into several intervals, which are connected together with the ends-beginning conditions.

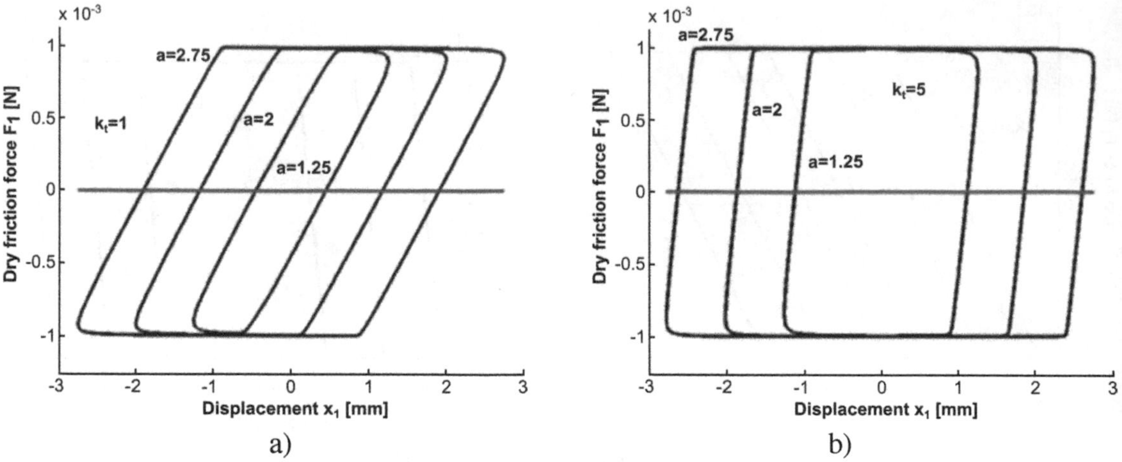

Fig. 3. Hysteresis loops of stick-slip "arc-tangent" friction element for a) weak tangential stiffness $k_t = 1\ \text{kg}\,\text{s}^{-2}$, b) strong stiffness $k_t = 5\ \text{kg}\,\text{s}^{-2}$

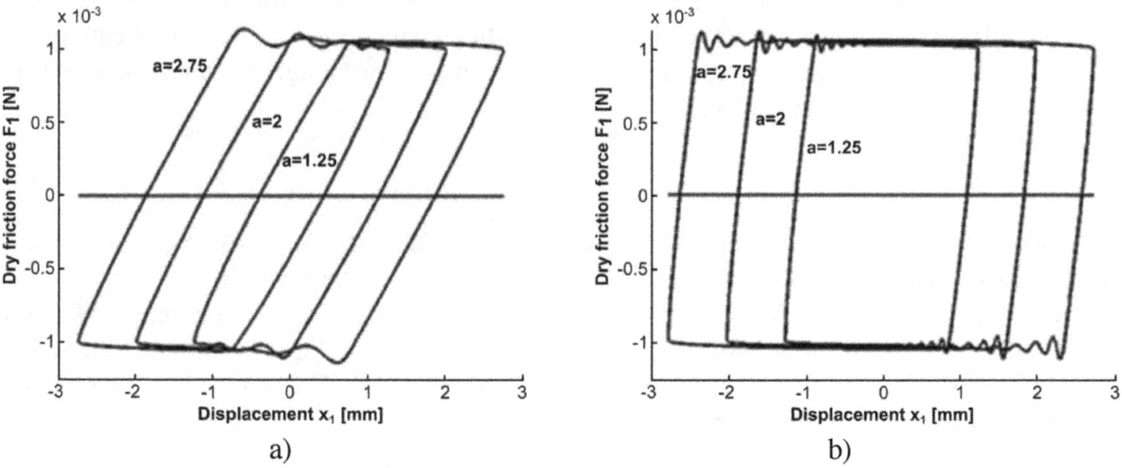

Fig. 4. Hysteresis loops of stick-slip Coulomb dry friction element for a) weak tangential stiffness $k_t = 1\ \text{kg}\,\text{s}^{-2}$, b) strong stiffness $k_t = 5\ \text{kg}\,\text{s}^{-2}$

4. Forced vibration of two-blades bunch with stick-slip damping connections

As it has been mentioned in the section Introduction, a part of research activities is oriented on the investigation of dynamics of turbine disk with blades connected by friction elements in shroud. The gained results of detail analysis of friction processes and their influence on blades vibrations were published in papers [5,6], where the simplest types of 2V Coulomb dry friction mathematical model and its modified version without any elastic deformation of contacting bodies were used.

In this section, the influence of the elastic micro-deformations in the contact surfaces will be analysed using more exact stick-slip friction mathematical models expressed by the 3V force-velocity-displacement characteristics. Laboratory experiments on the blades' models prepared by a turbine producer were the main motivation for this analysis. The scheme of one of these laboratory models is shown in Fig. 5a. The contact pressure on friction surfaces A was realized either by a spring, or by the torsion pre-stress applied to the oblique cut of shrouding.

The first simplified mathematical model of the blade couple shown in Fig. 5b consists of two identical 1 DOF slightly damped subsystems with the stiffness k and the damping coefficient b,

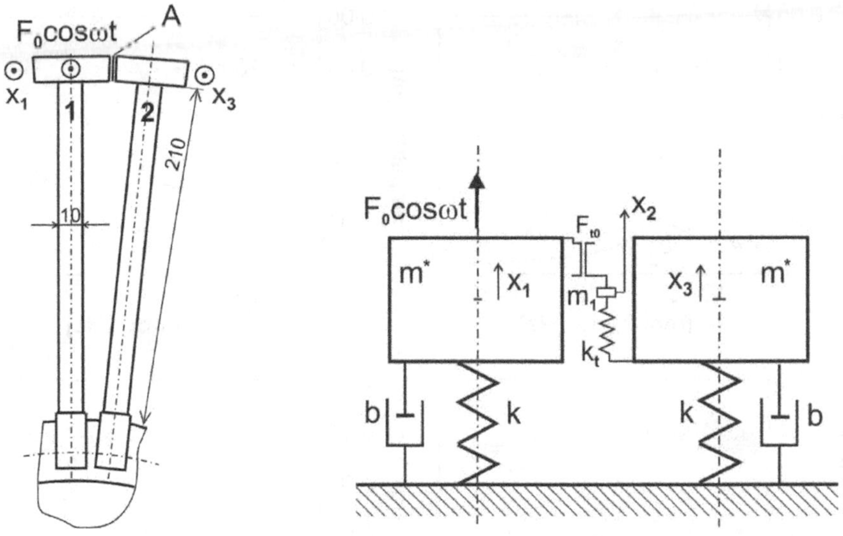

Fig. 5. Two-blades bunch with stick-slip dry friction connection: a) laboratory model, b) mathematical model

roughly corresponding to the experimentally ascertained values of real blades. The modelling of blades dynamic properties by 1 DOF subsystems is possible if the investigation is limited to the lowest resonance frequency range. Between these two subsystems is a small sprig-mass-friction system modelling stick-slip friction element.

Differential equations of motion of such a system with stick-slip dry friction contact, where due to computational reasons an arctg characteristic is applied, are:

$$\begin{aligned} m^*\ddot{x}_1 + b\dot{x}_1 + kx_1 + F_{t0}\tfrac{2}{\pi}\,\mathrm{arctg}[\alpha(\dot{x}_1 - \dot{x}_2)] &= F_0\cos\omega t, \\ m_1\ddot{x}_2 + k_t(x_2 - x_3) + F_{t0}\tfrac{2}{\pi}\,\mathrm{arctg}[\alpha(\dot{x}_2 - \dot{x}_1)] &= 0, \\ m^*\ddot{x}_3 + b\dot{x}_3 + kx_3 - k_t(x_2 - x_3) &= 0. \end{aligned} \tag{3}$$

Let the mass of a single blade be m. The mass m_1 of the elastically deformed part near the contact surface belongs to the blade's mass m, but during vibrations it moves separately, it is covered by own equation and therefore the masses of blades must be modelled by $m^* = m - m_1/2$ in the differential equations of motion.

In order to be ready for numerical solution, the equations (3) need to be rearranged into a set of equations of the first order:

$$\begin{aligned} \dot{x}_1 &= v_1, \\ \dot{v}_1 &= (-bv_1 - kx_1 - F_{t0}\tfrac{2}{\pi}\,\mathrm{arctg}[\alpha(v_1 - v_2)] + F_0\cos\omega t)/(m - m_1/2), \\ \dot{x}_2 &= v_2, \\ \dot{v}_3 &= \{-k_t(x_2 - x_3) + F_{t0}\tfrac{2}{\pi}\,\mathrm{arctg}[\alpha(v_1 - v_2)]\}/m_1, \\ \dot{x}_3 &= v_3, \\ \dot{v}_3 &= [-bv_3 - kx_3 - k_t(x_{31} - x_2)]/(m - m_1/2). \end{aligned} \tag{4}$$

The influence of the dry friction force F_{t_0} at constant contact stiffness $k_t = 10\,000\ \mathrm{kg\,s^{-2}}$ on response curves of both bodies is shown in Fig. 6 for the excitation amplitude $F_0 = 10$ N, and for four magnitudes of dry friction contact with the slipping force $F_{t_0} = 2; 4; 6; 8$ N.

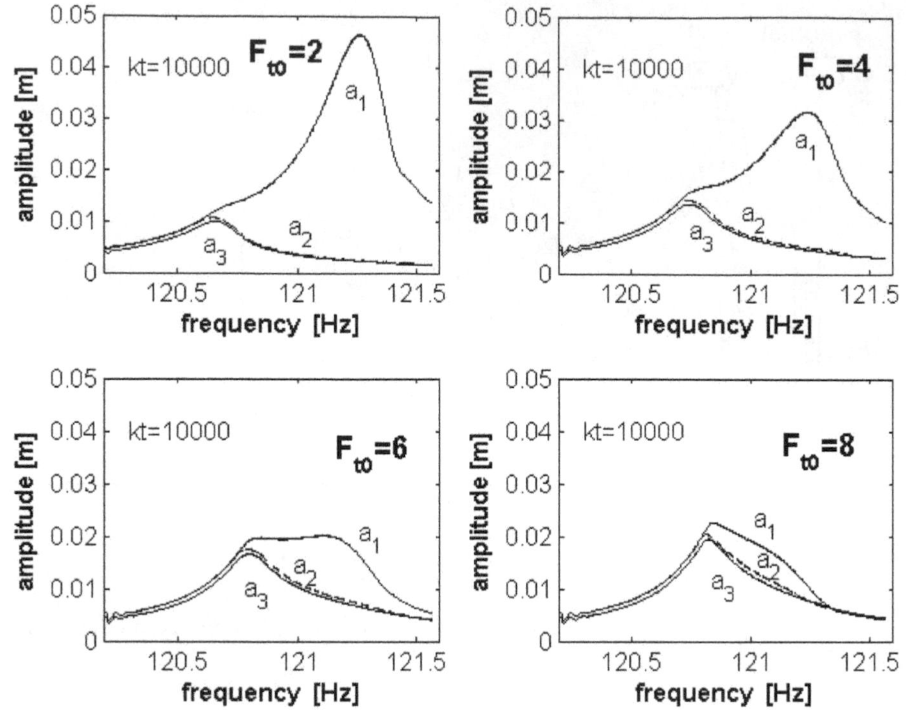

Fig. 6. Response curves of the blade couple model connected by stick-slip damping element. Mass of elastically deformed parts: $m_1 = m/100$

The function *arctg* was used for the description of the dependence of dry friction force F_{t_0} on relative slipping velocity $v = \dot{x}_2 - \dot{x}_1$ between the bodies with masses m, m_1, with sufficiently great parameter $\alpha = 25$ [s mm^{-1}] guaranteeing appropriate similarity to the exact Coulomb dry friction characteristic. The small body m_1, modelling the mass of elastically deformed parts of contacting bodies, is chosen relatively large in order to see its effect on the response curves: $m_1 = m/100 = 0.001\,82$ kg. Also the excitation amplitude $F_0 = 10$ N has been selected sufficiently high in these examples for similar reasons. For ten times lower forces (excitation, elastic, friction, etc.), the amplitudes reduce as well, but the character of the responses remain similar.

There are two response peaks, at the frequencies $f_r = 121.25$ Hz and $f_r = 120.75$ Hz. The first one corresponds to the eigenfrequency of the first excited body with mass $m - m_1/2$, the second one is given by connected bodies $m + m_1/2$. The stiffness k_t influences the difference between response curves $a_2(f_r)$ of the small middle body m_1 and $a_3(f_r)$ of the non-excited body with mass $m + m_1/2$. As the stiffness $k_t = 10\,000$ kg s^{-2} is very high in this case, both curves (the dashed and the solid lower one) lie near each other.

It is evident that the increasing value of the dry friction slipping force F_{t_0} causes a decrease of amplitudes a_1 of the excited blades to the values of amplitude a_2 which corresponds to the small body m_1 (dashed line). The peaks of these dashed lines together with the peaks of the lower response curves increase with higher dry friction slipping force F_{t_0}. For the highest friction force $F_{t_0} = 8$ N (Fig. 6), they overlap with the peak of curve $a_1(f_r)$ and the investigated system vibrates approximately with very similar amplitudes in the entire frequency resonance range.

If the elastically deformed parts of bodies near the contacting surfaces are twice smaller $m_1 = m/200$, then the response curves change. Again, there exist two response peaks, one at

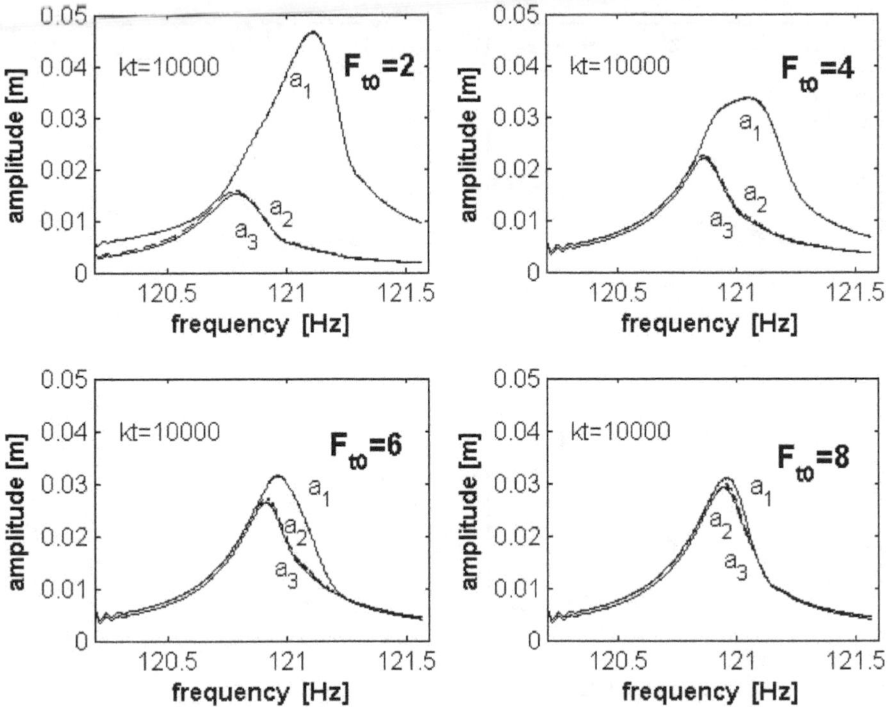

Fig. 7. Response curves of the blade couple model connected by stick-slip damping element. Mass of elastically deformed parts: $m_1 = m/200$

the response curve $a_1(f_r)$ of the excited body of mass $m - m_1/2$ and the second one of the closely joined bodies with response curves $a_2(f_r)$ and $a_3(f_r)$. Due to the smaller body mass m_1, the frequency difference between these peaks is smaller than in the previous case. The maximum amplitude a_1 of the excited body decreases with increasing dry friction force F_{t_0}, but the maximum of amplitudes a_2, a_3 increase simultaneously. All curves form a narrow bunch of response curves at the highest friction force $F_{t_0} = 8$ N.

The results of the numerical solution presented in Fig. 6 and Fig. 7 seem not to be realistic, as the gained amplitudes of vibrations up to 5 cm cannot occur in a real turbine. These high values are caused by the application of a large excitation force of the amplitude $F_0 = 10$ N acting on the relative small blade used by laboratory experiments ($m = 0.182$ kg). In order to show that these illustrative results can be successfully applied also to an analysis of real blade's responses, the next figure presented the results calculated for ten times lower excitation and for ten times lower friction forces ($F_0 = 1$ N, $F_{t_0} = 0.2; 0.4; 0.6; 0.8$ N) is added.

From the comparison of Fig. 6 and Fig. 8 it is evident that the qualitative properties of blade's responses are valid also for vibration with small amplitudes, of course when the conditions of physical similarity such as the ratio of forces F_0/F_{t_0} are fulfilled. The measurements of physical model of a blade couple (e.g. [9]) confirm the main properties of responses on harmonic excitations as well as that the form of hysteresis loops can be modelled by the stick-slip friction elements.

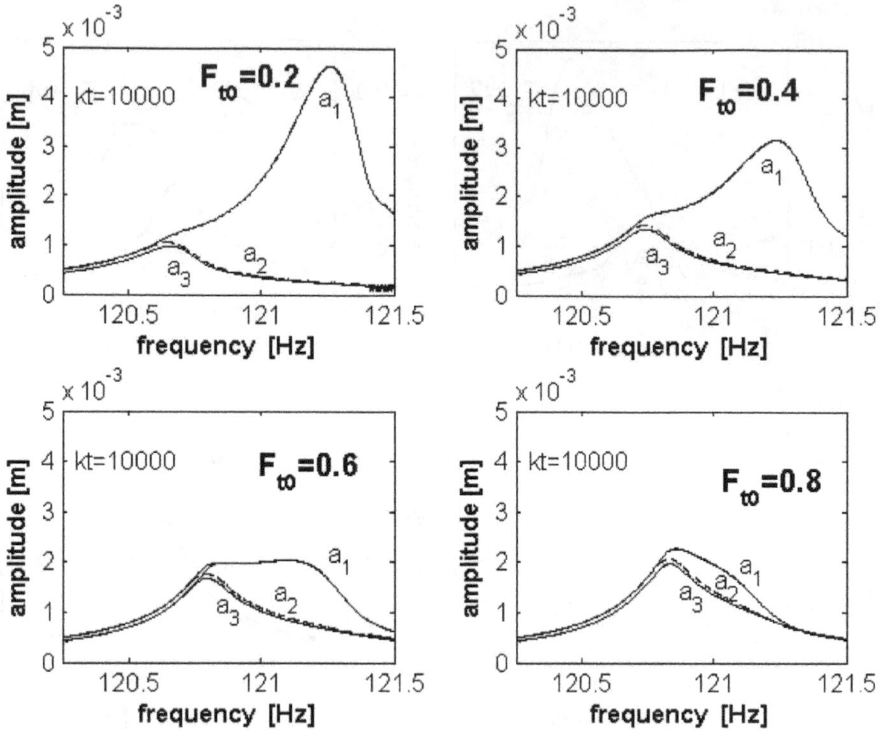

Fig. 8. Response curves of the blade couple model connected by stick-slip damping element. Mass of elastically deformed parts: $m_1 = m/100$, amplitude of excitation $F_0 = 1$ N

5. Symmetrical slip-stick friction element

The frequency shift between the peaks of response curves $a_1(f), a_2(f)$ in Figs. 6–8 are caused by the non-symmetry of the slip-stick friction element connected with a spring to the body 1 and with a dry friction contact to the body 2. This frequency shift can be removed by the application of a more exact slip-stick friction connection consisting of two springs and two mass bodies between which a dry friction element is placed — see Fig. 9.

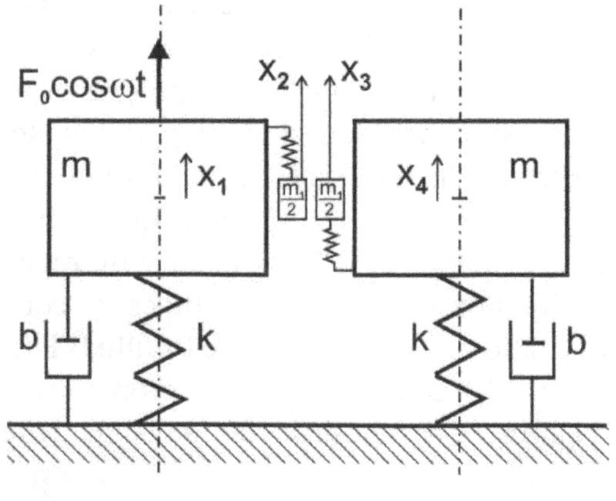

Fig. 9. Model of two-blades bunch with symmetrical stick-slip dry friction connection

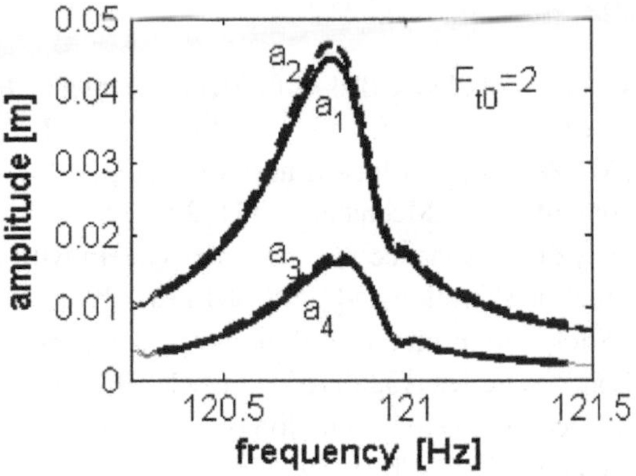

Fig. 10. Response curves of the blade couple model with symmetrical stick-slip damping elements

An example of response curves for the same parameters as used in Fig. 6 is presented in Fig. 10.

Comparing the last figure with the first subfigure in Fig. 6 it is obvious that the blade couple connected by the symmetric slip-stick friction element has the same resonant frequencies of both main bodies and that the peak amplitudes of response curves with the symmetric and non-symmetric friction elements are similar. It can be estimated that the courses of responses in both cases are different, but the damping effect of symmetric and non-symmetric structure are approximately the same.

The symmetrical model is applicable for modelling of a steel–steel friction contact. The non-symmetric model of slip-stick friction elements can be used when the contacting surfaces have different properties. Such cases were investigated, for example, in reports [9, 10], where various combinations of friction couples used in praxis were studied.

6. Conclusion

Two main types of stick-slip mathematical models with 3V "force-velocity-displacement" characteristics are presented and analysed by means of hysteresis loops at the harmonic excitation. It is shown that the approximate models with *arctg* description of Coulomb dry friction law as well as models including the inertia of elastically deformed parts near the contact surfaces have very similar rhomboid hysteresis loops as the exact piecewise solution.

The response curves of two-blades bunch with internal connection of blades by means of damping element with the inertia stick-slip characteristic and *arctg* description of dry friction have been investigated. The analysis of influence of different dry friction force and two values of elastically deformable parts near the contact surface on response curves showed the existence of two resonance peaks. Their heights and positions strongly depend on all parameters of 3V characteristic of connecting element. This frequency shift between the resonance peaks can be removed by the application of a more exact slip-stick friction connection consisting of two springs and two mass bodies between which a dry friction element is placed.

Acknowledgements

The work has been supported by the conceptual development fund RVO: 61388998 of the Institute of Thermomechanics AS CR, v.v.i.

Reference

[1] Brůha, J., Zeman, V., Vibration of bladed disk with friction elements, Proceedings of the conference Dynamics of Machines 2014, Prague, IT AS CR, 2014, pp. 15–22.

[2] Byrtus, M., Hajžman, M., Zeman, V., Linearization of friction effects in vibration of two rotating blades, Applied and Computational Mechanics 7 (1) (2013) 5–22.

[3] Ding, Q., Chen, Y., Analyzing resonance response of a system with dry friction damper using an analytical method, Journal of Vibration and Control 14 (8) (2008) 1 111–1 123.

[4] Pfeifer, F., Hájek, M., Stick-slip motion of turbine blade dampers, Philosophical Transactions of the Royal Society of London, Series A338 (1651) (1992) 503–517.

[5] Půst, L., Pešek, L., Influence of delayed excitation on vibrations of turbine blades couple, Applied and Computational Mechanics 7 (1) (2013) 39–52.

[6] Půst, L., Pešek, L., Vibration damping by dry friction with micro-slips, Proceedings of the conference Dynamics of Machines 2013, Prague, IT AS CR, 2013, pp. 93–102.

[7] Půst, L., Pešek, L., Radolfová, A., Various type of dry friction characteristics for vibration damping, Engineering Mechanics 18 (3-4) (2011) 203–224.

[8] Půst, L., Pešek, L., Radolfová, A., Vibrations of blades couple connected by slip-stick dry friction, Proceedings of the 30th conference Computational Mechanics, University of West Bohemia, Špičák, 2014, pp. 117–118.

[9] Půst, L., Veselý, J., Horáček, J., Radolfová, A., Effect of friction forces on dynamics of blades' model, Report Z-1412/07, IT ASCR, 2007, (in Czech).

[10] Půst, L., Veselý, J., Horáček, J., Radolfová, A., Investigation of friction effects on blades' model, Report Z-1422/10, IT ASCR, 2010, (in Czech).

[11] Sextro, W., Dynamical contact problems with friction, Springer-Verlag Berlin Heidelberg, 2007.

[12] Voldřich, J., Modelling of the three-dimensional friction contact of vibrating elastic bodies with rough sur-faces, Applied and Computational Mechanics 3 (1) (2009) 241–252.

3

Seismic response of nuclear fuel assembly

Z. Hlaváč[a,*], V. Zeman[a]

[a]*Faculty of Applied Sciences, University of West Bohemia, Univerzitní 22, 306 14 Plzeň, Czech Republic*

Abstract

The paper deals with mathematical modelling and computer simulation of the seismic response of fuel assembly components. The seismic response is investigated by numerical integration method in time domain. The seismic excitation is given by two horizontal and one vertical synthetic accelerograms at the level of the pressure vessel seating. Dynamic response of the hexagonal type nuclear fuel assembly is caused by spatial motion of the support plates in the reactor core investigated on the reactor global model. The modal synthesis method with condensation is used for calculation of the fuel assembly component displacements and speeds on the level of the spacer grid cells.

Keywords: seismic response, nuclear fuel assembly, modal synthesis method, condensation

1. Introduction

One of the basic requirements on operation conditions of the nuclear reactor is the feasible seismic response guarantee. Two basic approaches can be applied to seismic response determination. The stochastic approach [5] is based on statistical description of loading process and on the parameters of vibrating system. For the sake of simplicity, it is mostly supposed the stochasticity is solely due to the loading process, while the vibrating system is considered as a deterministic one. The deterministic approach is based on description of the seismic excitation in either analytical or digital form.

The seismic action is most often represented by the response spectrum in displacement, pseudo-velocity or pseudo-acceleration [1] expressed analytically as a function of the eigenfrequency and relative damping of a simple oscillator. The seismic response is calculated by the response spectrum method based on different combination of vibration mode contributions [8]. The specific method of response spectrum method, so called missing mass correction method, includes the high frequency rigid modes into the system response pseudostatically [4]. The seismic action in the digital form is represented by synthetic accelerograms corresponding to given response spectra generally for damping value 5 % for ground spectra and 2 % for floor spectra [1]. Both deterministic approaches require assemblage of the mathematical model of the reactor for frequency area up to about 50 Hz.

An assessment of nuclear fuel assemblies (FA) behaviour at standard and extreme operating conditions belongs to important safety and reliability audit. A significant part of FA assessment plays dynamic deformation and load of FA components especial of fuel rods (FR) and load-bearing skeleton (LS) (see Fig. 1). The beam type FA model used in seismic analyses of WWER type reactors [2] does not enable investigation of seismic deformations and load of FA

*Corresponding author. e-mail: hlavac@kme.zcu.cz.

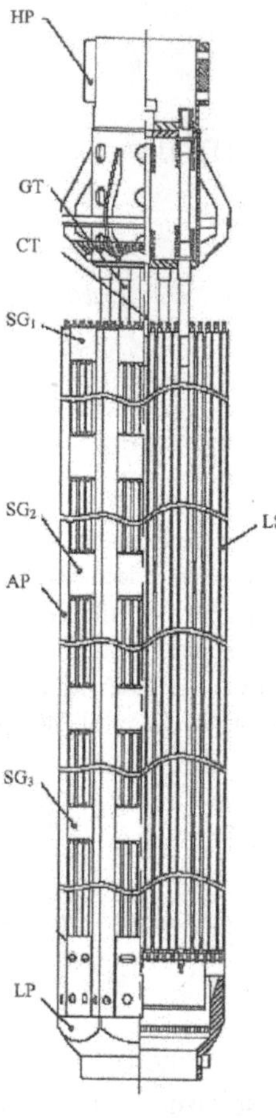

Fig. 1. Fuel assembly

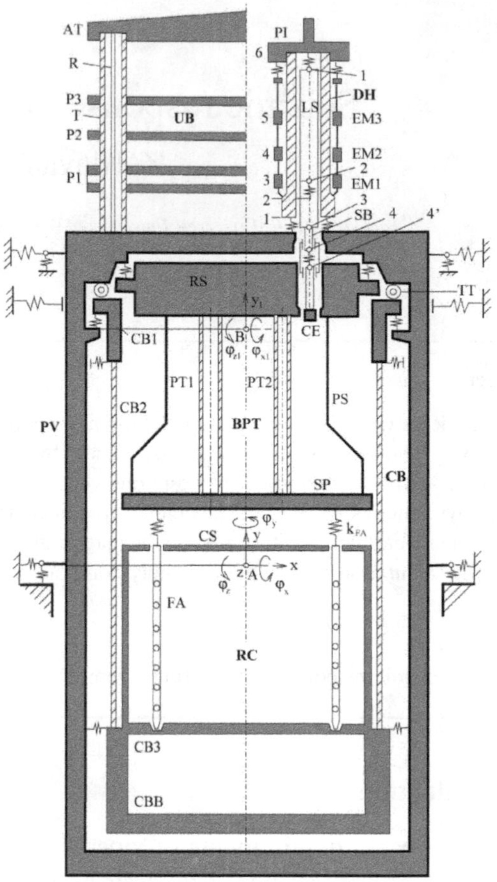

Fig. 2. Reactor

components. The goal of this contribution, in direct sequence at an interpretation of FA modelling, modal analysis and calculation of dynamic response caused by pressure pulsation [12], is a presentation of the newly developed method for seismic analysis of FA components. The seismic displacements, velocities and deformations of the FA components on the level of spacer grids (SG) can be used for their stress analysis.

2. The seismic motion of the supporting plates

The original linearized mathematical model of the WWER 1000/320 type reactor intended for seismic response calculation was derived on the basis of computational (physical) model, whose structure is shown in Fig. 2. It was derived using the decomposition method [11]. The reactor was decomposed into eight subsystems [3, 11] — pressure vessel (PV), core barrel (CB) composed from two rigid bodies which are connected by beam-type continuum (CB2), reactor core (RC) formed from 163 FA, block of protection tubes (BPT), upper block (UP), system of 61 control rod drive housing (DH), system of 61 electromagnet blocks (EM) and system of 61 drive assemblies composed from a lifting system mechanism (LS) which ensures a suspension bar (SB) motion with the control elements (CE). The mass and static stiffness of the primary coolant loops between a reactor pressure vesel nozzles and steam generators were approximately replaced by mass points and springs placed in gravity centers of the nozzles. The components marked by grey in Fig. 2 were reflected as rigid bodies with six degrees of freedom excepted tope part of core barrel (CB1) having only three degrees of freedom with respect to pressure vessel. Other components are modelled as one-dimensional continua of beam types.

The mathematical model of the reactor after discretization of the one-dimensional continuums and a completion of the damping approximated by the proportional damping matrix $\boldsymbol{B}$ and of the seismic excitation has the form [2]

$$\boldsymbol{M}\ddot{\boldsymbol{q}} + \boldsymbol{B}\dot{\boldsymbol{q}} + \boldsymbol{K}\boldsymbol{q} = -\boldsymbol{m}_1\ddot{u}_x(t) - \boldsymbol{m}_2\ddot{u}_y(t) - \boldsymbol{m}_3\ddot{u}_z(t)\,, \qquad (1)$$

where components of the vector generalized coordinates $\boldsymbol{q}$ are relative displacements of carried subsystems with respect to supporting subsystems. So, for example, the supporting subsystem for CB, BPT, UB, DH is the pressure vessel (PV). Pressure vessel generalized coordinates are relative displacements with respect to basis. The seismic excitation is expressed by the synthetic accelerograms $\ddot{u}_l$, $l = x, y, z$ of the reactor hall as basic in directions of axes x, y, z on the level of point A (see Fig. 2). The first three generalized coordinates of the pressure vessel and the whole reactor are relative translation displacements with respect to basis. That is why the vectors $\boldsymbol{m}_i, i = 1, 2, 3$ are the first three columns of the reactor mass matrix $\boldsymbol{M}$. The same horizontal accelerograms $\ddot{u}_x(t)$, $\ddot{u}_z(t)$ and one vertical accelerogram $\ddot{u}_y(t)$, given by Škoda Nuclear Machinery for NPP Temelín, are presented in Fig. 3 and Fig. 4, along with their power spectral densities.

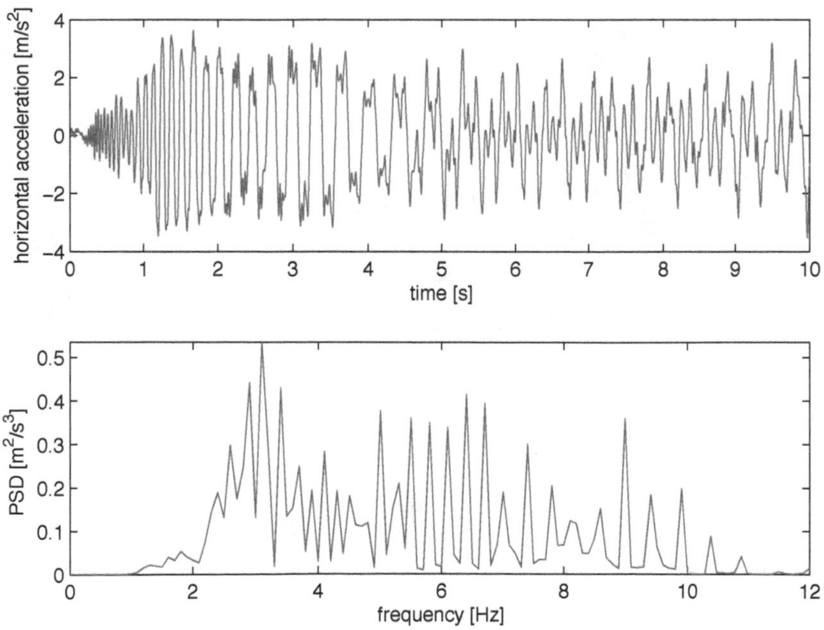

Fig. 3. Horizontal accelerogram

Each fuel assembly (see Fig. 1) is fixed by means of lower tailpiece (LP) into mounting plate in core barrel bottom and by means of head piece (HP) into lower supporting plate of the block of protection tubes. These support plates with pieces can be considered as rigid bodies.

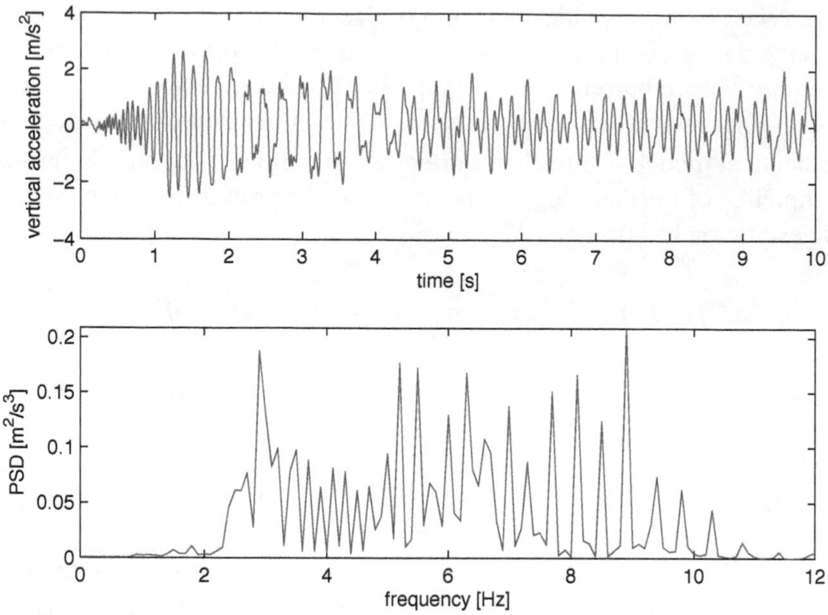

Fig. 4. Vertical accelerogram

Let use consider the spatial motion of the support plates described in coordinate systems x_X, y_X, z_X ($X = L, U$) with origins in plate gravity centres L, U by displacement vectors (see Fig. 5)

$$\boldsymbol{q}_X = [x_X, y_X, z_X, \varphi_{x,X}, \varphi_{y,X}, \varphi_{z,X}]^T, \; X = L, U. \tag{2}$$

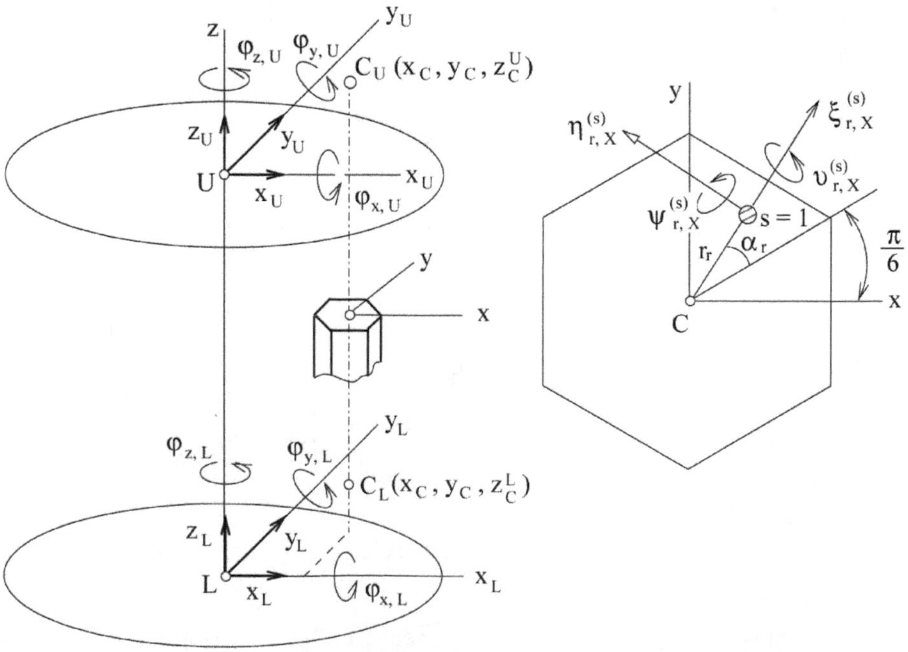

Fig. 5. Spatial motion of the FA support plates

The transformation relations between reactor generalized coordinates and absolute displacements of lower (L) and upper (U) FA supporting plates can be expressed in the global matrix form

$$\boldsymbol{q}_X = \boldsymbol{T}_{R,X}\boldsymbol{q} + \boldsymbol{u}(t)\,, \quad \boldsymbol{T}_{R,X} \in R^{6,n_R},\; X = L, U, \tag{3}$$

where n_R is reactor DOF number. The vector $\boldsymbol{u}(t)$ of basis translational motion, with respect to different reactor and plates coordinate systems (see Fig. 2 and Fig. 5), is

$$\boldsymbol{u}(t) = [u_x(t), -u_z(t), u_y(t), 0, 0, 0]^T. \tag{4}$$

3. Condensed mathematical model of the fuel assembly

In order to model, the hexagonal type FA (Fig. 6) is divided into subsystems-six identical rod segments $s = 1, \ldots, 6$, centre tube (CT) and load-bearing skeleton (LS). Each rod segment of the TVSA-T FA (on Fig. 6 drawn in lateral FA cross section and circumscribed by triangles) is composed of 52 fuel rods with fixed bottom ends in lower piece (LP) and 3 guide thimbles (GT) fully restrained in lower and head pieces (HP). The fuel rods and guide thimbles are linked by transverse spacer grids $g = 1, \ldots, 8$ of three types (SG1-SG3) inside the segments. All FA components are modelled as one dimensional continuum of beam type with nodal points in the gravity centres of their cross-section on the level of the spacer grids. Mathematical models of six segments $s = 1, \ldots, 6$ are identical in consequence of radial $\xi_{r,g}^{(s)}$ and orthogonal $\eta_{r,g}^{(s)}$ fuel rods and guide thimbles lateral displacements and bending angles $\vartheta_{r,g}^{(s)}, \psi_{r,g}^{(s)}$ around these lateral displacements on the level of spacer grid g (in the Fig. 5 on the level fixed ends).

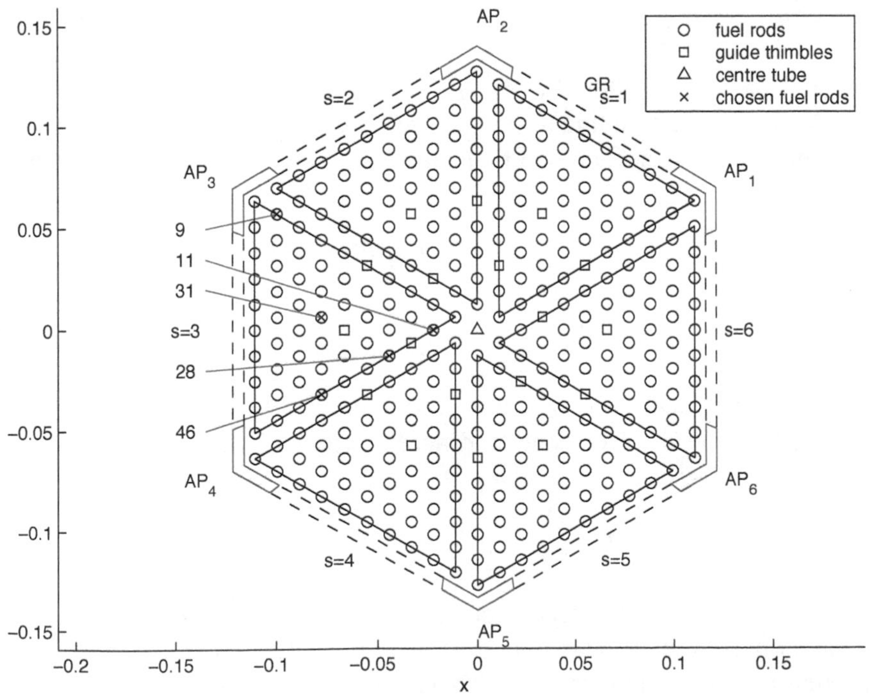

Fig. 6. The FA cross-section

The FA mathematical model was derived in the configuration space [3,11]

$$\boldsymbol{q} = [\boldsymbol{q}_1^T, \dots, \boldsymbol{q}_s^T, \dots, \boldsymbol{q}_6^T, \boldsymbol{q}_{CT}^T, \boldsymbol{q}_{LS}^T]^T \tag{5}$$

corresponding to FA decomposition. The vector s of generalized coordinates of each subsystem (rod segments, centre tube) losed in kinematically excited nodes fixed into lower and upper supporting plate can be partitioned in the form

$$\boldsymbol{q}_s = [(\boldsymbol{q}_L^{(s)})^T, (\boldsymbol{q}_F^{(s)})^T, (\boldsymbol{q}_U^{(s)})^T]^T, \; s = 1, \dots, 6, CT \tag{6}$$

and the skeleton $s = LS$ fixed in bottom ends only has the form

$$\boldsymbol{q}_{LS} = [(\boldsymbol{q}_L^{(LS)})^T, (\boldsymbol{q}_F^{(LS)})^T]^T \, . \tag{7}$$

The displacements of free system nodes (uncoupled with support plates) are integrated in vectors $\boldsymbol{q}_F^{(s)} \in R^{n_s}$. The conservative mathematical models of the loosed subsystems in the decomposed block form corresponding to partitioned vectors can be written as

$$\begin{bmatrix} \boldsymbol{M}_L^{(s)} & \boldsymbol{M}_{L,F}^{(s)} & \boldsymbol{0} \\ \boldsymbol{M}_{F,L}^{(s)} & \boldsymbol{M}_F^{(s)} & \boldsymbol{M}_{F,U}^{(s)} \\ \boldsymbol{0} & \boldsymbol{M}_{U,F}^{(s)} & \boldsymbol{M}_U^{(s)} \end{bmatrix} \begin{bmatrix} \ddot{\boldsymbol{q}}_L^{(s)} \\ \ddot{\boldsymbol{q}}_F^{(s)} \\ \ddot{\boldsymbol{q}}_U^{(s)} \end{bmatrix} + \begin{bmatrix} \boldsymbol{K}_L^{(s)} & \boldsymbol{K}_{L,F}^{(s)} & \boldsymbol{0} \\ \boldsymbol{K}_{F,L}^{(s)} & \boldsymbol{K}_F^{(s)} & \boldsymbol{K}_{F,U}^{(s)} \\ \boldsymbol{0} & \boldsymbol{K}_{U,F}^{(s)} & \boldsymbol{K}_U^{(s)} \end{bmatrix} \begin{bmatrix} \boldsymbol{q}_L^{(s)} \\ \boldsymbol{q}_F^{(s)} \\ \boldsymbol{q}_U^{(s)} \end{bmatrix} = \begin{bmatrix} \boldsymbol{f}_L^{(s)} \\ \boldsymbol{f}_C^{(s)} \\ \boldsymbol{f}_U^{(s)} \end{bmatrix} \tag{8}$$

for the $s = 1, \dots, 6, CT$ and for the skeleton as

$$\begin{bmatrix} \boldsymbol{M}_L^{(LS)} & \boldsymbol{M}_{L,F}^{(LS)} \\ \boldsymbol{M}_{F,L}^{(LS)} & \boldsymbol{M}_F^{(LS)} \end{bmatrix} \begin{bmatrix} \ddot{\boldsymbol{q}}_L^{(LS)} \\ \ddot{\boldsymbol{q}}_F^{(LS)} \end{bmatrix} + \begin{bmatrix} \boldsymbol{K}_L^{(LS)} & \boldsymbol{K}_{L,F}^{(LS)} \\ \boldsymbol{K}_{F,L}^{(LS)} & \boldsymbol{K}_F^{(LS)} \end{bmatrix} \begin{bmatrix} \boldsymbol{q}_L^{(LS)} \\ \boldsymbol{q}_F^{(LS)} \end{bmatrix} = \begin{bmatrix} \boldsymbol{f}_L^{(LS)} \\ \boldsymbol{f}_C^{(LS)} \end{bmatrix}, \tag{9}$$

where letters $\boldsymbol{M}$ ($\boldsymbol{K}$) correspond to mass (stiffness) submatrices of the subsystems. The force subvectors $\boldsymbol{f}_C^{(s)}$ express the coupling forces between subsystem s and adjacent subsystems transmitted by spacer grids. The second set of equations extracted from (8) and (9) for each subsystem $s = 1, \dots, 6, CT, LS$ is

$$\boldsymbol{M}_F^{(s)} \ddot{\boldsymbol{q}}_F^{(s)} + \boldsymbol{K}_F^{(s)} \boldsymbol{q}_F^{(s)} = -\boldsymbol{M}_{F,L}^{(s)} \ddot{\boldsymbol{q}}_L^{(s)} - \boldsymbol{M}_{F,U}^{(s)} \ddot{\boldsymbol{q}}_U^{(s)} - \boldsymbol{K}_{F,L}^{(s)} \boldsymbol{q}_L^{(s)} - \boldsymbol{K}_{F,U}^{(s)} \boldsymbol{q}_U^{(s)} + \boldsymbol{f}_C^{(s)}, \tag{10}$$

where for the skeleton (LS) is $\boldsymbol{M}_{F,U}^{(LS)} = \boldsymbol{0}, \boldsymbol{K}_{F,U}^{(LS)} = \boldsymbol{0}$ because the skeleton is fixed only with lower supporting plate.

Displacements and accelerations of the all kinematically excited nodes of the subsystems can be expressed by the displacements and accelerations of the lower ($X = L$) and upper ($X = U$) supporting plates as

$$\boldsymbol{q}_X^{(s)} = \boldsymbol{T}_X^{(s)} \boldsymbol{q}_X, \quad \ddot{\boldsymbol{q}}_X^{(s)} = \boldsymbol{T}_X^{(s)} \ddot{\boldsymbol{q}}_X, \quad X = L, U \, . \tag{11}$$

The transformation matrices $\boldsymbol{T}_X^{(s)}$ depend on the FA position in the reactor core.

The global model of the fuel assembly has too large DOF number for calculation of dynamic response excited by support plate motion. Therefore, we assemble the condensed model using the modal synthesis method presented in the paper [10]. Let the modal properties of the conservative models of the mutually uncoupled subsystems with the strengthened end-nodes coupled with immovable support plates be characterized by spectral $\boldsymbol{\Lambda}_s$ and modal $\boldsymbol{V}_s$ matrices of order n_s, suitable to orthonormality conditions

$$\boldsymbol{V}_s^T \boldsymbol{M}_F^{(s)} \boldsymbol{V}_s = \boldsymbol{E}, \quad \boldsymbol{V}_s^T \boldsymbol{K}_F^{(s)} \boldsymbol{V}_s = \boldsymbol{\Lambda}_s, \quad s = 1, \ldots, 6, CT, LS. \tag{12}$$

The vectors $\boldsymbol{q}_F^{(s)}$ of dimension n_s, corresponding to free nodes of subsystems, can be approximately transformed in the form

$$\boldsymbol{q}_F^{(s)} = {}^m\boldsymbol{V}_s \boldsymbol{x}_s, \quad \boldsymbol{x}_s \in R^{m_s}, \quad s = 1, \ldots, 6, CT, LS, \tag{13}$$

where ${}^m\boldsymbol{V}_s \in R^{n_s, m_s}$ are modal submatrices composed from chosen m_s master eigenvectors of fixed subsystems. The equations (10) can be rewritten using (11) and (13) in the form

$$\ddot{\boldsymbol{x}}_s(t) + {}^m\boldsymbol{\Lambda}_s \boldsymbol{x}_s(t) = -{}^m\boldsymbol{V}_s^T \sum_{X=L,U} [\boldsymbol{M}_{F,X}^{(s)} \boldsymbol{T}_X^{(s)} \ddot{q}_X + \boldsymbol{K}_{F,X}^{(s)} \boldsymbol{T}_X^{(s)} q_X] + {}^m\boldsymbol{V}_s^T \boldsymbol{f}_C^{(s)}, \tag{14}$$

where spectral submatrices ${}^m\boldsymbol{\Lambda}_s \in R^{m_s, m_s}$ correspond to chosen master eigenvectors in matrix ${}^m\boldsymbol{V}_s$. The models (14) of all subsystems can be written in the global configuration space $\boldsymbol{x} = [\boldsymbol{x}_s]$, $s = 1, \ldots, 6, CT, LS$ of dimension $m = \sum_s m_s = 6m_s + m_{CT} + m_{LS}$

$$\ddot{\boldsymbol{x}}(t) + (\boldsymbol{\Lambda} + \boldsymbol{V}^T \boldsymbol{K}_C \boldsymbol{V}) \boldsymbol{x}(t) = -\boldsymbol{V}^T \sum_{X=L,U} \left[\boldsymbol{M}_X \ddot{\boldsymbol{Q}}_X(t) + \boldsymbol{K}_X \boldsymbol{Q}_X(t) \right], \tag{15}$$

where global vector $\boldsymbol{f}_C = [\boldsymbol{f}_C^{(s)}]$ of coupling forces between subsystems was expressed by means of the stiffness matric $\boldsymbol{K}_C$ in the form $\boldsymbol{f}_C = -\boldsymbol{K}_C \boldsymbol{q}_F$, $\boldsymbol{q}_F = [\boldsymbol{q}_F^{(s)}] \in R^n$, $n = \sum n_s = 6n_s + n_{CT} + n_{LS}$ [3]. In the matrix equation (15), we introduced the block diagonal global matrices

$$\boldsymbol{\Lambda} = \text{diag}[{}^m\boldsymbol{\Lambda}_s] \in R^{m,m}, \ \boldsymbol{V} = \text{diag}[{}^m\boldsymbol{V}_s] \in R^{n,m},$$

$$\boldsymbol{M}_X = \text{diag}[\boldsymbol{M}_{F,X}^{(s)} \boldsymbol{T}_X^{(s)}], \ \boldsymbol{K}_X = \text{diag}[\boldsymbol{K}_{F,X} \boldsymbol{T}_X^{(s)}] \in R^{n,48}, \ X = L, U$$

and the global vectors

$$\boldsymbol{Q}_X(t) = [\boldsymbol{q}_X^T, \ldots, \boldsymbol{q}_X^T]^T, \ \ddot{\boldsymbol{Q}}_X(t) = [\ddot{\boldsymbol{q}}_X^T, \ldots, \ddot{\boldsymbol{q}}_X^T]^T \in R^{48}, \ X = L, U$$

describing the kinematical excitation given by FA supporting plates motion. These vectors $\boldsymbol{Q}_X(t)$, $\ddot{\boldsymbol{Q}}_X(t)$ are assembled, as a result of eight FA subsystems, from eight times repeating support plate displacement and acceleration vectors.

In consequence of slightly damped FA components we consider modal damping of the subsystems characterized in the space of modal coordinates $\boldsymbol{x}_s$ by diagonal matrices $\boldsymbol{D}_s = \text{diag}[2D_\nu^{(s)} \Omega_\nu^{(s)}]$, where $D_\nu^{(s)}$ are damping factors of natural modes and $\Omega_\nu^{(s)}$ are eigenfrequencies of the mutually uncoupled subsystems. The damping of spacer grids can be approximately expressed by damping matrix $\boldsymbol{B}_C = \beta \boldsymbol{K}_C$ proportional to stiffness matrix $\boldsymbol{K}_C$. The conservative condensed model (15) can be completed in the form

$$\ddot{\boldsymbol{x}}(t) + (\boldsymbol{D} + \beta \boldsymbol{V}^T \boldsymbol{K}_C \boldsymbol{V}) \dot{\boldsymbol{x}}(t) + (\boldsymbol{\Lambda} + \boldsymbol{V}^T \boldsymbol{K}_C \boldsymbol{V}) \boldsymbol{x}(t) = -\boldsymbol{V}^T \sum_{X=L,U} [\boldsymbol{M}_X \ddot{\boldsymbol{Q}}_X(t) + \boldsymbol{K}_X \boldsymbol{Q}_X(t)], \tag{16}$$

where $\boldsymbol{D} = \text{diag}[\boldsymbol{D}_s]$.

4. Seismic response of the fuel assembly components

The FA seismic response in modal coordinates $\boldsymbol{x}(t)$ can be investigated by integration of motion equations (16) in time domain transformed into $2m$ differential equations of the first order

$$\dot{\boldsymbol{z}}(t) = \begin{bmatrix} \dot{\boldsymbol{x}}(t) \\ \dot{\boldsymbol{y}}(t) \end{bmatrix} = \begin{bmatrix} \boldsymbol{y}(t) \\ -\widetilde{\boldsymbol{B}}\boldsymbol{y}(t) - \widetilde{\boldsymbol{K}}\boldsymbol{x}(t) + \widetilde{\boldsymbol{f}}(t) \end{bmatrix}, \tag{17}$$

where corresponding to (16)

$$\widetilde{\boldsymbol{B}} = \boldsymbol{D} + \beta \boldsymbol{V}^T \boldsymbol{K}_C \boldsymbol{V}, \quad \widetilde{\boldsymbol{K}} = \boldsymbol{\Lambda} + \boldsymbol{V}^T \boldsymbol{K}_C \boldsymbol{V}, \quad \widetilde{\boldsymbol{f}}(t) = -\boldsymbol{V}^T \sum_{X=L,U} [\boldsymbol{M}_X \ddot{\boldsymbol{Q}}_X(t) + \boldsymbol{K}_X \boldsymbol{Q}_X(t)].$$

These equations are solved at zero initial conditions $\boldsymbol{x}(0) = \boldsymbol{0}, \boldsymbol{y}(0) = \dot{\boldsymbol{x}}(0) = \boldsymbol{0}$ using standard software ODE in MATLAB code. We then calculate the displacements and velocities of the free subsystem nodes according to (13)

$$\boldsymbol{q}_F^{(s)}(t) = {}^m\boldsymbol{V}_s \boldsymbol{x}_s(t), \quad \dot{\boldsymbol{q}}_F^{(s)}(t) = {}^m\boldsymbol{V}_s \dot{\boldsymbol{x}}_s(t) \tag{18}$$

for selected subsystem $s = 1, \ldots, 6, CT, LS$.

The components of the rod segment vectors $\boldsymbol{q}_F^{(s)}$ $(s = 1, \ldots, 6)$ defined in (5) are absolute lateral displacements $\xi_{r,g}^{(s)}, \eta_{r,g}^{(s)}$ and bending angles $\vartheta_{r,g}^{(s)}, \psi_{r,g}^{(s)}$ of the fuel rod r cross-section [3] in segment s on the level of spacer grid g (see Fig. 1 and Fig. 5). Corresponding lateral displacements $\overline{\xi}_{r,g}^{(s)}, \overline{\eta}_{r,g}^{(s)}$ of the non-deformed fuel rods can be expressed by means of the lower supporting plate motion defined in (3) in the form

$$\begin{bmatrix} \overline{\xi}_{r,g}^{(s)} \\ \overline{\eta}_{r,g}^{(s)} \end{bmatrix} = \begin{bmatrix} C_r^{(s)} & S_r^{(s)} & 0 & -z_g S_r^{(s)} & z_g C_r^{(s)} & x_C S_r^{(s)} - y_C C_r^{(s)} \\ -S_R^{(s)} & C_r^{(s)} & 0 & -z_g C_r^{(s)} & -z_g C_R^{(s)} & x_C C_r^{(s)} + y_C S_r^{(s)} + r_r \end{bmatrix} \boldsymbol{q}_L, \tag{19}$$

where

$$C_r^{(s)} = \cos\left[\alpha_r + \frac{\pi}{6} + (s-1)\frac{\pi}{3}\right], \quad S_r^{(s)} = \sin\left[\alpha_r + \frac{\pi}{6} + (s-1)\frac{\pi}{3}\right],$$
$$r = 1, \ldots, 55; \quad s = 1, \ldots, 6,$$

x_C, y_C are coordinates of the FA centre C_L in the reactor core, z_g is vertical coordinate of the spacer grid g in x_L, y_L, z_L and r_r, α_r are polar coordinates of the selected fuel rod r in segment s. The fuel rod lateral deformations on the level of spacer grid g are

$$d_{r,g}^{(s)} = \sqrt{(\xi_{r,g}^{(s)} - \overline{\xi}_{r,g}^{(s)})^2 + (\eta_{r,g}^{(s)} - \overline{\eta}_{r,g}^{(s)})^2}, \quad r = 1, \ldots, 55; \; g = 1, \ldots, 8; \; s = 1, \ldots, 6. \tag{20}$$

As an illustration, the time behaviour of lateral deformations of the chosen fuel rod $r = 31$ in segment $s = 3$ on the level of the lower (for $g = 1$) and upper (for $g = 8$) spacer grid of the FA outside in the WWER 1000 reactor core ($x_C = 0.59, y_C = 1.431$ [m]) is presented in Fig. 7. The condensed FA model (16) with 960 DOF ($m_s = 150, m_{CT} = 20, m_{LS} = 40$) was used for numerical integration of equations (17).

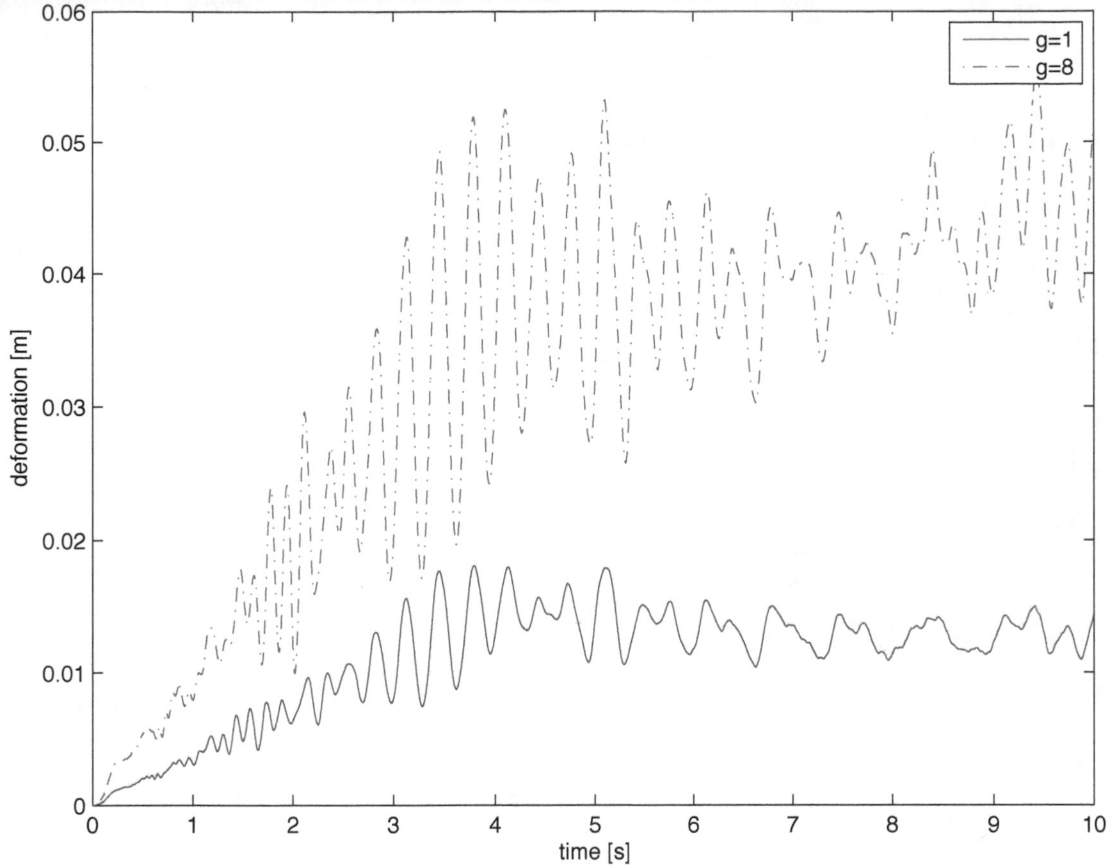

Fig. 7. Lateral seismic deformations of fuel rod

The components of the load-bearing skeleton vector $\boldsymbol{q}_{LS}$ defined in (5) are absolute lateral displacements $\xi_{AP,g}^{(s)}$, $\eta_{AP,g}^{(s)}$ of centre of gravity, torsional angles $\varphi_{AP,g}^{(s)}$ and bending angles $\vartheta_{AP,g}^{(s)}$, $\psi_{AP,g}^{(s)}$ of the angle pieces AP_s, $s = 1, \ldots, 6$ cross-section [3] on the level of spacer grid g (see Fig. 1 and Fig. 6). Corresponding lateral displacements $\overline{\xi}_{AP,g}^{(s)}$ in radial direction of the non-deformed angle pieces can be expressed similar as for fuel rods by means of lower supporting plate motion defined in (3) as

$$\overline{\xi}_{AP,g}^{(s)} = [C^{(s)}, S^{(s)}, 0, -z_g S^{(s)}, z_g C^{(s)}, x_C S^{(s)} - y_C C^{(s)}]\boldsymbol{q}_L\,, \tag{21}$$

where new $C^{(s)} = \cos\left[\frac{\pi}{6} + (s-1)\frac{\pi}{3}\right]$, $S^{(s)} = \sin\left[\frac{\pi}{6} + (s-1)\frac{\pi}{3}\right]$. The other quantity sense is same as in (19). The angle pieces lateral deformations in radial direction on the level of spacer grid g are

$$d_{AP,g}^{(s)} = |\xi_{AP,g}^{(s)} - \overline{\xi}_{AP,g}^{(s)}|, \quad s = 1, \ldots; \quad g = 1, \ldots, 8\,. \tag{22}$$

As an illustration, the time behaviour of lateral deformations of the chosen angle piece $s = 3$ on the level of spacer grids $g = 1, 8$ of the identical FA as in case of chosen fuel rod is presented in Fig. 8.

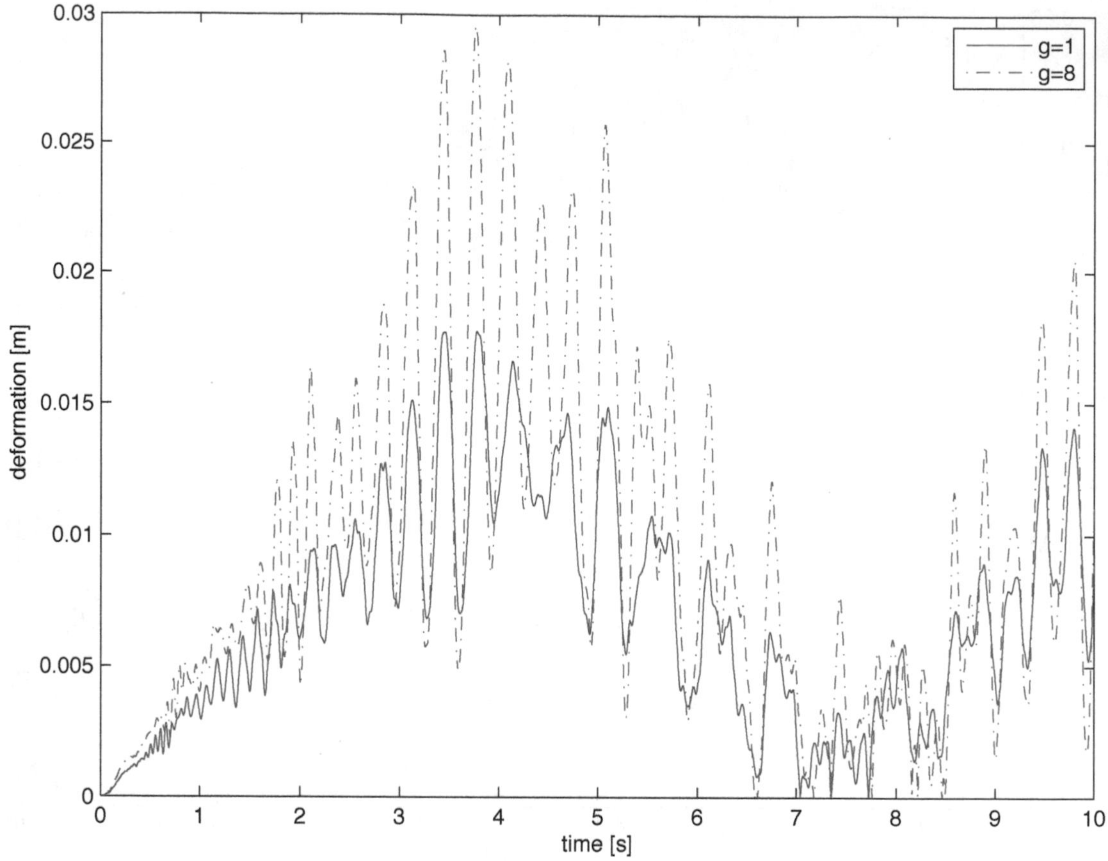

Fig. 8. Lateral seismic deformations of angle piece

Maximum lateral seismic deformations of fuel rods and angle pieces, in consequence of linking of these component with lower supporting plate, are on the highest level of spacer grids ($g = 8$). Outer radial deformations of the fuel assembly are determined by radial deformations of the angle pieces. The same condensed mathematical model was applied.

On the basis of lateral deformations on the level of all spacer grids we can relatively easily calculate the maximum stress of fuel assembly components excited by seismic events. The software developed in MATLAB code according to presented method makes possible to study an influence of the fuel assembly and reactor design parameters on seismic deformations of fuel assembly components. The publications dealing with seismic response of fuel assembly components have not been seen yet. The strength criterions of the fuel rods in a general form are presented in [7]. There are many procedures that can be used in the seismic engineering with the aim to mitigate the earthquake impacts [6]. Their description is outside a framework of this paper. The same condensed mathematical model was applied in this work.

5. Conclusion

The described method based on mathematical modelling and computer simulation of vibrations in time domain enables to investigate the seismic deformations of all nuclear fuel assembly components. The fuel assembly seismic vibrations are caused by spatial motion of the two horizontal supporting plates an the reactor core transformed into displacements and accelerations of the kinematically excited nodes of the fuel assembly components — fuel rods, guide thimbles, centre tube and skeleton angle pieces. The seismic motion of the supporting plates is investigated by numerical integration method in the previous analysis stage at the global reactor model whereof fuel assemblies are replaced by one dimensional continuums of beam type. Seismic excitation is described by synthetic accelerograms of the reactor hall translation motion on the level of the reactor pressure vessel seating.

The fuel assembly mathematical model has, in consequence of great number of fuel rods, too large DOF number for calculation of seismic response. Therefore, it is compiled fuel assembly condensed model based on reduction of the subsystems eigenvectors conducive to seismic response by modal synthesis method.

The developed software in MATLAB is conceived in such a way that enables to choose an arbitrary fuel assembly component — fuel rod, guide thimble, centre tube or angle piece of load-bearing skeleton — for calculation its deformation on the level of spacer grids. The presented method was applied for the Russian TVSA-T fuel assembly in the WWER 1000/320 type reactor in NPP Temelín.

Acknowledgements

This work was supported by the European Regional Development Fund (ERDF), project "NTIS", European Centre of Excellence, CZ.1.05/1.1.00/02.0090 within the research project "Fuel cycle of NPP" of the NRI Řež plc.

References

[1] Betbeder-Matibet, J., Seismic engineering. John Wiley and Sons Inc., London, 2008.

[2] Hlaváč, Z., Zeman, V., The seismic response affection of the nuclear reactor WWER 1000 by nuclear fuel assemblies, Engineering Mechanics 17 (3/4) (2010) 147–160.

[3] Hlaváč, Z., Zeman, V., Vibration of nuclear fuel assemblies. LAP LAMBERT Academic Publishing, Saarbrücken, 2013.

[4] Kubr, T., Methods of seismic resistance assessment of mechanical systems. Doctoral Thesis, University of West Bohemia, 2004. (in Czech)

[5] Manolis, G. D., Kiliopoulos, P. K., Stochastic structural dynamics in earthquake engineering. WIT Press, Southampton, Boston, 2001.

[6] Procházková, D., Demjančuková, K., Earthquakes, hazards and principles for trade-off with risks, University of West Bohemia, 2012.

[7] Sýkora, M., Reactor TVSA-T fuel assembly insertion, Part 4, Research Report PpBZ1, 2, ČEZ ETE, 2009. (in Czech)

[8] Zeman, V., Hlaváč, Z., Methodology and software creation for load calculation of WWER 1000 type reactor caused by seismic excitation. Research Report No. 52120-03-08, University of West Bohemia, 2008. (in Czech)

[9] Zeman, V., Hlaváč, Z., Modelling of WWER 1000 type reactor vibration by means of decomposition method. Proceedings of the 12th International Conference Engineering Mechanics 2006, Svratka, Institute of Theoretical and Applied Mechanics AS CR, 2006. (in Czech)

[10] Zeman, V., Hlaváč, Z., Vibration of the package of rods linked by spacer grids, Vibration Problems ICOVP 2011, Springer (2011), pp. 227–233.

[11] Zeman, V., Hlaváč, Z., Modelling and modal properties of the nuclear fuel assembly, Applied and Computational Mechanics 5 (2) (2011) 253–266.

[12] Zeman, V., Hlaváč, Z., Dynamic response of nuclear fuel assembly excited by pressure pulsations, Applied and Computational Mechanics 6 (2) (2012) 219–230.

4

A numerical study of planar discharge motion

F. Benkhaldoun[a], J. Fořt[b], K. Hassouni[c], J. Karel[a,b], D. Trdlička[b,*]

[a]*LAGA, University Paris 13, 99 Av. J. B. Clement, 93430 Villetaneuse, France*

[b]*Faculty of Mechanical Engineering, CTU in Prague, Karlovo namesti 13, 121 35 Prague, Czech Republic*

[c]*LSPM, University Paris 13, 99 Av. J. B. Clement, 93430 Villetaneuse, France*

Abstract

Presented paper describes a numerical study of discharge plasma motion. This non-stationary phenomenon with steep gradients and sharp peaks in unknowns is described as a coupled problem of convection-diffusion equation with source term for electron, ion densities and Poisson's equation for electric potential. The numerical method is 2nd order of accuracy in space and time and it uses dynamical adaptation of unstructured triangular mesh. Results of numerical studies included size of computational domain, type of boundary conditions and numerical convergence test are presented.

Keywords: numerical study, cold plasma, refinement, boundary conditions

1. Introduction

We deal with a numerical simulation of planar discharge motion. This phenomenon is also denoted as cold plasma motion, or avalanche of electrons or in 3D version as, e.g., streamer. The streamer motion simulation is subject of many scientific works, which consider either axi-symmetric or fully 3D configuration, e.g., [1–5,7]. Nevertheless, there is still a lot of challenging problems like, e.g., simulation of streamer in complicated geometries, streamer branching. This phenomenon can be used in different technical applications. In our case, we simulated the 2D discharge between two planar electrodes initializing by a narrow Gaussian pulse of charged particles of seven orders magnitude. The developed method will be next also extended into 3D.

2. Governing equations

We consider the so called minimal plasma model (see, e.g., [5] and [7]), which consists of two transport equations for electron and ion densities ($\mathbf{n_e}$, $\mathbf{n_i}$) coupled with Poisson's equation for the electric field potential V

$$\frac{\partial \mathbf{n_e}}{\partial t} + \operatorname{div}\left(\mathbf{n_e}\vec{v_e} - D_e\vec{\nabla}\mathbf{n_e}\right) = S_e, \tag{1}$$

$$\frac{\partial \mathbf{n_i}}{\partial t} = S_i^+, \tag{2}$$

$$\Delta V = -\frac{e}{\epsilon}\frac{(\mathbf{n_i} - \mathbf{n_e})}{10^4}, \tag{3}$$

*Corresponding author. e mail: david.trdlicka@fs.cvut.cz.

where $\vec{v_e}$ is the electron drift velocity, S_e is the electron source term, the positive ion source term is simplified as $S_i^+ = S_e$, the diffusive coefficient D_e, the unit electron charge e, the dielectric constant ϵ and the intensity of electric field is computed by $\vec{E} = -\vec{\nabla} V$.

The source term and diffusive coefficient are computed by the following formulas (see, e.g., [6]):

$$S_e = \frac{\alpha}{N} \cdot \|\vec{v}_e\| \cdot \mathbf{n_e}, \tag{4}$$

$$\text{if } \frac{\|\vec{E}\|}{N} > 1.5 \cdot 10^{-15} \qquad \frac{\alpha}{N} = 2 \cdot 10^{-16} \cdot \exp\left(\frac{-7.248 \cdot 10^{-15}}{\|\vec{E}\|/N}\right), \tag{5}$$

$$\text{else} \qquad \frac{\alpha}{N} = 6.619 \cdot 10^{-17} \cdot \exp\left(\frac{-5.593 \cdot 10^{-15}}{\|\vec{E}\|/N}\right), \tag{6}$$

$$D_e = \left[0.3341 \cdot 10^9 \cdot \left(\frac{\|\vec{E}\|}{N}\right)^{0.54069}\right] \cdot \frac{\|\vec{v}_e\|}{\|\vec{E}\|}, \tag{7}$$

$$N = 2.5 \cdot 10^{19}\,\text{cm}^{-3}, \tag{8}$$

where α is the ionization coefficient and N is the neutral gas density.

3. Numerical method

The sufficiently accurate computation inside moving narrow region of discharge head (characterized by very steep gradients of unknowns) is the crucial point of numerical simulation. Therefore, our numerical method is based on the unstructured dynamically adapted grid. We use the following algorithm for computation of unknowns in the new time (t^{k+1}) level:

1. Computation of $\mathbf{n_e}^{k+1}, \mathbf{n_i}^{k+1}$ from evolution equations (1), (2).
2. Computation of V with results of step 1 on the right hand side of Poisson's equation (3).
3. Computation of $\vec{E}, \vec{v_e}, D_e$ and S_e from equations (4)–(8).

We approximate the equation for the electron density (1) by the finite volume method. It leads to

$$\mathbf{n_{e_i}}^{k+1} = \mathbf{n_{e_i}}^{k} - \frac{\Delta t}{\mu_i} Rez_{e_i}^k, \tag{9}$$

$$Rez_{e_i}^k = \sum_{j=1}^{m} \left(\mathbf{n_{e_{ij}}}^k \vec{v}_{e_{ij}} \vec{n_{ij}} ds_{ij} - D_{e_{ij}} \left((\mathbf{n_{e_{ij}}})_x, (\mathbf{n_{e_{ij}}})_y\right) \vec{n_{ij}} ds_{ij}\right) + S_{e_i}, \tag{10}$$

where Δt is a time step, m is the number of faces of the cell i, $\vec{n}_{ij}$ is a unit normal vector of the face ij (between volumes i and j) and ds_{ij} is its length. Other variables denoted by the subscripts ij represent variables on the face ij. The convective flux is computed by the upwind scheme

$$\begin{aligned} \mathbf{n_{e_{ij}}} &= \mathbf{n_{e_i}} \ \text{ if } (\vec{v}_{e_{ij}} \cdot \vec{n_{ij}}) \geq 0, \\ \mathbf{n_{e_{ij}}} &= \mathbf{n_{e_j}} \ \text{ in other case,} \end{aligned} \tag{11}$$

the normal vector $\vec{n_{ij}}$ is oriented from the cell i to the cell j. The gradient components $(\mathbf{n_{e_{ij}}})_x$, $(\mathbf{n_{e_{ij}}})_y$ in the center of edge (face) ij are computed on a dual cell with a central diamond scheme. Unknowns in the vertices of the edge are computed by the least square method.

Discretization of the equation for the ion density is based also on a forward difference for the time derivative

$$\mathbf{n_{i_i}}^{n+1} = \mathbf{n}_{ii}^{n} + \Delta t\, S_i^{+}. \tag{12}$$

The Poisson's equation for the electric potential is discretized by a central finite volume approximation. We use a similar finite volume approximation as for diffusive terms in the equation for electron density. It leads to a system of linear equations

$$A\vec{V}^{n+1} = \vec{b}^{n+1}, \tag{13}$$

where A is a matrix of coefficients, $\vec{V}$ is a vector of unknowns and $\vec{b}$ is a vector of the right hand side. An i-th row in the matrix A corresponds to the cell i.

The system of equations (13) is solved directly by LU decomposition in an implementation for sparse matrices (with Intel MKL library). Numerical scheme used for evolution equation is first order of accuracy in time (explicit Euler's scheme) and space (approximation of convective term). The scheme for the Poisson's equation is second order of accuracy. Next the scheme for evolution equations has been extended into the theoretically second order of accuracy. The three stage Runge-Kutta scheme has been used for approximation in time

$$\begin{aligned}
\mathbf{n_e}^{(0)} &= \mathbf{n_e}^{(k)}, \\
\mathbf{n_e}^{(1)} &= \mathbf{n_e}^{(0)} - \frac{1}{2}\Delta t\; Rez_e^{(0)}, \\
\mathbf{n_e}^{(2)} &= \mathbf{n_e}^{(0)} - \frac{1}{2}\Delta t\; Rez_e^{(1)}, \\
\mathbf{n_e}^{(k+1)} &= \mathbf{n_e}^{(2)}
\end{aligned} \tag{14}$$

and second order approximation of convective term in (10) has been achieved by linear reconstruction and Barth-Jesperson limiter.

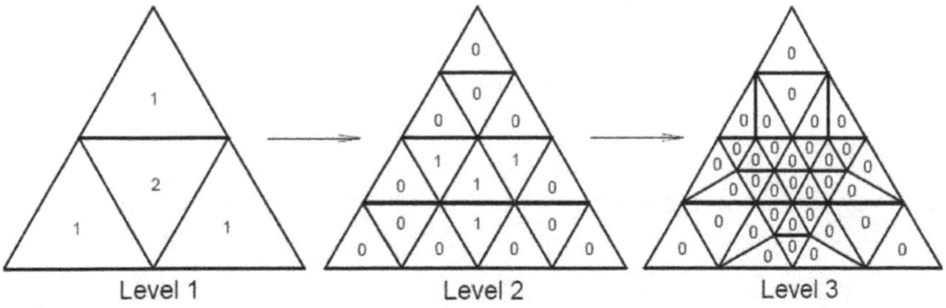

Fig. 1. Mechanism of multi-level mesh refinement

The dynamic mesh adaptation is used to capture sharp peaks and steep gradients of unknowns. The triangle of (starting) reference grid can be divided to 4^r triangles according to a refinement criterion (where r is the refinement level), see Fig. 1. The criterion is considered as a grid function, which has some non-zero value in each triangle of reference grid. These values are firstly smoothed by computation of several time steps of the simple diffusion equation. This is approximated by the explicit scheme and criterion function is used as the initial condition.

The values are then scaled into interval $\langle 0, 1\rangle$. The interval $\langle 0, 1\rangle$ is split into sub-intervals with prescribed level of adaptation. When the level of adaptation is evaluated for each triangle, the grid conformity procedure is applied. The new adapted grid is then generated and values of unknowns are conservatively interpolated from the previous adapted grid into the new one. The adapted grid is fixed for a prescribed number of time steps and then the grid adaptation procedure is repeated. The criterion consists of two parameters: the electron density gradient and the source term magnitude.

4. Problem formulation

We simulated planar discharge motion in homogenous electric field described by equations (1)–(7). This field is generated in a gaseous gap between two planar electrodes with high electric potential. We simulated this phenomenon in rectangular domain $x \times y = \langle 0; 1\rangle \times \langle 0; 0.5\rangle$ [cm] (standard calculation). Left ($x = 0$ cm) and right ($x = 1$ cm) boundaries are the plane anode and cathode, respectively. Upper ($y = 0.5$ cm) and lower ($y = 0$ cm) boundaries are artificial. They made the computational domain finite and sufficiently small.

We used the common initial distribution of charged particles for considered shape of electrodes, which represents a small electrically neutral region with high density of particles and proper background density of particles, which substitute neglected effect of photo-ionization. The corresponding electric potential in initial time is a linear function of the x coordinate

$$\begin{aligned} \mathbf{n_e}(x, y, 0) &= 10^{16} \exp\left(\frac{-(x - x_0)^2 + (y - y_0)^2}{0.01^2}\right) + 10^9 \ [\text{cm}^{-3}], \\ \mathbf{n_i}(x, y, 0) &= \mathbf{n_e}(x, y, 0) \ [\text{cm}^{-3}], \\ V(x, y, 0) &= 25\,000\,(1 - x) \ [\text{V}]. \end{aligned} \tag{15}$$

Initial Gaussian pulse is placed in $[x_0; y_0] = [0.2; 0.25]$ cm. Neumann's boundary condition is prescribed on the anode and cathode for electron and ion densities. Electric potential is 0 V on the cathode and 25 000 V on the anode

$$\begin{aligned} \frac{\partial \mathbf{n_e}}{\partial \vec{n}}(0, y, t) &= 0, \\ \frac{\partial \mathbf{n_e}}{\partial \vec{n}}(1, y, t) &= 0, \\ V(0, y, t) &= 25\,000 \ [\text{V}], \\ V(1, y, t) &= 0 \ [\text{V}]. \end{aligned} \tag{16}$$

We used either Newton's homogeneous or periodical boundary condition for upper and lower boundary.

5. Results

A typical example of discharge motion is shown in Fig. 2. Electron density is plotted in Fig. 2a at time $t = 5.25 \cdot 10^{-8}$ s. We can see an initial pulse in position of $[x_0; y_0]$ and avalanche of electrons parallel with the x coordinate, discharge head is appropriately in the position $[0.4; 0.25]$ cm. We can observe charge with high magnitude — red color in Fig. 2b. This Figure shows isolines of the net-charge density

$$\rho = -\frac{e}{\epsilon}\frac{(\mathbf{n_i} - \mathbf{n_e})}{10^4}, \tag{17}$$

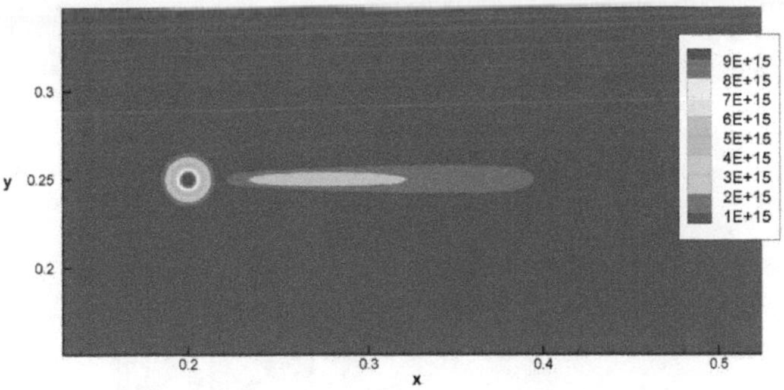

(a) Electron density

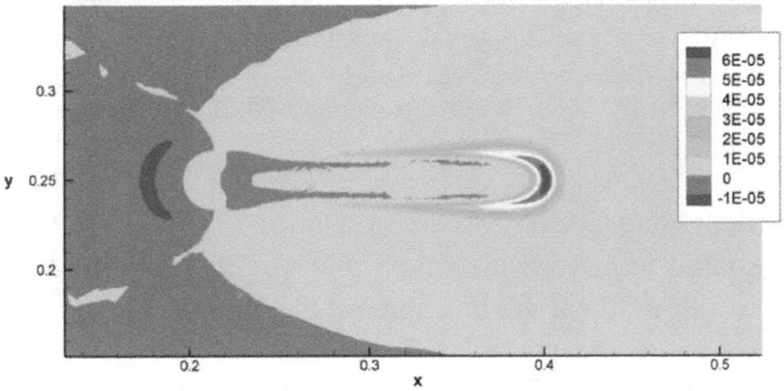

(b) Net-charge density

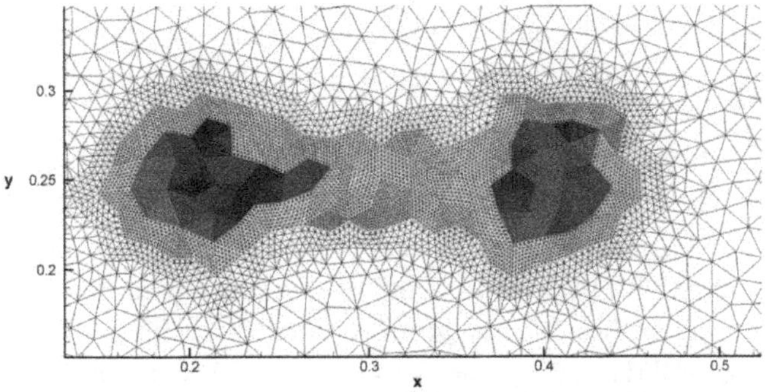

(c) Adapted grid

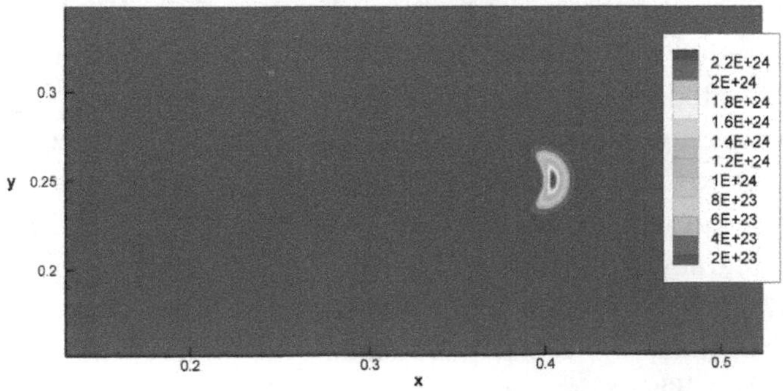

(d) Source term

Fig. 2. Results in time $t = 5.25 \cdot 10^{-8}$ s

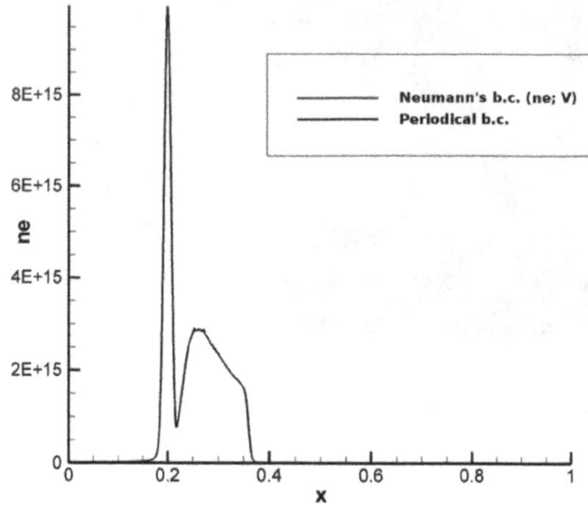

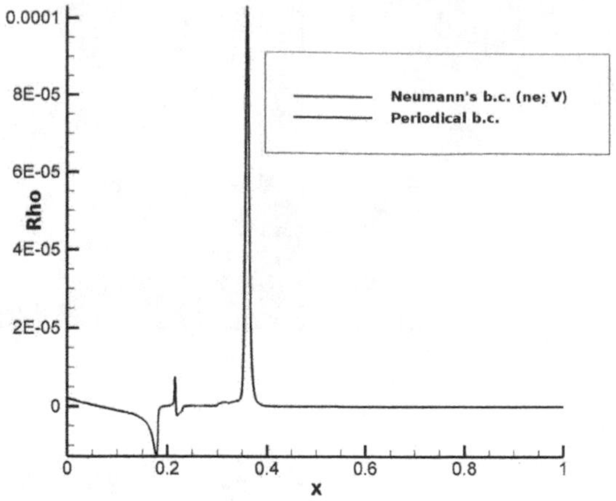

Fig. 3. Electron density (left), net-charge density (right) at time $t = 5.25 \cdot 10^{-8}$ s

corresponding to the right hand side of Poisson's equation (3). This relates to the narrow region with high values of source term (S_e) displayed in Fig. 2d. Therefore, we need grid refinement corresponding to the gradient $\mathbf{n_e}$ and magnitude of S_e. Our refined grid is plotted in Fig. 2c.

The comparison of the results of computations with Neumann's and periodical boundary conditions for the upper and the lower boundaries is shown in Fig. 3. There is electron density (left) and net-charge density (right) along line $y = 0.25$ cm. The figures confirm that results are almost identical and do not depend on type of boundary condition prescribed on the upper and lower boundary of the computational domain. Therefore, we use further only the Newton's boundary condition.

Figs. 4 and 5 illustrate dependence of the results on the width of the computational domain. Electron density computed by the first (left) and the second (right) order of accuracy along line $y = 0.25$ cm is shown in Fig. 4, net-charge density is in Fig. 5. These figures document increase of discharge head velocity with increasing width of the computational domain. Discharge head is shown (for electron density) as nearly vertical line which has the farthest position on the x coordinate, or in case of the net-charge density, as peek. Results computed on a domain with width 1.5 cm and higher are almost the same and we can take them as independent on the domain width.

The results of numerical convergence tests are shown in Figs. 6 and 7. We plot the electron density and the distribution of net-charge density along line $y = 0.25$ cm in the same time calculated on series of adapted grids with increasing maximal level of adaptation. We can observe the significant influence on grid density as well as much better convergence of second order approximation of convective term. Computation with the first order of accuracy shows decrease in velocity of discharge head with increasing number of adaptation levels. But computation with the second order of accuracy shows increase in velocity of discharge head with increasing number of adaptation levels. Solutions obtained from both orders of accuracy are moving towards each other. Results calculated by second order scheme are almost without differences for adaptation level 4 and higher. Therefore, we take 4th level of adaptation as satisfying compromise between accuracy and CPU time.

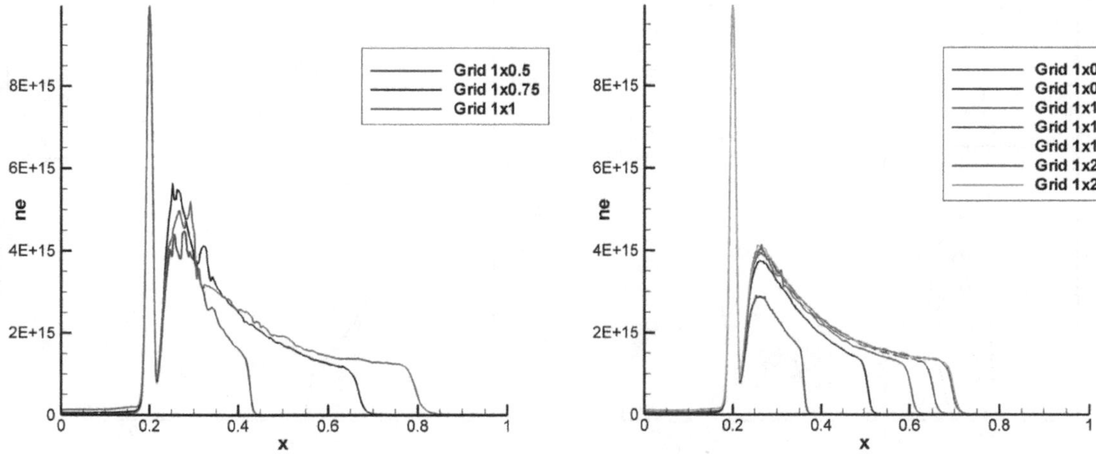

Fig. 4. Electron density at time $t = 5.25 \cdot 10^{-8}$ s and its dependence on domain width, 1st order (left) and 2nd order (right)

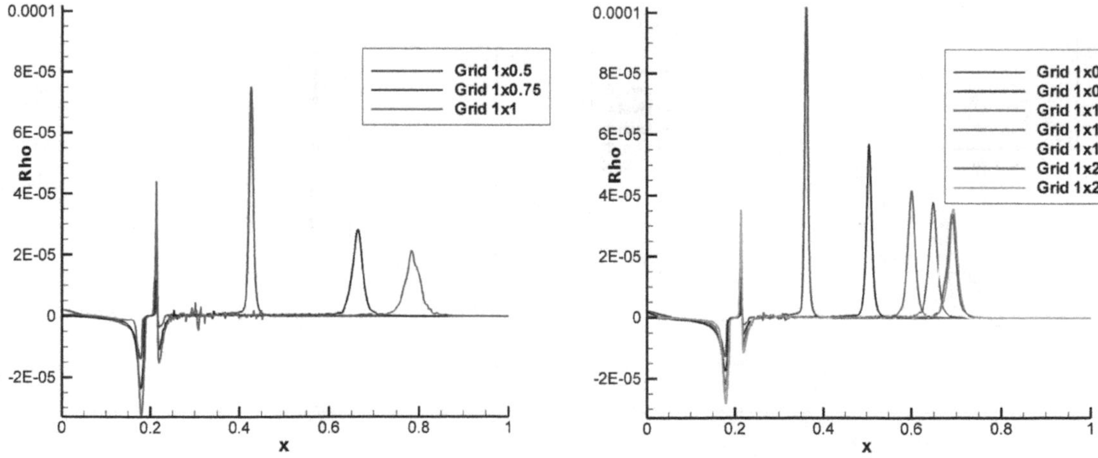

Fig. 5. Net-charge density at time $t = 5.25 \cdot 10^{-8}$ s and its dependence on domain width, 1st order (left) and 2nd order (right)

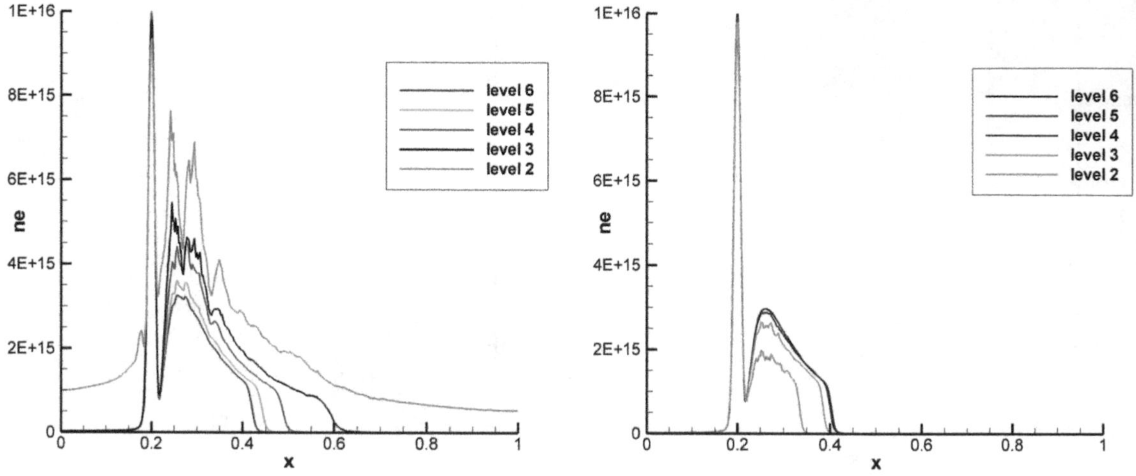

Fig. 6. Electron density at time $t = 5.25 \cdot 10^{-8}$ s and its dependence on refinement level, 1st order (left) and 2nd order (right)

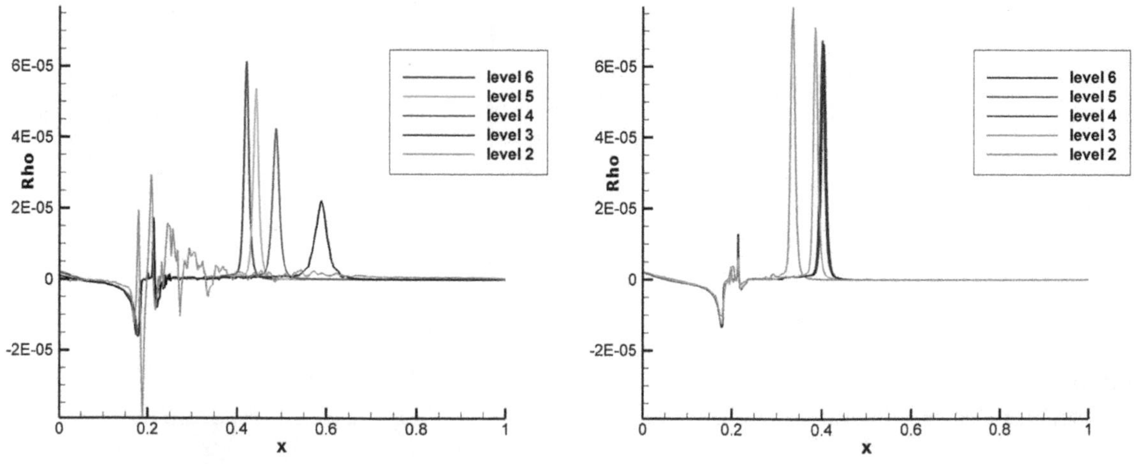

Fig. 7. Net-charge density at time $t = 5.25 \cdot 10^{-8}$ s and its dependence on refinement level, 1st order (left) and 2nd order (right)

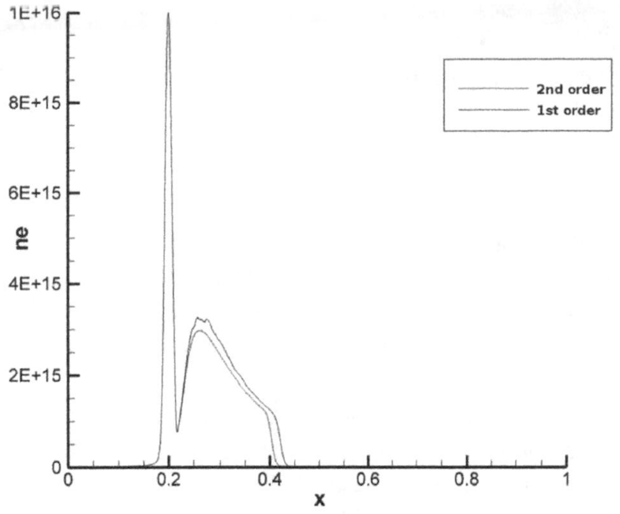

Fig. 8. Electron density for the 6th level of refinement, time $t = 5.25 \cdot 10^{-8}$ s, 1st and 2nd order

The proximity of the results for 6th level of refinement computed by 1st and 2nd order of accuracy illustrates Fig. 8. We can see only small difference in magnitude of the electron density and in the position of discharge head.

Table 1. The dependence of CPU time on adaptation levels

Level	Number of cells	Number of time steps	CPU time
2	5 024	2 395	0.009 9
3	10 592	10 607	0.089 6
4	41 082	24 552	1.000 0
5	62 392	56 629	3.840 0
6	186 802	153 090	31.800 0

Table 1 shows computational consumption in dependence on the maximal adaptation level. The table also contains numbers of cells and numbers of time steps for appropriate adaptation levels. Values of CPU time are scaled to the one with regard to the 4th level of adaptation, which is used as the reference value.

6. Conclusion

Presented results confirm ability of the numerical method and the grid adaptation strategy to capture the main features of planar discharge motion. They also show that for sufficiently wide (1.5 or 2 times distance of electrodes) computational domain can be prescribed relatively simple boundary conditions (homogenous Neumann or periodical condition) along artificial side boundaries without influence on the discharge propagation. Numerical convergence tests have been presented with conclusions that behavior of the first and the second order scheme differs with increased grid refinement level. Four levels of adaptation seems to be an optimal compromise between accuracy and efficiency of computation.

Acknowledgements

This work was supported by the Grant Agency of the Czech Technical University in Prague, grant No. SGS13/174/OHK2/3T/12.

References

[1] Bartoš, P., Hrach, R., Jelínek, P., Multidimensional fluid-particle modelling technique in low-temperature argon plasma at low pressure, Vacuum 82 (2) (2007) 220–223.

[2] Benkhaldoun, F., Fořt, J., Hassouni, K., Karel, J., Simulation of planar ionization wave front propagation on an unstructured adaptive grid, Journal of Computational and Applied Mathematics 236 (2012) 4 623–4 634.

[3] Bourdon, A., Bessieres, D., Paillol, J., Michau, A., Hassouni, K., Marode, E., Ségur, P., Influence of numerical schemes on positive streamer propagation, Proceedings of the 27th International Conference on Phenomena in Ionized Gases (ICPIG), Eindhoven, Netherlands, 2005, pp. 17 422.

[4] Bourdon, A., Pasko, V. P., Liu, N. Y., Célestin, S., Ségur, P., Marode, E., Efficient models for photoionization produced by non-thermal gas discharges in air based on radiative transfer and the Helmholtz equations, Plasma Sources Science and Technology 16 (3) (2007) 656–678.

[5] Montijn, C., Hundsdorfer, W., Ebert, U., An adaptive grid refinement strategy for the simulation of negative streamers, Journal of Computational Physics 219 (2) (2006) 801–835.

[6] Morrow, R., Lowke, J. J., Streamer propagation in air, Journal of Physics D: Applied Physics 30 (4) (1997) 614–627.

[7] Nikandrov, D. S., Arslanbekov, R. R., Kolobov, V. I., Streamer simulations with dynamically adaptive cartesian mesh, IEEE Transactions on Plasma Science 36 (4) (2008) 932–933.

Three-scale model of single bone osteon modelled as double-porous fluid saturated body: Study of influence of micro/meso-structure

J. Turjanicová[a,*], E. Rohan[a], S. Naili[b]

[a] *Faculty of Applied Sciences, University of West Bohemia, Univerzitní 22, 306 14 Plzeň, Czech Republic*
[b] *Laboratoir modelisation et simulation multi échelle, Université Paris-est, 61 avenue du Général de Gaulle, Créteil cedex, France*

Abstract

This paper deals with the multiscale description of a single osteon of cortical bones. The cortical bone tissue is modeled as a double-porous medium decomposed into the solid matrix and the fluid saturated canals. The resulting homogenized model describes deformation of such medium in response to a static loading by external forces and to an injection of slightly compressible fluid. Three numerical examples are presented, showing the influence of selected lower-scales geometrical features on the macroscopic body behavior.

Keywords: homogenization, poroelasticity, osteon, cortical bone

---

1. Introduction

This paper is devoted to the macroscopic behavior of the poroelastic materials with the double porosity. Such materials consist of two very distinct porous systems so that their interaction has a strong influence on their macroscopic behavior. The two interacting systems featured by strongly different pore sizes are arranged hierarchically, one is embedded in the other. This characterizes materials which are abundant in nature, such as rocks, or biological tissues, as well as engineered materials comprising certain types of foams. The model reported in this paper can be applied to describe the cortical bone tissue, where the pores filled with bone fluid can be found on multiple scales. The material properties of the bone can serve as a basis for the development of new biomaterials or for a better understanding of the processes in the bone tissue such as remodeling.

The concept of double porosity was first presented in context of geomechanics in work [1], which studied flows in cracked rocks. This work considered only the fluid exchange between two porous systems. The model was simplified few years later in work [10], where authors considered situation such that the flow in the porous matrix can be neglected. Under the same assumption, in the recent work [6] we presented a model of the double-porous medium derived using unfolding homogenization method applied at two levels of the heterogeneity coresspond-ing to a hierarchical arrangement of the bone sructure. This model can be used to compute the effective poroelasticity coefficients for a given geometry and topology of the micro- and mesoscopic levels.

The aim of this paper is to perform a computational study which explains how the micro- and mesostructure influence the macroscopic body response. We focus on the poroelastic properties

*Corresponding author. e-mail: turjani@kme.zcu.cz.

which are independent of the pore fluid flow. The solid-fluid interaction is studied under the static loading whereby any fluid flow is excluded. It should be emphasized that the poroelastic material constants derived and computed under such thermodynamic steady state conditions are valid also in quasistatic situations featured by slow flows; this has been justified, e.g., in [3,8].

The paper is composed from five sections. In the Section 2 the definition and the geometry description of porosity levels is given. The following Section 3 deals with the upscaling procedure of equations describing poroelasticity. The Section 4 introduces the set of numerical simulations with their results discussed. Finally, a conclusion is drawn in Section 5.

2. Geometrical configuration

The cortical bone is a hierarchical system with a complex structure on different scale levels. From the macroscopic point of view, the cortical bone tissue is made of a system of approximately cylindrical mineralized structures called osteons, each with an exterior radius of roughly 100–150 μm, [11]. These structures are hollow; the center of each osteon is perforated by the so-called Haversian canal, which contains blood vessels, nerves, and is saturated by the bone fluid. In addition to these canals, other smaller pathways exist within the bone tissue. The aim of this section is to describe this multiscale arrangement of the bone tissue and, in particular, to explain geometry of the structure.

The walls of the Haversian canal are covered with bone cells and behind this bone cell layer, there are entrances to multiple small tunnels [11]. These tunnels with the diameters of a few hundreds of nanometers are called canaliculi and create the network connecting Haversian canal and the small ellipsoidal pores. These are called lacunae with dimension about a few tens of micrometers and each of them contains one osteocyte, the bone cell responsible for the tissue growth processes. Canaliculi and lacunae create one system of a mutually connected network filled with bone fluid.

In literature, [2, 4, 11], there are commonly distinguished three porosity levels with the biggest importance in the bone structure. The uppermost level is the so-called vascular porosity level involving the Haversian canals, the mean one, the lacunar-canalicular porosity, is associated with lacunae and canaliculi, and lowest one is the collagen-apatite porosity associated with the space between collagen and the crystallites of the mineral apatite.

The present work is focused on describing the mechanical properties of the single bone osteon which is relevant to the macroscopic scale. Therefore, we shall neglect the vascular porosity (or osteonal porosity). Considering the difference in dimensions of lacunae and canaliculi, the lacuno-canalicular porosity is split into two levels; the lacunar porosity further referred to as a mesoscopic porosity level or β-level and canalicular porosity referred to as microscopic porosity or α-level. Porosity of the collagen-apatite matrix is neglected in this work, being replaced by a solid elastic material.

In the numerical examples reported in Section 4 we shall represent the hierarchical structure of the osteon using a simple geometrical models which, however, enable us to introduce geometrical parameters with a clear interpretation and to assess their influence on the effective material properties.

3. Homogenized model

In this section we summarize the homogenization results witch was obtained in [6] by unfolding homogenization method of the poroelasticity problem in a dual porous medium saturated by a slightly compressible fluid. In throughout the text, we use labelling by two superscripts: α for microscopic (or α-) level and β for mesoscopic (β-) level. First, we recall a few general

Table 1. Basic nomenclature

Superscript	Describtion	Superscript	Describtion
$\square^{\varepsilon}$	quantities dependent on the scale-parameter	$\square^{m}$	quantities defined on matrix (solid)
$\square^{\alpha}$	quantities relevant to the microscopic level	$\square^{c}$	quantities defined on canals (fluid)
$\square^{\beta}$	quantities relevant to the mesoscopic level		
Quantity	**Describtion**	**Quantity**	**Describtion**
$\mathbf{u}$	displacement	$\mathbf{v}$	admissible displacement
$\tilde{\mathbf{u}}$	matrix-to-canal extension of $\mathbf{u}$	$\bar{p}$	fluid pressure
$\mathbf{f}$	volume forces	$\mathbf{g}$	traction forces
γ	fluid compressibility	J	fluid volume (injected)
ω^{ij}	characteristic responses relevant to displacement	ϕ	porosity
ω^{P}	characteristic responses relevant to pressure	$\mathbb{D} = (D_{ijkl})$	elasticity tensor
$\mathbf{\Pi}^{ij} = (\Pi_k^{ij})$	transformation vectors, $k = 1, 2, 3$	$\mathbf{n}$	outer unit normal vector
$\mathbb{A} = (A_{ijkl})$	effective elasticity tensor	$\mathbf{B} = (B_{ij})$	effective tensor of Biot coefficients
M	effective Biot modulus of compressibility	E	Young's modulus
G	shear modulus	ν	Poisson's ratio

principles of homogenization and then we focus on the microscopic scale problem. It is worth noting, that for the better orientation throught text, the used notation is summarized in Table 1.

3.1. Scale parameter

We consider a poroelastic medium whose material properties vary periodically with position; the period length is proportional to a small parameter ε which is the so called scale-parameter defined as the ratio between the characteristic lengths of the micro- and macroscopic levels, thus, $\varepsilon = L_{micro}/L_{macro}$.

Through the text all quantities varying with this periodicity are denoted with superscript $^{\varepsilon}$. In our case we have three scales, micro-, meso- and macro-scale, thus it is more accurate to define scale-parameter as ratio of characteristic length of lower and upper scale.

3.2. Material periodicity

For simplicity we consider material with the periodic geometrical decomposition on both the micro- and mesoscale. The following general definition applies to both α and β levels and through the text all quantities belonging to those levels are labelled by superscripts $^{\alpha}$ and $^{\beta}$.

We consider an open bounded domain $\Omega \in \mathbb{R}^3$ with boundary $\partial\Omega$. Further we consider domain $Y = \prod_{i=1}^{3}(0, \hat{y}_i)$ which is called the representative periodic cell (RPC), since it generates the domain Ω as a periodic lattice. The material point positions are defined in the coordinate system $(0, x_1, x_2, x_3)$. The rescaled material point positions in the RPC are defined in the coordinate system $(\hat{0}, y_1, y_2, y_3)$, where $\hat{y}_i$ is the length of the i-th side of Y.

Application of this definition to the three scale model is illustrated in the Fig. 1. Thus, the domain Ω is generated (as periodic lattice) by using mesoscopic RPC Y^{β}. The matrix part

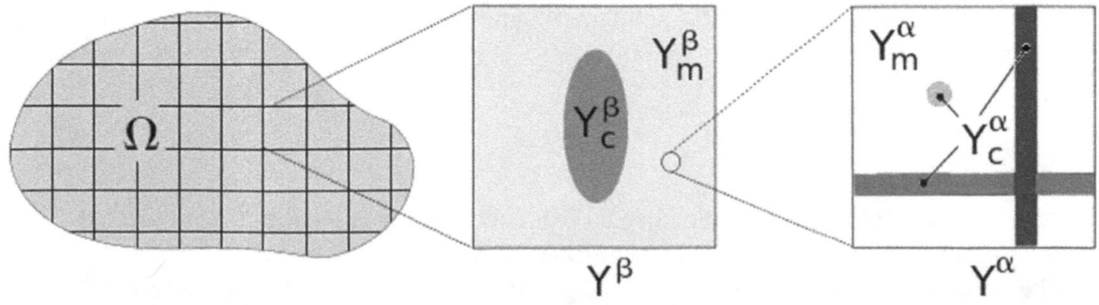

Fig. 1. Three scales of homogenization: Macroscopic body Ω, mesoscopic periodic cell Y^{β}, microscopic cell Y^{α}

within Y^β is also generated as periodic structure by using a microscopic RPC Y^α. Details are given below.

3.3. Microscopic problem

Let us consider $\Omega^\alpha \subset \mathbb{R}^3$, which represents the microscopic scale of the osteon. With respect to the material structure, the domain $\Omega^\alpha \subset \mathbb{R}^3$ is decomposed into two non-overlapping parts; the (solid) matrix $\Omega_m^{\alpha,\varepsilon}$, canals $\Omega_c^{\alpha,\varepsilon}$ and their interface $\Gamma^{\alpha,\varepsilon}$ are defined, as follows:

$$\Omega^\alpha = \Omega_m^{\alpha,\varepsilon} \cup \Omega_c^{\alpha,\varepsilon} \cup \Gamma^{\alpha,\varepsilon}, \qquad \Omega_c^{\alpha,\varepsilon} = \Omega^{\alpha,\varepsilon} \setminus \bar{\Omega}_m^{\alpha,\varepsilon}, \quad \Gamma^{\alpha,\varepsilon} = \bar{\Omega}_m^{\alpha,\varepsilon} \cap \bar{\Omega}_c^{\alpha,\varepsilon}\,. \tag{1}$$

The outer boundaries $\partial_{ext}\Omega_c^{\alpha,\varepsilon} \subset \partial\Omega^\alpha$ and $\partial_{ext}\Omega_m^{\alpha,\varepsilon} \subset \partial\Omega^\alpha$, are defined as follows:

$$\partial_{ext}\Omega_m^{\alpha,\varepsilon} = \partial\Omega_m^{\alpha,\varepsilon} \setminus \Gamma^{\alpha,\varepsilon} = \partial\Omega_m^{\alpha,\varepsilon} \cap \partial\Omega^\alpha, \qquad \partial_{ext}\Omega_c^{\alpha,\varepsilon} = \partial\Omega_c^{\alpha,\varepsilon} \setminus \Gamma^{\alpha,\varepsilon} = \partial\Omega_c^{\alpha,\varepsilon} \cap \partial\Omega^\alpha. \tag{2}$$

The outer unit normal vectors of the boundaries $\partial\Omega_m^{\alpha,\varepsilon}$ and $\partial\Omega_c^{\alpha,\varepsilon}$ are denoted by $\mathbf{n}^m$ and $\mathbf{n}^c$, respectively. In this section, the superscript $^\varepsilon$ refers to the scale-dependence, where $\varepsilon = L_\alpha / L_\beta$ describes the ration of the characteristic sizes of the micro- and mesoscopic levels.

In what follows, we will use the same notation as in [6]. By symbols ∇ and $\nabla\cdot$ the gradient and divergence operators are denoted, respectively. The scalar product is denoted by "$\cdot$" and the symbol ":" between tensors of any order denotes their double contraction. The superscript s in the gradient operator ∇^s denotes its symmetric part.

The linear poroelastic body characterized by its elasticity tensor D_{ijkl}^ε, occupying the domain $\Omega^\alpha \subset \mathbb{R}^3$, is loaded by volume forces $\mathbf{f}^{\alpha,\varepsilon}$ in $\Omega_m^{\alpha,\varepsilon}$ and surface traction forces $\mathbf{g}^{\alpha,\varepsilon}$ on $\partial_{ext}\Omega_m^{\alpha,\varepsilon}$.

Since the objective here is to characterize the effective poroelastic properties of the bone tissue, we restrict ourselves to the stationary problems only and neglect the fluid flow. Therefore, the fluid in the deforming pores is characterized by its compressibility γ only, whereas its viscosity does not interfere in the problem. Further, mass conservation has to be considered: The pore deformation resulting in a decrease of the porosity is compensated by an increased pressure acting on the interface Γ_{mc} and by an amount of the fluid leaking from $\Omega_c^{\alpha,\varepsilon}$ into the outer space through $\partial_{ext}\Omega_c^{\alpha,\varepsilon}$, cf. [6]. Using the tensorial notation, the linear poroelasticity problem is defined as follows

$$\begin{aligned} -\nabla(\mathbb{D}^{\alpha,\varepsilon}\nabla^S \mathbf{u}^{\alpha,\varepsilon}) &= \mathbf{f}^{\alpha,\varepsilon} && \text{in} \quad \Omega_m^{\alpha,\varepsilon}, \\ \mathbf{n}^m \cdot \mathbb{D}^{\alpha,\varepsilon}\nabla^S \mathbf{u}^{\alpha,\varepsilon} &= \mathbf{g}^{\alpha,\varepsilon} && \text{on} \quad \partial_{ext}\Omega_m^{\alpha,\varepsilon}, \\ \mathbf{n}^m \cdot \mathbb{D}^{\alpha,\varepsilon}\nabla^S \mathbf{u}^{\alpha,\varepsilon} &= -\bar{p}^{\alpha,\varepsilon} && \text{on} \quad \Gamma^{\alpha,\varepsilon}, \end{aligned} \tag{3}$$

where $\mathbf{u}^{\alpha,\varepsilon}$ is the displacement of the solid matrix and $\bar{p}^{\alpha,\varepsilon}$ is the fluid pressure. The announced mass conservation attains the following form

$$\int_{\partial\Omega_c^{\alpha,\varepsilon}} \tilde{\mathbf{u}}^{\alpha,\varepsilon} \cdot \mathbf{n}^c \, \mathrm{d}S_x + \gamma\bar{p}^{\alpha,\varepsilon}|\Omega_c^{\alpha,\varepsilon}| = -J^{\alpha,\varepsilon}, \tag{4}$$

where $-J^{\alpha,\varepsilon}$ denotes a fluid volume injected from outside through $\partial_{ext}\Omega_c^{\alpha,\varepsilon}$ into $\Omega_c^{\alpha,\varepsilon}$. By $\tilde{\mathbf{u}}$ we mean a matrix-to-canal differentiable extension of $\mathbf{u}$ from Ω_m^ε to whole Ω.

The solvability condition yields $\int_{\partial_{ext}\Omega_m^{\alpha,\varepsilon}} \mathbf{g}^{\alpha,\varepsilon} \, \mathrm{d}S_x + \int_{\Omega_m^{\alpha,\varepsilon}} \mathbf{f}^{\alpha,\varepsilon} \, \mathrm{d}V_x = \mathbf{0}$, where $\mathrm{d}S_x$ is the differential element of the surface and $\mathrm{d}V_x$ of the volume.

Weak formulation We introduce set of admissible displacements V, which due to assumed boundary conditions $(3)_{2,3}$ coincide with the space of virtual displacement $V = V_0 = \{\mathbf{v} \in \mathbf{H}^1(\Omega_m^{\alpha,\varepsilon}) | \mathbf{v} = \bar{\mathbf{u}} \text{ on } \partial_{ext}\Omega_m^{\alpha,\varepsilon}\}$, where $\mathbf{H}^1$ is the Sobolev space $[W^{1,2}(\Omega^\alpha)]^3$ of vector functions. The problem (3)–(4) can be rewritten in the weak form[1]: Find $(\boldsymbol{u}^{\alpha,\varepsilon}, \bar{p}^{\alpha,\varepsilon}) \in \mathbf{H}^1(\Omega_m^{\alpha,\varepsilon}) \times \mathbb{R}$ such that

$$\int_{\Omega_m^{\alpha,\varepsilon}} (\mathbf{D}^{\alpha,\varepsilon}\nabla^s \mathbf{u}^{\alpha,\varepsilon}) : \nabla^s \mathbf{v} \, \mathrm{dV}_x + \bar{p}^{\alpha,\varepsilon} \int_{\Gamma_m^{\alpha,\varepsilon}} \mathbf{n}^m \cdot \mathbf{v} \, \mathrm{dS}_x = \int_{\partial_{ext}\Omega_m^{\alpha,\varepsilon}} \mathbf{g}^{\alpha,\varepsilon} \cdot \mathbf{v} \, \mathrm{dS}_x + \int \mathbf{f}^{\alpha,\varepsilon} \cdot \mathbf{v} \, \mathrm{dV}_x,$$

$$\int_{\partial\Omega_c^{\alpha,\varepsilon}} \tilde{\mathbf{u}}^{\alpha,\varepsilon} \cdot \mathbf{n}^c \, \mathrm{d}S_x + \gamma \bar{p}^{\alpha,\varepsilon} |\Omega_c^{\alpha,\varepsilon}| = -J^{\alpha,\varepsilon} \qquad \forall \mathbf{v} \in V_0. \tag{5}$$

3.3.1. Representative periodic cell (RPC)

The domain Ω^α representing the periodic microstructure is generated by representative cell Y^α which adheres the decomposition introduced in (1), thus,

$$Y^\alpha = Y_m^\alpha \cup Y_c^\alpha \cup \Gamma_{mc}^\alpha, \qquad Y_c^\alpha = Y^\alpha \setminus \bar{Y}_m^\alpha, \qquad \Gamma_{mc}^\alpha = \bar{Y}_m^\alpha \cap \bar{Y}_c^\alpha. \tag{6}$$

We assume that the elastic tensor parameters $D_{ijkl}^{\alpha,\varepsilon}$ obey the standard positive definiteness and symmetries; in our study we consider an isotropic material[2] $D_{ijkl}^{\alpha,\varepsilon}(x) = \lambda^\varepsilon(x)\delta_{kl}\delta_{ij} + \mu^\varepsilon(x)(\delta_{ik}\delta_{jl} + \delta_{il}\delta_{jk})$ with the Lame coefficients $\lambda^\varepsilon, \mu^\varepsilon > 0$ which by virtue of the unfolding are defined with respect to material points in Y_m^α. In Section 4 we use a constant tensor in Y_m^α, thus $\lambda^\varepsilon, \mu^\varepsilon$ are constants.

We shall record the homogenized mesoscopic problem reported in [6], see [7] for the proofs. It is expressed using homogenized material coefficients defined through the solutions of auxiliary microscopic problem known as the corrector basis functions. In the following text we present all microscopic problems constituting the mesoscopic homogenized medium.

3.3.2. Microscopic problems and corrector basis functions

In what follows we shall use the following bilinear form:

$$a_y^d(\mathbf{w}, \mathbf{v}) = \oint_{Y_d} (\mathbf{D}\nabla_y^S \mathbf{w}) : \nabla_y^S \mathbf{v}, \tag{7}$$

where $\oint_{Y_d} = |Y|^{-1} \int_{Y_d}$ for $d = m, c$. We define the transformation vectors $\boldsymbol{\Pi}^{ij} = (\Pi_k^{ij})$, $i, j, k = 1, 2, 3$, whose components $\Pi_k^{ij} = y_j \delta_{ik}$ are constituted by coordinates y_j. In following text we denote the Sobolev space $[W^{1,2}(Y^\alpha)]^3$ of Y-periodic vector functions by $\mathbf{H}_\#^1(Y^\alpha)$.

We shall now define $\boldsymbol{\omega}^{ij}, \boldsymbol{\omega}^P$, the characteristic responses of the microstructure, which are solutions of the following microscopic problems: Find $\boldsymbol{\omega}^{ij} \in \mathbf{H}_\#^1(Y^\alpha)$ and $\boldsymbol{\omega}^P \in \mathbf{H}_\#^1(Y^\alpha)$ satisfying

$$a_y^m(\boldsymbol{\omega}^{ij} + \boldsymbol{\Pi}^{ij}, \mathbf{v}) = 0, \quad i, j = 1, \ldots, 3,$$
$$a_y^m(\boldsymbol{\omega}^P, \mathbf{v}) = \oint_{\Gamma_m} \mathbf{v} \cdot \mathbf{n}^m \mathrm{d}S_y. \tag{8}$$

---

[1]The convergence of the weak formulation can be found in the [6].

[2]δ_{ik} is the Kronecker symbol.

3.3.3. Homogenized coefficients

The material obtained by upscaling from the microscopic scale to the mesoscopic one is characterized by three effective coefficients, $\mathbb{A}^\alpha, \mathbf{B}^\alpha$ and M^α, which are expressed using the computed corrector functions $\boldsymbol{\omega}^{ij}, \boldsymbol{\omega}^P$ as follows:

$$\begin{aligned} A^\alpha_{ijkl} &= a^m_y(\boldsymbol{\omega}^{ij} + \mathbf{\Pi}^{ij}, \boldsymbol{\omega}^{kl} + \mathbf{\Pi}^{kl}), \\ B^\alpha_{ij} &= - \fint_{Y_m} \mathrm{div}_y \boldsymbol{\omega}^{ij}, \\ M^\alpha &= a^m_y(\boldsymbol{\omega}^P, \boldsymbol{\omega}^P). \end{aligned} \tag{9}$$

The tensors $\mathbb{A}^\alpha$ and $\mathbf{B}^\alpha$ are symmetric, i.e. $\mathbb{A} = (A_{ijkl})$ satisfies $A^\alpha_{ijkl} = A^\alpha_{klij} = A^\alpha_{jikl}$ and $\mathbf{B}^\alpha = (B^\alpha_{ij})$ satisfies $B^\alpha_{ij} = B^\alpha_{ji}$. It can be shown, [6], that $M^\alpha > 0$. Further we may introduce

$$\hat{\boldsymbol{B}}^\alpha = \boldsymbol{B}^\alpha + \phi \boldsymbol{I}, \qquad \hat{M}^\alpha = M^\alpha + \phi\gamma, \tag{10}$$

where $\hat{\mathbf{B}}^\alpha$ is the tensor of the Biot coefficients and $\hat{M}^\alpha$ is the effective Biot modulus of compressibility, which expresses the complete compressibility of the fluid and of the matrix deformed by the pore fluid pressure, [6]. The tensor $\mathbb{A}^\alpha$ is the effective elasticity tensor of the drained skeleton.

3.4. Mesoscopic problem

This section summarizes the upscaling procedure to obtain effective material properties relevant to the macroscopic scale. Thus, we shall introduce the poroelastic problem on the mesoscopic scale, using effective coefficients $\mathbb{A}^\alpha, \hat{\mathbf{B}}^\alpha$ and M^α obtained by upscaling the microstructure. This poroelastic model describes the behavior of the matrix part on the mesoscopic scale.

The domain Ω^β is decomposed as in (1)–(2) introduced for the heterogeneous microstructure, here representing the heterogeneous mesoscopic structure. We recall that the matrix $\Omega^{\beta,\varepsilon}_m$ is now formed by the porous medium associated with the upscaled microstructure of the α-level, and "canals". Canals $\Omega^{\beta,\varepsilon}_c$ are filled with the fluid and connected with pores of the α-level.

Since the α- and β-level porosities are mutually connected, the pressure $\bar{p}$ is evenly distributed through the fluid on both levels and can be characterized only by one scalar value only. The mesoscale matrix is loaded by the volume-force field $\hat{\boldsymbol{f}}^\alpha = (1 - \phi^\alpha)\boldsymbol{f}^\alpha$ and mean surface traction $\bar{\boldsymbol{g}}^\alpha := (1 - \phi^\alpha_s)\boldsymbol{g} + \phi^\alpha_s(-\bar{p})\boldsymbol{n}$ on $\partial_{ext}\Omega^{\beta,\varepsilon}_m$. The pores $\Omega^{\beta,\varepsilon}_c$ are drained out through $\partial_{ext}\Omega^{\beta,\varepsilon}_c$. The total outflow from Ω^β is denoted as $J^{\beta,\varepsilon}$ and it incorporates also the flux from α-level pores and from the β-level pores, [6]. The displacement $\boldsymbol{u}^{\beta,\varepsilon} \in \mathbf{H}^1(\Omega^{\beta,\varepsilon}_m)$ and the pressure $\bar{p}^\varepsilon \in \mathbf{H}^1(\Omega^{\beta,\varepsilon}_m)$ satisfy the equation

$$\begin{aligned} &\int_{\Omega^{\beta,\varepsilon}_m} (\mathbb{A}^\alpha \nabla^S \mathbf{u}^{\beta,\varepsilon} - \bar{p}^\varepsilon \hat{\mathbf{B}}^\alpha) : \nabla^S \mathbf{v} + \bar{p}^\varepsilon \int_{\Gamma^{\beta,\varepsilon}} \mathbf{v} \cdot \mathbf{n}^m \, \mathrm{d}S_x = \\ &\int_{\partial_{ext}\Omega^{\beta,\varepsilon}_m} \bar{\mathbf{g}}^\alpha \cdot \mathbf{v} \, \mathrm{d}S_x + \int_{\Omega^{\beta,\varepsilon}_m} \hat{\mathbf{f}}^\alpha \cdot \mathbf{v}, \quad \forall \mathbf{v} \in V(\Omega^{\beta,\varepsilon}_m) \end{aligned} \tag{11}$$

and the volume conservation

$$\int_{\Omega^{\beta,\varepsilon}_m} (\hat{\mathbf{B}}^\alpha : \nabla^S \mathbf{u}^{\beta,\varepsilon} + \int_{\partial\Omega^{\beta,\varepsilon}_c} \tilde{\mathbf{u}}^{\beta,\varepsilon} \cdot \mathbf{n}^c \, \mathrm{d}S_x + \bar{p}^\varepsilon \left[(M^\alpha) + \gamma\bar{\phi}^\alpha |\Omega^{\beta,\varepsilon}_m| + \gamma |\Omega^{\beta,\varepsilon}_c| \right] = -J^{\beta,\varepsilon}, \tag{12}$$

where $\tilde{\boldsymbol{u}}^{\beta,\varepsilon}$ is the displacement extension to canals $\Omega^{\beta,\varepsilon}_c$ and $\bar{\phi}^\alpha = |\Omega^\alpha|^{-1} \int_{\Omega^\alpha} \phi^\alpha$ is the mean porosity[3].

[3]However, in this work we consider constant porosity, thus, $\phi^\alpha = \bar{\phi}^\alpha$.

Analogically to previous section, the domain Ω^β is generated as periodic lattice using RPC Y^β. Note that like domain Ω^β, RPC Y^β has the decomposition analogical to Y^α, see Eq. (6). The upscaling procedure described in [6, 7] leads to the mesoscopic auxiliary problems for corrector basis functions of the β-level. These are employed to express the effective coefficients relevant to the macroscopic level of the osteon.

3.4.1. Mesoscopic problems and corrector basis functions

In what follows, we record the mesoscopic problems in Y^β for the corrector basis functions of displacement, $\boldsymbol{\omega}_\beta^{ij}$ and $\boldsymbol{\omega}_\beta^P$. Find $\boldsymbol{\omega}_\beta^{ij} \in \mathbf{H}^1_\#(Y_m^\beta)$ and $\boldsymbol{\omega}_\beta^P \in \mathbf{H}^1_\#(Y_m^\beta)$ satisfying

$$\begin{aligned}
&\int_{Y_m^\beta} \left[\mathbb{A}^\alpha \nabla_y^S(\boldsymbol{\omega}_\beta^{ij} + \boldsymbol{\Pi}_\beta^{ij})\right] : \nabla_y^S \mathbf{v} = 0 \qquad \forall \boldsymbol{v} \in \mathbf{H}^1_\#(Y_m^\beta), \\
&\int_{Y_m^\beta} \left[\mathbb{A}^\alpha \nabla_y^S \boldsymbol{\omega}_\beta^P\right] : \nabla_y^S \mathbf{v} = -\int_{Y_m^\beta} \hat{\mathbf{B}}^\alpha : \nabla_y^S \mathbf{v} + \int_{\Gamma_Y^\beta} \mathbf{v}\cdot \mathbf{n}^m \, \mathrm{d}S_y \qquad \forall \boldsymbol{v} \in \mathbf{H}^1_\#(Y_m^\beta),
\end{aligned} \tag{13}$$

where transformation vector $\boldsymbol{\Pi}_\beta^{ij} = (\Pi_{k,\beta}^{ij})$, $i, j, k = 1, 2, 3$ have components $\Pi_{k,\beta}^{ij} = y_j^\beta \delta_{ik}$ are constituted by coordinates y_j^β of RPC Y^β.

While the solutions $\boldsymbol{\omega}_\beta^{ij}$ express fluctuations with respect to the local macroscopic unite strain, the $\boldsymbol{\omega}_\beta^P$ expresses the local response with respect to the unit pore pressure, [6].

3.4.2. Homogenized coefficients

The corrector functions specified in the previous section are involved in the expressions of the effective material coefficients relevant to the macroscopic body with porosities on micro- and mesoscale and are denoted by $\mathbb{A}, \boldsymbol{B}, M$ (on the macroscale the superscript β is not necessary and has been released). These expressions are as follows

$$A_{ijkl} = \fint_{Y_m^\beta} \left[\mathbb{A}^\alpha \nabla_y^S(\boldsymbol{\omega}_\beta^{kl} + \boldsymbol{\Pi}_\beta^{kl})\right] : \nabla_y^S \mathbf{v}(\boldsymbol{\omega}_\beta^{ij} + \boldsymbol{\Pi}_\beta^{ij}), \tag{14}$$

$$B_{ij} = \fint_{Y_m^\beta} \hat{\mathbf{B}}^\alpha : \nabla_y^S \mathbf{v}(\boldsymbol{\omega}_\beta^{ij} + \boldsymbol{\Pi}_\beta^{ij}) - \fint_{Y_m^\beta} \mathrm{div}_y \boldsymbol{\omega}_\beta^{ij}, \tag{15}$$

$$M = \fint_{Y_m^\beta} \left[\mathbb{A}^\alpha \nabla_y^S(\boldsymbol{\omega}_\beta^p)\right] : \nabla_y^S \mathbf{v}(\boldsymbol{\omega}_\beta^p). \tag{16}$$

Further, we introduce

$$\hat{\mathbf{B}} := \mathbf{B} + \phi^\beta \mathbf{I}, \qquad \hat{M} := M + \gamma\bar{\phi}^\beta + (M^\alpha + \gamma\bar{\phi}^\alpha)(1 - \bar{\phi}^\beta)\,. \tag{17}$$

These effective coefficients are analogical to the classical Biot model coefficients. Advantageously, the given formulae enable to compute them for a given representative cells at both the micro- and mesoscopic levels.

3.4.3. Macroscopic problem

In the previous section, we derive the equations for effective coefficients relevant to the macroscopic poroelastic body with porosities distributed at the two scales. Responses of such a body are described by the displacement $\mathbf{u}$ and by the pressure $\bar{p}$ satisfying

$$\begin{aligned}\int_{\Omega^\beta} (\mathbb{A}\nabla_x^S \mathbf{u} - \bar{p}\hat{\mathbf{B}}) : \nabla_x^S \mathbf{v} &= \int_{\Omega^\beta} (1-\phi^\beta)\mathbf{f}^\alpha \cdot \mathbf{v} + \int_{\partial\Omega^\beta} \bar{\mathbf{g}}^\beta \cdot \mathbf{v}\, \mathrm{d}S_x \quad \forall \boldsymbol{v} \in \mathbf{H}^1(\Omega^\beta),\\ \int_{\Omega^\beta} \hat{\mathbf{B}}\nabla_x^S \mathbf{u} + \bar{p}\hat{M}|\Omega^\beta| &= -J^\beta \qquad \forall \boldsymbol{v} \in \mathbf{H}^1(\Omega^\beta),\end{aligned} \tag{18}$$

where in analogy with $\bar{\mathbf{g}}^\beta$ introduced above at the mesoscopic scale,

$$\bar{\mathbf{g}}^\beta := (1-\phi_S^\beta)\bar{\mathbf{g}}^\alpha + \phi_S^\beta(-\bar{p})\mathbf{n} \tag{19}$$

is the mean surface stress and $\bar{\phi}_S^\beta$ is the mean surface β-level porosity.

Now we have a complete set of equations for obtaining the poroelastic properties of one single osteon in cortical bone tissue. The effective coefficients can be used for further applications and can be also associated with the Biot poroelasticity model, from which the increase of fluid content J^β can be computed.

In the section below, we assume that the interface on both levels $\Gamma^{\alpha,\varepsilon}$ and $\Gamma^{\beta,\varepsilon}$ is impermeable, but porosities on α- and β- beta level are mutually connected; this configuration results in one hydrostatic pressure $\bar{p}$ in all porosity levels, see section 3.4.3. In addition, the matrix cannot be drained throught $\partial_{ext}\Omega_m^{\beta,\varepsilon}$, thus $\phi_S^\beta = 0$ on the exterior surface. In this case $\bar{\mathbf{g}}^\beta(\bar{p}) = \bar{\mathbf{g}}^\beta$ is independent of the pore pressure and also external flux $J^\beta = 0$.

4. Two-level computational homogenization

In this section we provide illustrations to the two-level computational homogenization. We consider an idealized structure of the cortical bone tissue characterized at the micro- and mesoscopic scales using the representative periodic cells Y^α and Y^β, respectively; the details will be given in Section 4.1. We use the finite element method (FEM) to discretize in space the weak formulations presented in previous sections. As explained above, the linearity of the multiscale problem leads to a computational algorithm consisting of the following steps:

1. Solve the auxiliary corrector problems in the RPC Y^α defined by Eq. (8) to obtain the local response functions $\omega^{ij}|^\alpha, \omega^p|^\alpha$.

2. Using $\omega^{ij}|^\alpha, \omega^p|^\alpha$ substituted in formulae (9) and (10), compute effective coefficients $\mathbb{A}^\alpha$, $\hat{\mathbf{B}}^\alpha$ and $\hat{M}^\alpha$ which characterize the matrix phase of the β-level structure.

3. Solve the auxiliary corrector problems in the RPC Y^β given by Eq. (13) to obtain local response functions $\omega^{ij}|^\beta, \omega^p|^\beta$.

4. Using $\omega^{ij}|^\beta, \omega^p|^\beta$ substituted in formulae (15) and (17), compute effective coefficients $\mathbb{A}^\beta$, $\hat{\mathbf{B}}^\beta$ and $\hat{M}^\beta$ involved in the macroscopic model.

5. Using the effective coefficients $\mathbb{A}^\beta$, $\hat{\mathbf{B}}^\beta$ and $\hat{M}^\beta$, solve the macroscopic problem (18) imposed in domain Ω.

Although the Neumann boundary conditions are assumed in Eq. (18), other boundary conditions can be prescribed. The classical formulation yields the equilibrium condition

$$-\nabla \cdot (\mathbb{A}\nabla_x^S \mathbf{u} - \bar{p}\hat{\mathbf{B}}) = (1-\phi^\beta)\mathbf{f}^\alpha \quad \text{in } \Omega\,, \tag{20}$$

which is accompanied by the mass balance $(18)_2$. In general, the boundary conditions may involve the Dirichlet type boundary conditions imposed on $\partial_u\Omega \subset \partial\Omega$, whereas the traction (Neumann) conditions are prescribed on the rest of the boundary, $\partial_\sigma\Omega \subset \partial\Omega \subset \partial_u\Omega$, thus,

$$\boldsymbol{n} \cdot (\mathbb{A}\nabla_x^S \mathbf{u} - \bar{p}\hat{\mathbf{B}}) = \bar{\mathbf{g}}^\beta \qquad \text{on } \partial_\sigma\Omega\,, \tag{21}$$

where $\boldsymbol{n}$ is the unit normal vector on $\partial_\sigma\Omega$. Recalling definition (19) of $\bar{\mathbf{g}}^\beta$, it should be noted, that (21) involves $\bar{p}$ through $\bar{\mathbf{g}}^\beta$ and $\bar{\mathbf{g}}^\alpha$ if the surface porosities ϕ_S^β and ϕ_S^α do not vanish. For discussion of possible cases of the boundary conditions and their consequences on the symmetry of problem (18) we refer to [7].

4.1. *Geometrical representation*

As discussed in the introduction, the cortical bone tissue can be represented by the hierarchical porous structure which is defined at the two levels by the RPCs Y^α and Y^β, see Fig. 2. The two-level upscaling procedure yields the effective material properties of bone osteon which constitutes the macroscopic body. All dimensional parameters describing employed geometrical features[4] are listed in Table 2. In this study we consider an artificial, simplified geometry of the canalicular and lacunar porosities which can easily be parameterized. This allows us to describe qualitatively influences of the selected structural features on the effective material properties of the upscaled bone.

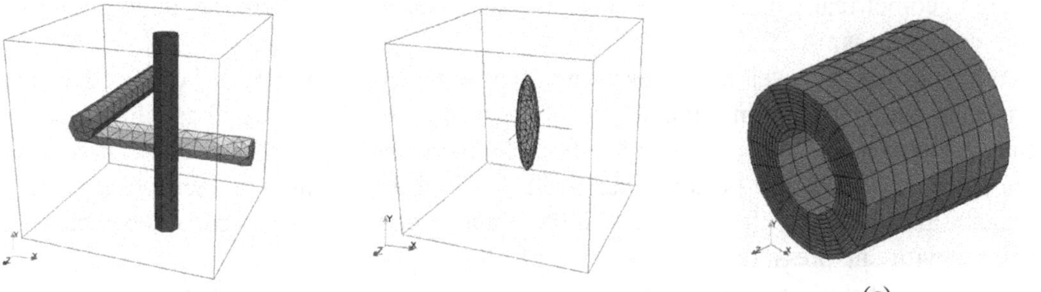

(a) (b) (c)

Fig. 2. Geometry representations of (a) RPC Y^α, (b) RPC Y^β and (c) macroscopic level body

Table 2. Parameters of geometrical configuration

Symbol	Parameter	Unit	Value
L_α	characteristic length of RPC on micro-level	μm	4.3
L_β	characteristic length of RPC on meso-level	μm	43.0
$r_x = d_x/2$	radius of canaliculi in x-direction	μm	0.6
$r_y = d_y/2$	radius of canaliculi in y-direction	μm	0.6
$r_z = d_z/2$	radius of canaliculi in z-direction	μm	0.6
a_0	semi-axis of ellipsoid in x-direction	μm	2.5
b_0	semi-axis of ellipsoid in y-direction	μm	12.5
c_0	semi-axis of ellipsoid in z-direction	μm	5.0

[4]In this section we use labelling of coordinate axes by x, y, z rather than $1, 2, 3$.

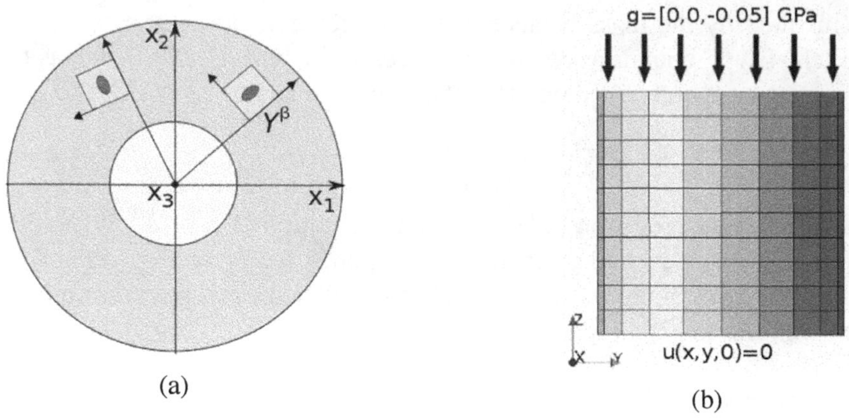

Fig. 3. (a) Circumferential orientation of RPC Y^β in cylindrical macroscopic body, (b) Loading and boundary conditions of macroscopic body

At the microscopic level, the bone matrix is represented by the cubic RPC Y^α of the edge size L_α. The solid phase corresponding to the bone matrix occupies domain Y_m^α. The canalicular pores (domain Y_c^α) are represented by three cylindrical channels defined by radii r_x, r_y and r_z. We choose them in such a way that the cross-section area of each channel is approximately equal to the total cross-section area of all actual canaliculi which lead in the corresponding direction. Although geometrically disconnected, in a "physical sense" we adhere the assumption of their mutual cennectness.

At the mesoscopic level, the lacunar porosity is represented by RPC Y^β of the edge size L_β. The porous matrix occupying domain Y_m^β is constituted by the upscaled canalicular porosity, so that it represents a network of orthogonal channels leading in x-, y- and z-directions. A single ellipsoidal inclusion occupies domain Y_c^β, which represents one lacunae, is defined by three semi-axes a_0, b_0, c_0. Recall that both the α and β porosities form one connected porosity associated with one pressure $\bar{p}$.

In the present study, the macroscopic level is represented by a single bone osteon occupying domain Ω, although at this level the bone tissue can also be described approximatively as a heterogeneous structure constituted by periodically distributed osteons, as considerd, e.g., in [5]. Thus, Ω is shaped as a hollow cylinder, see Fig. 2c. The material model obtained using the two-level homogenization can be associated with a local coordinate system, in general. Below we consider that the material structure is locally periodic, generated by cell Y^β which is aligned with the radial, tangential and axial macroscopic coordinate axes related naturally to the cylindrical geometry, see Fig. 3a. Obviously, the orientation of the α-level anisotropy is fixed within the mesoscopic RPC Y^β.

4.2. Numerical examples

The purpose of this example is to show how different structures at micro- and mesoscopic level influence the homogenized coefficients and the macroscopic response. For better understanding of how each level influence the final effective coefficients, we present following three numerical examples.

Example 1 We modify the canalicular porosity ϕ^α by changing the diameter of one of the three cylindrical channels aligned with the x-axis. The geometry representing RPC Y^β is preserved, see Fig. 4a.

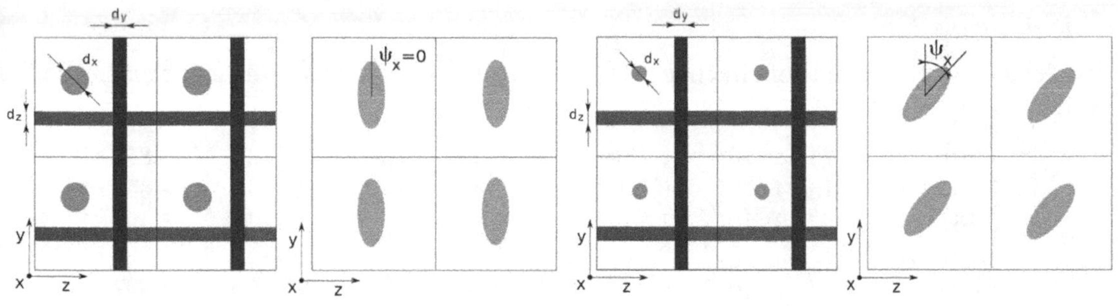

Fig. 4. Numerical example schemes: (a) Example 1 – changing diameter d_x on the RPCs Y^α geometry while preserving lacunae position in RPC Y^β; (b) Example 2 – while preserving RPC Y^α ($d_x = d_y = d_z$), rotating ellipsoid conclusion in RPC Y^β on axes x, y and z (rotation angle $\psi_i \in \langle 0, \pi/2 \rangle, i = x, y, z$)

Example 2 While preserving the geometry representation of RPC Y^α, we change orientation of ellipsoidal inclusion in RPC Y^β by rotating it subsequently on one of its axes by angles ψ_i, $i = x, y, z$, see Fig. 4b.

Example 3 We solve the macroscopic problems with effective material coefficients obtained according to Example 1 and Example 2, as described above.

In the following text, we report numerical results obtained for these three numerical examples. First, we compute final effective coefficients for models Example 1 and Example 2 and then use them to solve the macroscopic problems, as specified in Example 3. The two-scale, two-level homogenization was implemented in our in-house developed FEM software *SfePy*, for more information see the website *sfepy.org*.

4.3. *Effective coefficients*

We use an isotropic material with Young's modulus $E = 18\,\mathrm{GPa}$ and Poisson's ratio $\nu = 0.3$ to describe the solid matrix representing bone tissue in RPC Y^α, while the fluid is characterized by its compressibility $\gamma = 0.9$. These values were choosen as our estimation of the cortical bone material properties relevant to the microstructure (the canalicular porosity level) on the basis of information found in the literature [11], see also [9].

The resulting homogenized material is orthotropic, whereby the effective compliance tensor $\mathbb{A}^{-1}$ is symetric, attaining the following form

$$\mathbb{A}^{-1} = \begin{bmatrix} \frac{1}{E_1} & -\frac{\nu_{21}}{E_2} & -\frac{\nu_{31}}{E_3} & 0 & 0 & 0 \\ -\frac{\nu_{12}}{E_1} & \frac{1}{E_2} & -\frac{\nu_{32}}{E_3} & 0 & 0 & 0 \\ -\frac{\nu_{13}}{E_1} & -\frac{\nu_{23}}{E_2} & \frac{1}{E_3} & 0 & 0 & 0 \\ 0 & 0 & 0 & \frac{1}{G_{12}} & 0 & 0 \\ 0 & 0 & 0 & 0 & \frac{1}{G_{23}} & 0 \\ 0 & 0 & 0 & 0 & 0 & \frac{1}{G_{31}} \end{bmatrix}, \quad (22)$$

where E_1, E_2, E_3 are Young's moduli in directions $1, 2, 3$. ν_{ij} represents the Poisson's ratio for the strain in direction j while loaded in direction i. G_{12}, G_{23}, G_{31} are shear moduli in 1-2, 2-3 a 3-1. Due to symmetry $\frac{\nu_{ij}}{E_i} = \frac{\nu_{ji}}{E_j}$, the orthotropic material can be described by only 12 material constants stated above.

4.3.1. Example 1

First, let us give an example of effective coefficients resulting from upscaling to the macroscopic level. The coefficients

$$\mathbb{A}^\beta = \begin{bmatrix} 1.801 \cdot 10^{1} & 7.064 \cdot 10^{0} & 6.731 \cdot 10^{0} & 1.501 \cdot 10^{-4} & -1.404 \cdot 10^{-4} & -1.851 \cdot 10^{-4} \\ 7.064 \cdot 10^{0} & 1.804 \cdot 10^{1} & 6.730 \cdot 10^{0} & 1.446 \cdot 10^{-4} & -1.798 \cdot 10^{-4} & 8.559 \cdot 10^{-6} \\ 6.731 \cdot 10^{0} & 6.730 \cdot 10^{0} & 1.806 \cdot 10^{1} & 2.876 \cdot 10^{-6} & -5.283 \cdot 10^{-4} & 5.491 \cdot 10^{-6} \\ 1.501 \cdot 10^{-4} & 1.445 \cdot 10^{-4} & 2.723 \cdot 10^{-6} & 5.549 \cdot 10^{0} & 2.040 \cdot 10^{-6} & -7.826 \cdot 10^{-5} \\ -1.408 \cdot 10^{-4} & -2.031 \cdot 10^{-4} & -5.507 \cdot 10^{-4} & 2.308 \cdot 10^{-6} & 5.205 \cdot 10^{0} & 1.637 \cdot 10^{-4} \\ -1.852 \cdot 10^{-4} & 7.385 \cdot 10^{-6} & 4.367 \cdot 10^{-6} & -7.824 \cdot 10^{-5} & 1.648 \cdot 10^{-4} & 5.205 \cdot 10^{0} \end{bmatrix},$$

$$\hat{\mathbf{B}}^\beta = \left[1.096 \cdot 10^{0}, 1.096 \cdot 10^{0}, 1.097 \cdot 10^{0}, -1.134 \cdot 10^{-5}, -3.161 \cdot 10^{-6}, -3.177 \cdot 10^{-7}\right],$$
$$\hat{M}^\beta = 1.004 \cdot 10^{-4}$$

were computed for the start configuration, i.e., $r_x = r_y = r_z = 0.6\,\mu$m and $\psi_x = \psi_y = \psi_z = 0$. When commponents smaller than 10^{-3} are neglected, stiffness $\mathbb{A}^\beta$ attain the form such of orthotropic material.

A shown in Fig. 5a, the macroscopic effective coefficients vary with the total porosity $\phi_\gamma = \phi_\alpha + \phi_\beta - \phi_\alpha\phi_\beta$ since porosity ϕ_α is being changed. For better understanding how variation of porosity on α-level influence the characteristics of macroscopic body, we transformed effective stiffness tensor into components of Young's modulus E_i, Poisson's ratio $\nu_{ij}, i, j = 1, 2, 3, i \neq j$

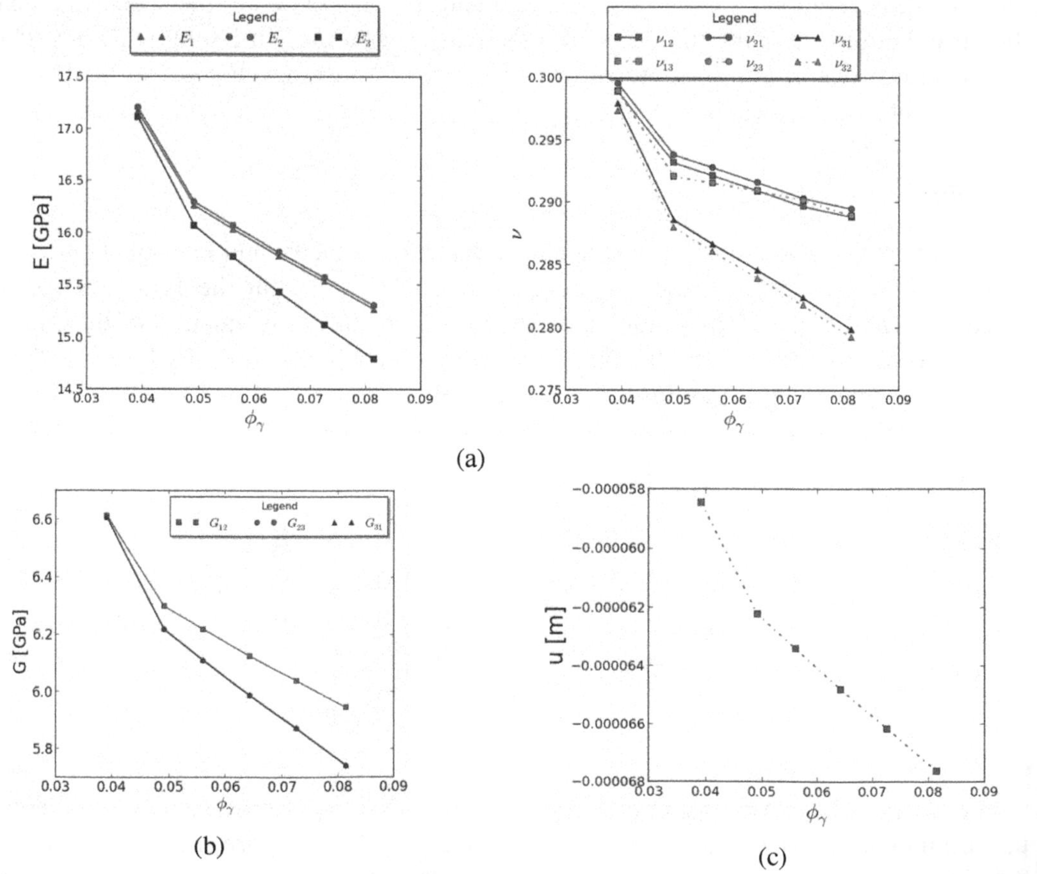

Fig. 5. Example 1 – Effective coefficients influenced by varying x-direction canaliculi radius r_x

and shear modulus G_{12}, G_{23}, G_{31}. It is clear, that as porosity rises, the solid phase is being replaced by the fluid part in the RPC Y^α, thus the curves of components of elastic coefficients decrease.

4.3.2. Example 2

A more complex influence of the orientation of the lacunae on the elastic coefficients is apparent in Figs. 6a, 6b and 6c, where we display the dependence of Young's modulus and Poisson's ratio on rotation angle $\psi_i, i = x, y, z$. The Dependence of sheare modulus G on ψ_i is then shown in Figs. 7a, 7c and 7e. In this study, the varying mutual positions of the ellipsoidal inclusions of the neighbouring cells (copies of Y^β), have to be taken into account. As the ellipsoids rotate, the cross-sectional areas of the solid phase in the planes perpendicular to main axes vary, which results into components of effective elastic coefficients having different trends.

4.4. Macroscopic problem solution

Using effective coefficients obtained above for various volume fractions, or varying orientations of the lacunae at the mesoscopic level, we now solve the macroscopic problem given by Eq. (18). In order to enable that to the two parameteric studies considered in the Example 1 and Example 2 can be compared, we use for both the same geometry representation of macroscopic body (see Fig. 2c) and the same boundary conditions. The single bone osteon is loaded by a uniform traction force $g = [0, 0, -5]$ MPa on the upper face $z = l$, where l is the cylinder length aligned with the z-coordinate, and stuck to the rigid frame on the bottom face $z = 0$, see Fig. 3b. Note, that the RPC Y^β representing mesoscopic level structure are circumferentially arranged in the macroscopic body, see Fig. 3a, so that the ellipsoid orientation with respect to the local tangential and radial axes is fixed.

Responses to the static loading of the macroscopic body for materials studied in Example 1 are shown in Fig. 5c. We display only the z-component of displacement, u_z, because in this direction maximal displacements appear. As expected, we may see that the curve of u_z follows the trend of E_3. Similarily, curves of u_z displacement for materials studied in Example 2 follow the effective elastic coefficients, see Figs. 7b, 7d and 7f.

5. Conclusion

In this paper we presented the three-scale model of cortical bone tissue obtained by homogenization procedure based on [6]. Such model was motivated by a strictly hierarchical structure of a cortical bone saturated by a bone fluid. We briefly described the computational steps leading to evaluation of the effective coefficients on the micro- and mesoscopic scale levels, the latter being involved in the macroscopic problem formulation. The resulting homogenized model describes deformation of the fluid saturated double-porous body in response to a static loading by external forces and to an injection of slightly compressible fluid.

The solutions of this model was implemented in the software *SfePy* and two numerical examples were presented, showing the influence of selected micro- and mesostructure geometrical features on the effective material poroelastic properties apparent at the macroscopic scale. As an advantage, this two-scale hierarchical approach allows for geometrically based identification of the Biot material model. Therefore, this model can be used as a basis for further research not only in bone tissue mechanics, but also in tissue biomechanics in general, as well as in other fields of engineering sciences related to porous media.

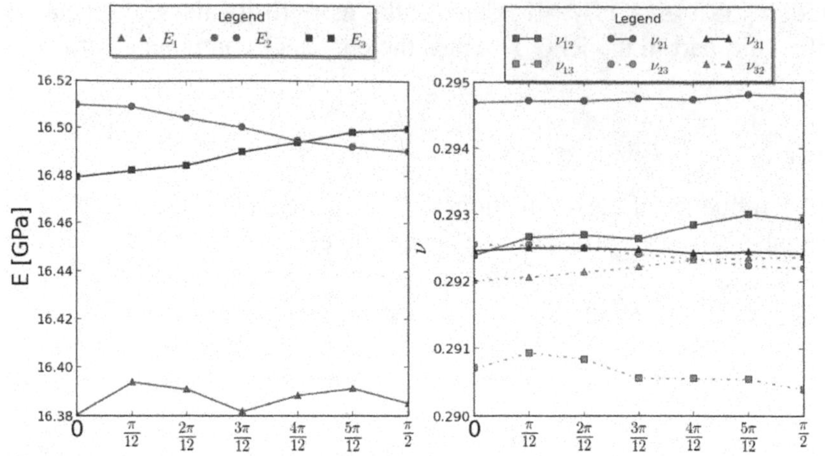

(a) Dependence of Young's modulus and Poisson's ratio on rotation angle ψ_x

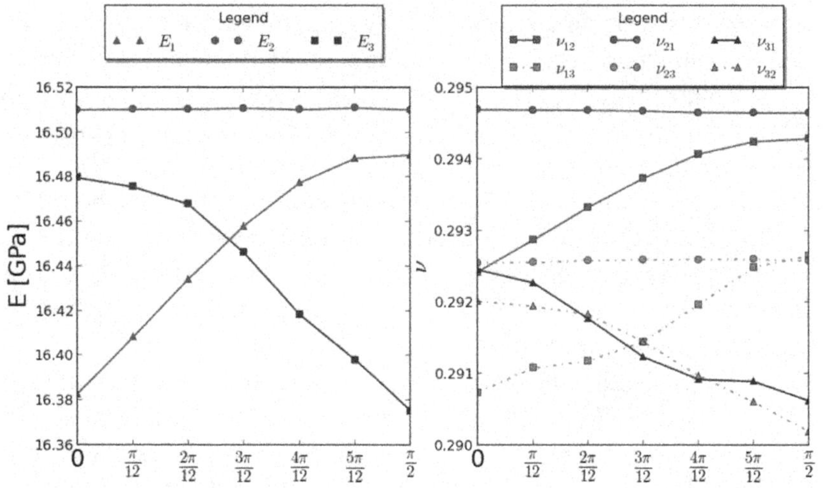

(b) Dependence of Young's modulus and Poisson's ratio on rotation angle ψ_y

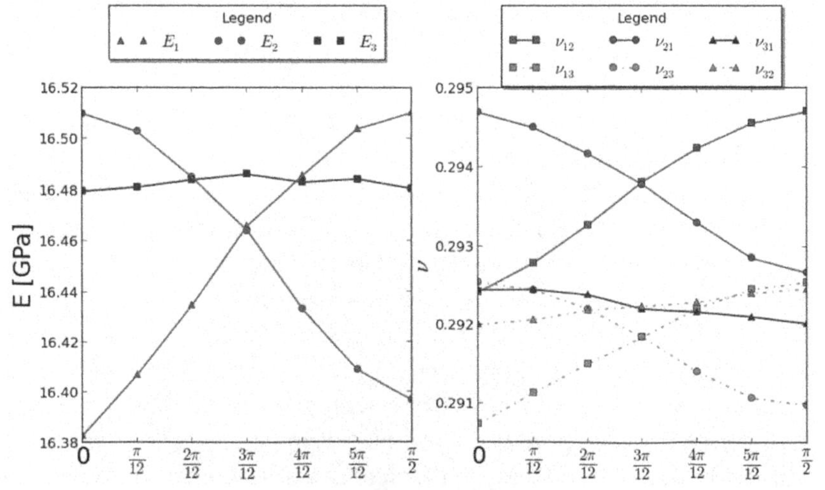

(c) Dependence of Young's modulus and Poisson's ratio on rotation angle ψ_z

Fig. 6. Example 2 – Effective coefficients influenced by rotation angle ψ_i of lacuna

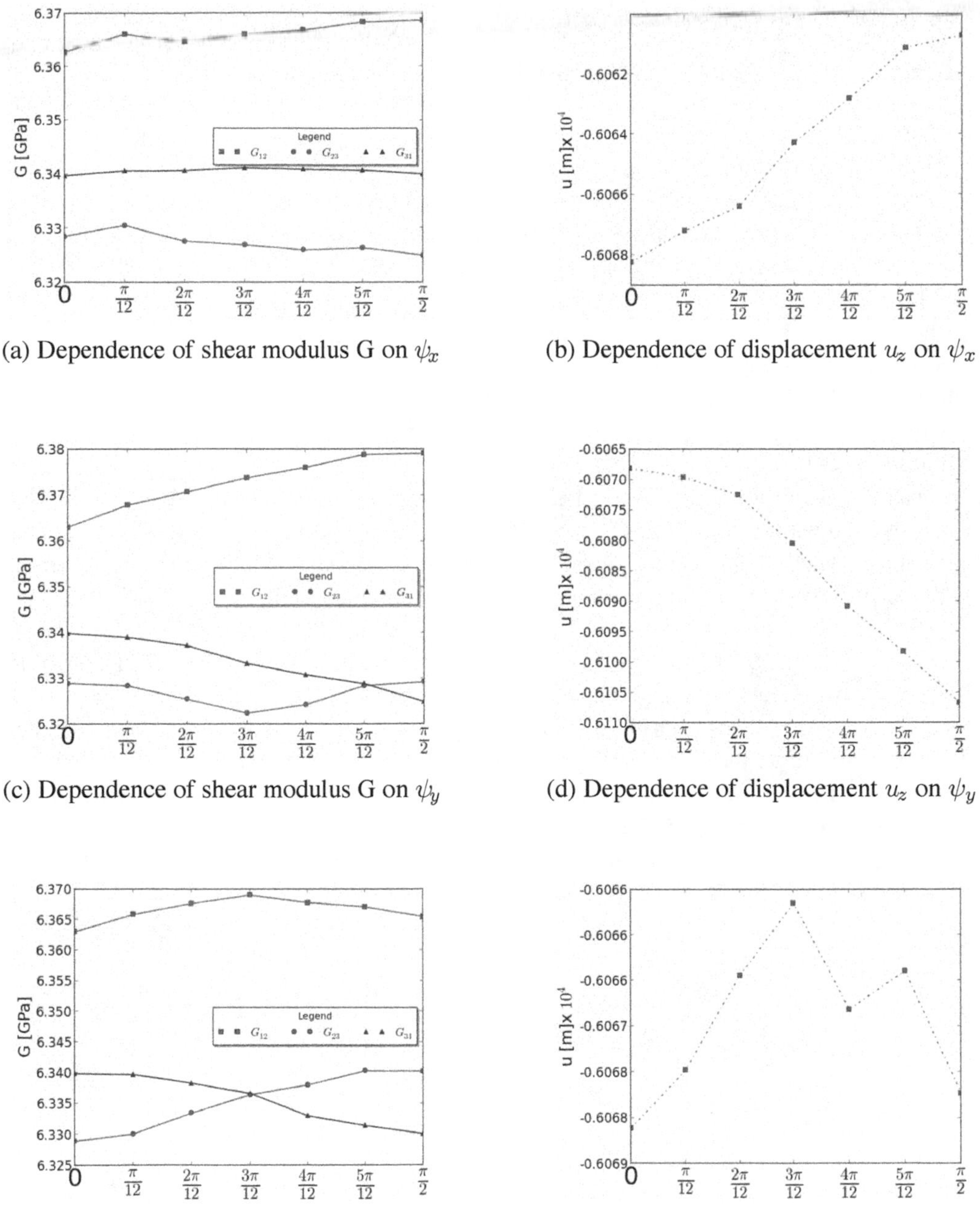

(a) Dependence of shear modulus G on ψ_x

(b) Dependence of displacement u_z on ψ_x

(c) Dependence of shear modulus G on ψ_y

(d) Dependence of displacement u_z on ψ_y

(e) Dependence of shear modulus G on ψ_z

(f) Dependence of displacement u_z on ψ_z

Fig. 7. Example 2 – (Left) Components of shear moduli influenced by rotation angle ψ_i of lacuna; (Right) z-component of displacement u_z influenced by rotation angle ψ_i of lacuna

Acknowledgements

The research of was supported in part by the project NT 13326 of the Ministry of Health of the Czech Republic and by the European Regional Development Fund (ERDF), project 'NTIS – New Technologies for Information Society', European Centre of Excellence, CZ.1.05/1.1.00/02.0090. Jana Turjanicová is also grateful for the support of her work by the project SGS-2013-026.

References

[1] Barenblatt, G. I., Zheltov, Iu P., Kochina, I. N., Basic concepts in the theory of seepage of homogeneous liquids in fissured rocks, Journal of Applied Mathematics and Mechanics 24 (5) (1960) 1 286–1 303.

[2] Beno, T., Yoon, Y.-J., Cowin, S. C., et al., Estimation of bone permeability using accurate microstructural measurements, Journal of Biomechanics 39 (13) (2006) 2 378–2 387.

[3] Mikelic, A., Wheeler, M., On the interface law between a deformable porous medium containing a viscous fluid and an elastic body, Mathematical Models and Methods in Applied Sciences 22 (2012) 1–32.

[4] Rho, J. Y., Kuhn-Spearing, L., Zioupos, P., Mechanical properties and the hierarchical structure of bone, Medical Engineering & Physics 20 (2) (1998) 92–102.

[5] Rohan, E., Naili, S., Cimrman, R., Lemaire, T., Multiscale modeling of a fluid saturated medium with double porosity: Relevance to the compact bone, Journal of the Mechanics and Physics of Solids 60 (2012) 857–881.

[6] Rohan, E., Naili, S., Cimrman, R., Lemaire, T., Hierarchical homogenization of fluid saturated porous solid with multiple porosity scales, Comptes Rendus Mecanique 340 (10) (2012) 688–694.

[7] Rohan, E., Naili, S., Lemaire, T., Double porosity in fluid-saturated elastic media — deriving effective parameters by hierarchical homogenization, Continuum Mechanics and Thermodynamics (2014). (submitted)

[8] Rohan, E., Shaw, S., Whiteman, J. R., Poro-viscoelasticity modelling based on upscaling quasistatic fluid-saturated solids, Computational Geosciences 18 (2014) 883–895.

[9] Turjanicová, J., Electro-mechanical coupling in porous bone structure — homogenization method application, Master thesis, University of West Bohemia, Pilsen, 2013.

[10] Warren, J. E., Root, P. J., The behavior of naturally fractured reservoirs, Society of Petroleum Engineers Journal 3 (3) (1963) 245–255.

[11] Yoon, Y. J., Cowin, S. C., An estimate of anisotropic poroelastic constants of an osteon, Biomechanics and Modeling in Mechanobiology 7 (1) (2008) 13–26.

Double pendulum contact problem

J. Špička[a,*], L. Hynčík[a], M. Hajžman[a]

[a] *Faculty of Applied Sciences, University of West Bohemia in Pilsen, Univerzitní 8, 306 14 Plzeň, Czech Republic*

Abstract

The work concerns contact problems focused on biomechanical systems modelled by a multibody approach. The example is modelling of impact between a body and an infrastructure. The paper firstly presents algorithm for minimum distance calculation. An analytical approach using a tangential plain perpendicular to an initial one is applied. Contact force generated during impact is compared by three different continuous force models, namely the Hertz's model, the spring-dashpot model and the non-linear damping model. In order to identify contact parameters of these particular models, the method of numerical optimization is used. Purpose of this method is to find the most corresponding results of numerical simulation to the original experiment. Numerical optimization principle is put upon a bouncing ball example for the purpose of evaluation of desirable contact force parameters. The contact modelling is applied to a double pendulum problem. The equation of motion of the double pendulum system is derived using Lagrange equation of the second kind with multipliers, respecting the contact phenomena. Applications in biomechanical research are hinted at arm gravity motion and a double pendulum impact example.

Keywords: contact, continuous contact model, minimum distance calculation, contact force parameters

1. Introduction

Contact or impact is a very frequent phenomenon that occurs when two or more bodies undergo a collision. A contact problem arises in numerous engineering applications, such as multibody dynamics, robotics, biomechanics and many others. Impact in biomechanical research studies the consequences to the human body impact like a car crash, pedestrian impact, falls and sports injuries or contact in forensic applications. This field motivates engineers and designers to develop better safety systems for people exposed to impact injuries. Virtual human body models start to play an important role in the impact biomechanics. Multibody models can evaluate human body kinematics under external loading quickly. Detailed deformable models can then simulate tissue injuries, however these models spend a lot of computational time. Thus articulated rigid bodies can be sufficient tool for the first approximation and they might predict long duration global human body behaviour in very short time. For such models, contact modelling and contact parameters evaluating are crucial aspects of a successful description.

This work describes double pendulum as a simple articulated rigid body system based on multibody approach, e.g. [10] or [12]. Author uses Lagrange equation with multipliers to evaluate equations of motion. Derivation of an impact algorithm using various contact force models is demonstrated. The solution of contact problems is very complex as is shown e.g. in [8]. The three implemented contact force models are Hertz model, spring-dashpot model and non-linear damping model [5], respectively. This work also presents an algorithm for minimum distance calculation between a body and a plain using analytical approach based on the plain tangential

*Corresponding author. e-mail: spicka@ntc.zcu.cz.

to the body. Principle of numerical optimization is applied on a simple mechanics example of a bouncing ball in order to evaluate contact parameters according to a real system. Optimized values of contact force parameters are used in a bouncing ball example and the results of simulations are presented. The double pendulum system is assumed to be a model of a human arm. The results were compared with 2D approach model of a human arm and also with an experiment. Possibilities of further biomechanics applications are demonstrated using the double pendulum contacting a plain example.

2. The method

2.1. Double pendulum model

The double pendulum is assumed to be composed by two ellipsoids constrained together. Both ellipsoids have semi-principal axes a_{ij}, mass m_i and moments of inertia I_{ij}, $i \in \{1,2\}$ and $j \in \{1,2,3\}$. The global coordinate system $\boldsymbol{x}_1 = [x_1, y_1, z_1]$ is defined to be a Cartesian right handed coordinate system with an origin at frame point of the first pendulum (joint). While $\boldsymbol{x}_2$ and $\boldsymbol{x}_3$ are local coordinate systems of particular bodies with origin located at the centre of the bodies. The two bodies are linked with a spherical kinematic joint together and the first one is linked to a rigid frame also with spherical joint. Whole system of bodies is shown in Fig. 1 and the coordinate systems are displayed.

2.1.1. Spherical joint

A spherical joint is a type of a primitive kinematic constraint with three rotational degrees of freedom.

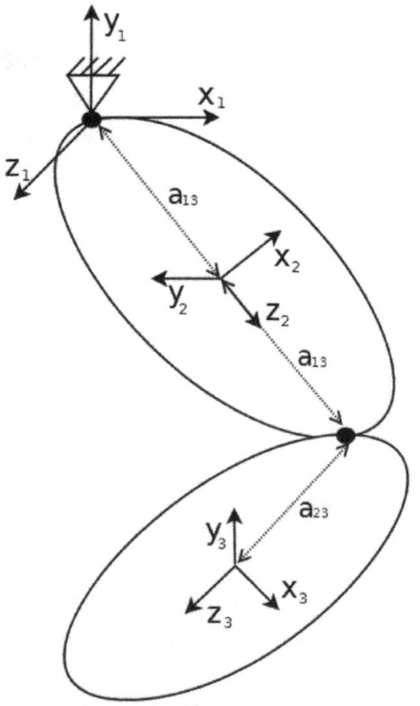

Fig. 1. Double pendulum

Spherical motion can be considered as three independent rotations, namely precession around the z axis represented with angle ψ, nutation around the actual x axis represented with ϑ and rotation around the actual z axis represented by angle φ. The three independent spatial motions can be described by transformation matrices, namely the

$$\boldsymbol{S}_{pre}(\psi) = \begin{bmatrix} \cos(\psi) & -\sin(\psi) & 0 \\ \sin(\psi) & \cos(\psi) & 0 \\ 0 & 0 & 1 \end{bmatrix}, \tag{1}$$

$$\boldsymbol{S}_{nut}(\vartheta) = \begin{bmatrix} 1 & 0 & 0 \\ 0 & \cos(\vartheta) & -\sin(\vartheta) \\ 0 & \sin(\vartheta) & \cos(\vartheta) \end{bmatrix}, \tag{2}$$

$$\boldsymbol{S}_{rot}(\varphi) = \begin{bmatrix} \cos(\varphi) & -\sin(\varphi) & 0 \\ \sin(\varphi) & \cos(\varphi) & 0 \\ 0 & 0 & 1 \end{bmatrix}. \tag{3}$$

Transformation formula in case of a spherical motion can be written using coordinates of centre of gravity and multiplication of precession, nutation and rotation matrices. Thus the general transformation of any point from local to a global coordinate system is described as

$$\boldsymbol{x}_1 = \boldsymbol{x_s} + \boldsymbol{S}_{pre}(\psi)\ \boldsymbol{S}_{nut}(\vartheta)\ \boldsymbol{S}_{rot}(\varphi)\ \boldsymbol{x}_2, \tag{4}$$

where $\boldsymbol{S}_{pre}(\psi)$, $\boldsymbol{S}_{nut}(\vartheta)$ and $\boldsymbol{S}_{rot}(\varphi)$ are transformation matrices of precession, nutation and rotation, respectively. $\boldsymbol{x}_2$ are coordinates of a point at the body expressed in local coordinate system, $\boldsymbol{x}_1$ represents coordinates of this point in the global coordinate system and $\boldsymbol{x}_s$ are coordinates of centre of gravity of body expressed in the global coordinate system. Equation (4) can be rewritten as

$$\boldsymbol{x}_1 = \boldsymbol{x}_{s_i} + \boldsymbol{S}_{1i}(\psi_i, \vartheta_i, \varphi_i)\ \boldsymbol{x}_i, \quad i \in \mathbb{N}, \tag{5}$$

where $\boldsymbol{S}_{1i}$ is a transformation matrix between local body-fixed coordinate system i and global coordinate system 1.

Since the system here considers two bodies, two local body fixed coordinate systems are required, and the global coordinates $\boldsymbol{x}_1$ of any point can be defined:

- The first body global coordinates: $i = 2$

$$\boldsymbol{x}_1 = \boldsymbol{x}_{s_2} + \boldsymbol{S}_{12}(\psi_2, \vartheta_2, \varphi_2)\boldsymbol{x}_2. \tag{6}$$

- The second body global coordinates: $i = 3$

$$\boldsymbol{x}_1 = \boldsymbol{x}_{s_3} + \boldsymbol{S}_{13}(\psi_3, \vartheta_3, \varphi_3)\boldsymbol{x}_3, \tag{7}$$

where $\boldsymbol{x}_{s_2} = [x_{s_2}, y_{s_2}, z_{s_2}]^T$ represents coordinates of centre of gravity of the first body and $\boldsymbol{x}_2$ are coordinates of the particular point in the local coordinate system of the first body. $\boldsymbol{x}_{s_3} = [x_{s_3}, y_{s_3}, z_{s_3}]^T$ are coordinates of centre of gravity of the second body and $\boldsymbol{x}_3$ are coordinates of a point in the local coordinate system of the second body. Variables $\psi_i, \vartheta_i, \varphi_i$, $i \in 2, 3$ are known as Euler's angles [8].

Vector of generalized coordinates of the whole system is defined as

$$\boldsymbol{q} = [x_{s_2}, y_{s_2}, z_{s_2}, \psi_2, \vartheta_2, \varphi_2, x_{s_3}, y_{s_3}, z_{s_3}, \psi_3, \vartheta_3, \varphi_3]^T.$$

The set of kinematics constraint equations can be defined as

$$\boldsymbol{\Phi} = \begin{bmatrix} \boldsymbol{\Phi}_1 \\ \boldsymbol{\Phi}_2 \end{bmatrix} = \begin{bmatrix} \Phi_1 \\ \Phi_2 \\ \Phi_3 \\ \Phi_4 \\ \Phi_5 \\ \Phi_6 \end{bmatrix} = \begin{bmatrix} \boldsymbol{x}_{s2} + \boldsymbol{S}_{12} \begin{bmatrix} 0 \\ 0 \\ -a_{13} \end{bmatrix} \\ \boldsymbol{x}_{s2} + \boldsymbol{S}_{12} \begin{bmatrix} 0 \\ 0 \\ a_{13} \end{bmatrix} - \boldsymbol{x}_{s3} - \boldsymbol{S}_{13} \begin{bmatrix} 0 \\ 0 \\ -a_{23} \end{bmatrix} \end{bmatrix} = \begin{bmatrix} 0 \\ 0 \\ 0 \\ 0 \\ 0 \\ 0 \end{bmatrix}. \tag{8}$$

This generates six equations of the kinematics constraint in term of the local coordinates

$$\boldsymbol{\Phi}(\boldsymbol{q}, t) = \begin{bmatrix} \mathrm{x_{s2}} - \mathrm{a_{13}} \sin(\vartheta_2) \sin(\psi_2) \\ \mathrm{y_{s2}} + \mathrm{a_{13}} \cos(\psi_2) \sin(\vartheta_2) \\ \mathrm{z_{s2}} - \mathrm{a_{13}} \cos(\vartheta_2) \\ \mathrm{x_{s2}} - \mathrm{x_{s3}} + \mathrm{a_{13}} \sin(\vartheta_2) \sin(\psi_2) + \mathrm{a_{23}} \sin(\vartheta_3) \sin(\psi_3) \\ \mathrm{y_{s2}} - \mathrm{y_{s3}} - \mathrm{a_{13}} \cos(\psi_2) \sin(\vartheta_2) - \mathrm{a_{23}} \cos(\psi_3) \sin(\vartheta_3) \\ \mathrm{z_{s2}} - \mathrm{z_{s3}} + \mathrm{a_{13}} \cos(\vartheta_2) + \mathrm{a_{23}} \cos(\vartheta_3) \end{bmatrix} = \boldsymbol{0}, \tag{9}$$

where a_{ij} represent a length of semi-principal axes.

2.2. Equations of motion

Equations of motion are derived using Lagrange equations of second kind with multipliers. Second derivatives on the kinematics constraints were added to the system and these formulate equation of motion of the double pendulum system

$$\begin{bmatrix} \boldsymbol{M} & \Phi_{\mathrm{q}}^T \\ \Phi_{\mathrm{q}} & \boldsymbol{0} \end{bmatrix} \begin{bmatrix} \ddot{\boldsymbol{q}} \\ -\boldsymbol{\lambda} \end{bmatrix} = \begin{bmatrix} \boldsymbol{f}(\boldsymbol{q}, \dot{\boldsymbol{q}}, t) \\ \boldsymbol{\gamma}(\boldsymbol{q}, \dot{\boldsymbol{q}}, t) \end{bmatrix}, \tag{10}$$

where $\boldsymbol{M}$ is mass matrix, $\ddot{\boldsymbol{q}}$ represents generalized accelerations vector, $\boldsymbol{\lambda}$ is vector of Lagrange's multipliers, Φ_{q} is the Jacobian of the vector of constraints, $\boldsymbol{f}$ and γ are vectors of external forces (including contact force), and rest after derivation, respectively. Equation (10) is a differential-algebraic equation of second order. An important classification of differential equations is whether it is a stiff or a non-stiff problem, associated with eigenfrequency distribution [4]. The example here is considered to be stiff problem and this can cause difficulties during numerical integration. Thus the special numerical solvers are implemented.

To express accelerations $\ddot{q}$ and solve the equation by numerical integration, the approach called elimination of the Lagrange multipliers is applied. Using this technique, following system is obtained

$$\begin{bmatrix} \dot{\boldsymbol{u}} \\ \dot{\boldsymbol{v}} \end{bmatrix} = \begin{bmatrix} \boldsymbol{v} \\ \ddot{\boldsymbol{q}} \end{bmatrix} = \begin{bmatrix} \dot{\boldsymbol{q}} \\ \boldsymbol{M}^{-1}\{\boldsymbol{f} + \Phi_{\mathrm{q}}^T(\Phi_{\mathrm{q}}\boldsymbol{M}^{-1}\Phi_{\mathrm{q}}^T)^{-1}(\gamma - \Phi_{\mathrm{q}}\boldsymbol{M}^{-1}\boldsymbol{f})\} \end{bmatrix}. \tag{11}$$

Equation (11) can be solved using standard techniques of numerical integration, however it has some undesirable troubles. It might be numerically unstable for a certain properties, thus Baumgarte's stabilization method solving bad stability is applied [4].

This brings new formulation of the first order differential-algebraic equation, which can be numerically solved

$$\begin{bmatrix} \dot{\boldsymbol{u}} \\ \dot{\boldsymbol{v}} \end{bmatrix} = \begin{bmatrix} \boldsymbol{v} \\ \ddot{\boldsymbol{q}} \end{bmatrix} = \begin{bmatrix} \dot{\boldsymbol{q}} \\ \boldsymbol{M}^{-1}\{\boldsymbol{f} + \Phi_{\mathrm{q}}^T(\Phi_{\mathrm{q}}\boldsymbol{M}^{-1}\Phi_{\mathrm{q}}^T)^{-1}(\gamma - 2\alpha\dot{\boldsymbol{\Phi}} - \beta^2\boldsymbol{\Phi} - \Phi_{\mathrm{q}}\boldsymbol{M}^{-1}\boldsymbol{f})\} \end{bmatrix}. \tag{12}$$

Constants α and β were chosen based on literature [4]. MATLAB [6] software is applied to calculate numerical solution. There are some suitable numerical ODE solvers for the stiff problems implemented in MATLAB, such as ODE15s, ODE23t, ODE23tb.

2.3. Arm gravity motion

Double pendulum system is used as an approximation of a human arm. Valdmanová in [13] established a 2D model of an arm based on multibody approach. This model represents the main parts of a human arm, namely the upper arm, the forearm and the hand. Later on, the model is simplified into a two bodies system only, since the motion between forearm and hand can be neglected. Valdmanová compared in her work a simulation with a result of an experiment. Joints between bodies are modelled to be joints with an internal stiffness. Thus the bodies load with torques representing rigidity of a shoulder and of an elbow, respectively. Geometric properties of the bodies are set of from [13]. Passive bending moments of joints are defined by curves based on [9].

Initial position of the arm corresponding to an experiment is based on anthropometric data, namely the driver's position while holding a steering wheel. Initial conditions of the arm are shown in Fig. 2, where angles $\varphi_1 = -45°$ and $\varphi_2 = 23°$.

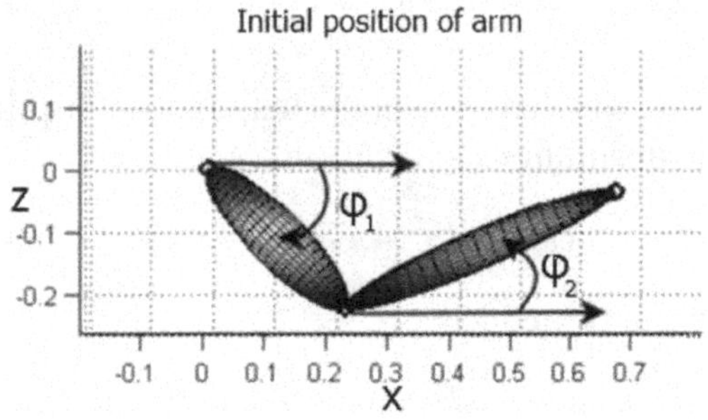

Fig. 2. Initial position of arm

2.4. Contact calculation

This work concerns possible impact between any ellipsoid of the double pendulum and a plain. If the bodies get into a collision, the crucial question is to evaluate impact performance of a contact force. Several approaches for a contact force expression were developed. The concept of this work is to use a continuous contact force model, where the contact force is a function of local penetration δ and local penetration velocity $\dot{\delta}$, respectively. Three contact force models are presented here, namely Hertz model, spring-dashpot model and non-linear damping model, respectively. To capture the effect of contact force in case of interaction bodies, the penetration depth is calculated. To identify whether the bodies are getting into a collision, the minimum distance between them is required. As long as the distance is positive, the bodies are disjointed. Change of the sign indicates a collision and negative distance magnitude is equal to penetration δ. Several algorithms for minimum distance calculation were published [1,3,11,14]. This study is focused on the analytical approach of minimum distance problem [2]. Idea of this method is to create a new plain, parallel to an initial one and tangential to the ellipsoid. When the common point, marked as C, of a new plain and of an ellipsoid is detected, distance between this point and the plain can be calculated, using adequate equation from analytical geometry. There always exist two such parallel plains, as is shown in Fig. 3.

2.5. Minimum distance problem application

Let us show analytical solution of the contact problem between an ellipsoid and a plain. The standard equation of an ellipsoid centred at the origin of a Cartesian coordinate system and

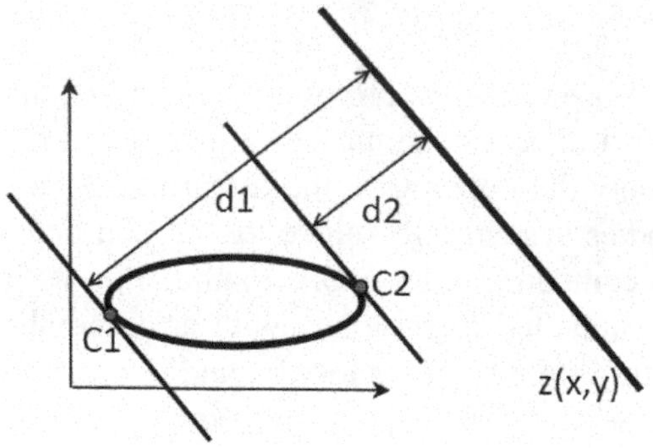

Fig. 3. Ellipsoid and two parallel plains

aligned with the axes is

$$\frac{x^2}{a^2} + \frac{y^2}{b^2} + \frac{z^2}{c^2} = 1, \tag{13}$$

where a, b and c are constants, which represent the length of semi-principal axes. Equation (13) can be rearranged by set of substitutions to the form

$$Ax^2 + By^2 + Cz^2 + D = 0. \tag{14}$$

A general equation of plain can be defined as

$$kx + ly + mz + n = 0. \tag{15}$$

An ellipsoid is a type of a quadric surface in coordinates $\{x, y, z\}$. Thus equation (15) can be rearranged to be a function $z = z(x, y)$ as

$$z(x, y) = -\frac{k}{m}x - \frac{l}{m}y - \frac{n}{m}. \tag{16}$$

A plain can be defined using one point and two vectors. To ensure new plain being parallel with the initial one, at least two gradient vectors of both plains have to be the same. The two gradient vectors together with one point on the surface of the ellipsoid, can identify the required tangential plain. Hence the gradients $\frac{\partial z}{\partial x}$ and $\frac{\partial z}{\partial y}$ of the plain $z = z(x, y)$ are evaluated as

$$\frac{\partial z(x,y)}{\partial x} = -\frac{k}{m} \tag{17}$$

and

$$\frac{\partial z(x,y)}{\partial y} = -\frac{l}{m}. \tag{18}$$

Equation (14) is differentiated with respect to variables x and y as

$$\frac{\partial}{\partial x}: \qquad 2Ax + 2By\underbrace{\frac{\partial y}{\partial x}}_{0} + 2Cz\underbrace{\frac{\partial z}{\partial x}}_{-\frac{k}{m}} = 0 \tag{19}$$

and

$$\frac{\partial}{\partial y}: \qquad 2Ax\underbrace{\frac{\partial x}{\partial y}}_{0} + 2By + 2Cz\underbrace{\frac{\partial z}{\partial y}}_{-\frac{l}{m}} = 0. \tag{20}$$

Since x and y are independent variables, mixed derivatives are equal to zero. If equation (17) and equation (18) are substituted into equation (19) and equation (20) together with general equation of ellipsoid (14) generate the system of three equations for unknown variables x, y and z as

$$2Ax - 2Cz\frac{k}{m} = 0, \tag{21}$$

$$2By - 2Cz\frac{l}{m} = 0 \tag{22}$$

and

$$f(x,y,z) = Ax^2 + By^2 + Cz^2 + D = 0. \tag{23}$$

Solution of this system of equations generates two points, namely point $C_1 = [x_{10}, y_{10}, z_{10}]$ and point $C_2 = [x_{20}, y_{20}, z_{20}]$, which are mutual points of the body and the new tangential plain. These points are also the points of extrema distance (minimum and maximum) between plain and body, see Fig. 3. When coordinates of these points are known, it is very straightforward to calculate distance between these points and plain. The distance between point $X_0 = [x_0, y_0, z_0]$ and plain $kx + ly + mz + n = 0$ is given by

$$d = \frac{kx_0 + ly_0 + mz_0 + n}{\sqrt{k^2 + l^2 + m^2}}. \tag{24}$$

Equation (24) results two extrema distances between the ellipsoid and the plain, so minimum one is required. However, this method is working only for the ellipsoid, whose semi principal axes are parallel with coordinate axes. Both of the entities (the body and the plain) have to be expressed in the identical coordinate system to applied the method defined above. Here, the equation of the plain in form (15) is expressed in the global, frame-fixed, coordinate system, but equation of ellipsoid (13) is evaluated in the local body fixed coordinate system.

2.5.1. Transformation

Actual position of any point of ellipsoid is defined by 6 independent coordinates x_s, y_s, z_s, ψ, ϑ, φ, where x_s, y_s and z_s are coordinates of centre of gravity and ψ, ϑ and φ are Euler's angles. The principle applied here is based on the transformation of a plain equation from a global coordinate system, marked as $\boldsymbol{x}_1$ into a local body fixed coordinate system, marked as $\boldsymbol{x}_2$. For the transformation, it is useful to write the plain and the ellipsoid equations in a matrix form using homogeneous coordinates. Thus the plain equation is

$$\begin{bmatrix} k & l & m & n \end{bmatrix} \begin{bmatrix} x_1 \\ y_1 \\ z_1 \\ 1 \end{bmatrix} = 0, \tag{25}$$

or in compact matrix form

$$\boldsymbol{r}^T \boldsymbol{x}_1 = 0. \tag{26}$$

The ellipsoid equation comes to

$$\begin{bmatrix} x_2 & y_2 & z_2 & 1 \end{bmatrix} \begin{bmatrix} A & 0 & 0 & 0 \\ 0 & B & 0 & 0 \\ 0 & 0 & C & 0 \\ 0 & 0 & 0 & D \end{bmatrix} \begin{bmatrix} x_2 \\ y_2 \\ z_2 \\ 1 \end{bmatrix} = 0, \tag{27}$$

or in matrix form

$$(\boldsymbol{x}_2)^T \boldsymbol{A} \boldsymbol{x}_2 = 0. \tag{28}$$

As is mentioned above, both the entities need to be expressed in the same coordinate system. The very crucial question is how to transform those equations to be expressed in the same coordinate system. A purpose of the transformation is to obtain equations in such a form, which the method of minimum distance calculation can be applied on. There are two possibilities how to assure this condition:

- The first option is using matrix $\boldsymbol{T}$ to transform ellipsoid equation (28) from the local coordinate system to the global one (where the plain is defined) as

$$\boldsymbol{x}_1^T \boldsymbol{T}^T \boldsymbol{A} \boldsymbol{T} \boldsymbol{x}_1 = 0. \tag{29}$$

- The second one is to use matrix $\boldsymbol{T}^{-1}$ to transform plain equation (26) from the global coordinate system to the local one (where the ellipsoid is defined) as

$$\boldsymbol{r} \boldsymbol{T}^{-1} \boldsymbol{x}_2 = 0, \tag{30}$$

  in which $\boldsymbol{T}$ is a transformation matrix between the local and the global coordinate system and obviously $\boldsymbol{T}^{-1}$ is a transformation matrix from the global to the local coordinate system.

- The first option results a scalar equation, but it is highly non-linear and it is not possible to arrange that in a form

$$\tilde{A}x_1^2 + \tilde{B}y_1^2 + \tilde{C}z_1^2 + \tilde{D} = 0, \tag{31}$$

  where $\tilde{A}$, $\tilde{B}$, $\tilde{C}$ and $\tilde{D}$ can be any arbitrary constants. So, this option is not suitable for this purpose.

- The second option also results a scalar equation, but this can be written in the same form as the original one as

$$\tilde{k}x_2 + \tilde{l}y_2 + \tilde{m}z_2 + \tilde{n} = 0, \tag{32}$$

  where $\tilde{k}, \tilde{l}, \tilde{m}, \tilde{n}$ are constants defined by particular transformations.

Now both (the plain and the ellipsoid) equations are expressed in the same coordinate system (local body-fixed) and the standard distance calculation method described above can be used.

Equation of a transformed plain equation (32) together with original equation of ellipsoid (14) are satisfactory inputs to the minimum distance calculation method. By solving system of equations, two points of extreme distance C_1 and C_2 are evaluated and two extreme distances can be calculated and obviously minimum one is required

$$d = \min_i(d_i) = \frac{\tilde{k}x_{i0} + \tilde{l}y_{i0} + \tilde{m}z_{i0} + \tilde{n}}{\sqrt{\tilde{k}^2 + \tilde{l}^2 + \tilde{m}^2}}, \qquad i \in \{1, 2\}. \tag{33}$$

2.6. *Contact force*

This study is focused on a continuous models implementation, in which impact force is defined to be a function of local penetration δ and local penetration velocity $\dot{\delta}$, respectively. Relative normal contact force acts at the contact point and can be defined as

$$\boldsymbol{f}_n = \boldsymbol{f}_n(\delta, \dot{\delta}). \tag{34}$$

Normal vector of the ellipsoid expressed at point C is then

$${}^c\tilde{\boldsymbol{n}} = \left[\frac{\partial f(x,y,z)}{\partial x}\bigg|_C, \frac{\partial f(x,y,z)}{\partial y}\bigg|_C, \frac{\partial f(x,y,z)}{\partial z}\bigg|_C \right]^T, \tag{35}$$

where f is the smooth regular surface, defined by (23). For the purpose of defining contact force, normal vector is normalized to have a unit length

$${}^c\boldsymbol{n} = \frac{{}^c\tilde{\boldsymbol{n}}}{\|{}^c\,\tilde{\boldsymbol{n}}\,\|}. \tag{36}$$

Calculation of the relative normal contact velocity (penetration velocity) is done by differentiating equation (24)

$$\dot{\delta} = \frac{\mathrm{d}}{\mathrm{d}t}\delta = \frac{\mathrm{d}}{\mathrm{d}t}\left\{ \frac{kx_0 + ly_0 + mz_0 + n}{\sqrt{k^2 + l^2 + m^2}} \right\}. \tag{37}$$

Vector of contact force $\boldsymbol{f}_n$ can be evaluated using entities above, regarding adequate contact force models:

- Hertz model

$$\boldsymbol{f}_n = f_n\,{}^c\boldsymbol{n} = k_h\,\delta^n\,{}^c\boldsymbol{n}. \tag{38}$$

- Spring dashpot model

$$\boldsymbol{f}_n = f_n\,{}^c\boldsymbol{n} = (k_{sd}\,\delta + b_{sd}\,\dot{\delta})^c\boldsymbol{n}. \tag{39}$$

- Non-linear damping model

$$\boldsymbol{f}_n = f_n\,{}^c\boldsymbol{n} = (k_{nl}\,\delta^n + b_{nl}\,\delta^p\dot{\delta}^q)^c\boldsymbol{n}. \tag{40}$$

Parameters k, b, p, q, n are constants and it is common to set them $p = n$ and $q = 1$. Parameter k represents artificial spring stiffness and b is artificial damping coefficient. Constants k and b depend on various aspects, such as material and geometric properties of contacting bodies. Acting force $\boldsymbol{f}_n$ is then translated to the centre of gravity of the body including a torque $\boldsymbol{m}$ caused by the translation. Fig. 4 shows two equivalent systems, first one with contact force acting at the contact point and second system loaded with moment and force acting at the centre of gravity.

Moment is then defined as

$$\boldsymbol{m} = \boldsymbol{r} \times \boldsymbol{f}_n, \tag{41}$$

where vector $\boldsymbol{r}$ can be expressed using coordinates of a contact point.

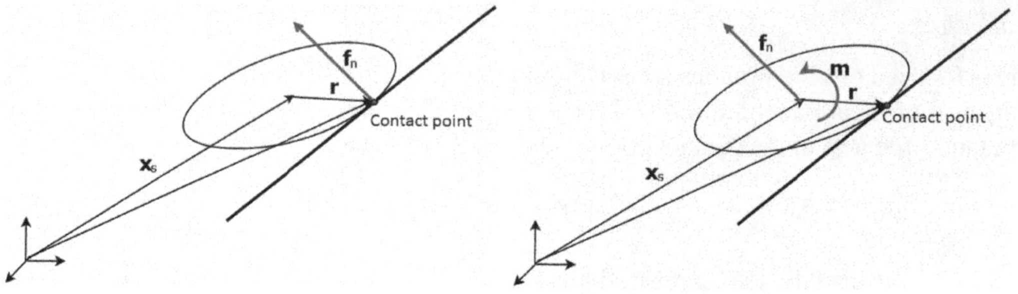

(a) Original position of contact force (b) Translated force and a moment

Fig. 4. Two equivalent systems

2.6.1. Force implementation

In case of contacting bodies right hand side of equation of motion (10) includes contact force $\boldsymbol{f}_n$ and torque $\boldsymbol{m}$ and comes to a following form

$$\boldsymbol{f} = \begin{bmatrix} F_{nx} \\ F_{ny} \\ F_{nz} - mg \\ M_x \\ M_y \\ M_z \end{bmatrix} . \tag{42}$$

For separated bodies, $\boldsymbol{f}_n = \boldsymbol{0}$ and thus vector $\boldsymbol{f}$ comes to a simple form of unconstrained model loaded only with gravity

$$\boldsymbol{f} = \begin{bmatrix} 0 \\ 0 \\ -mg \\ 0 \\ 0 \\ 0 \end{bmatrix} . \tag{43}$$

2.7. Contact parameters optimization

Continuous model defines contact force to be a function of penetration δ between contacting bodies and penetration velocity $\dot{\delta}$ and parameters k and b, respectively. Experimental results of simple impact example are used in comparison with numerical simulations to obtain appropriate values of parameters k and b for each model. By varying the theoretical quantities, the most corresponding results of simulation to an original experiment can be achieved. Stiffness and damping parameters are so called optimization parameters and difference between experimental and calculated results is an objective function, which is desirable to be minimised. Th method of numerical optimization is introduced here. An example of application considered here is bouncing ball, published in [5]. An elastic ball with an initial height equalling to 1.0 m, mass of 1 kg, moment of inertia equalling to 0.1 kg $\cdot$ m^2 and radius equalling to 0.1 m, are released from initial position under action only with acceleration of gravity g equalling to 9.81 m $\cdot$ s^{-2}, see Fig. 5. The ball is falling down until it collides with a rigid ground. When the ball collides a contact takes place and the ball rebounds, producing a jump, which height depends on parameters of the contact force.

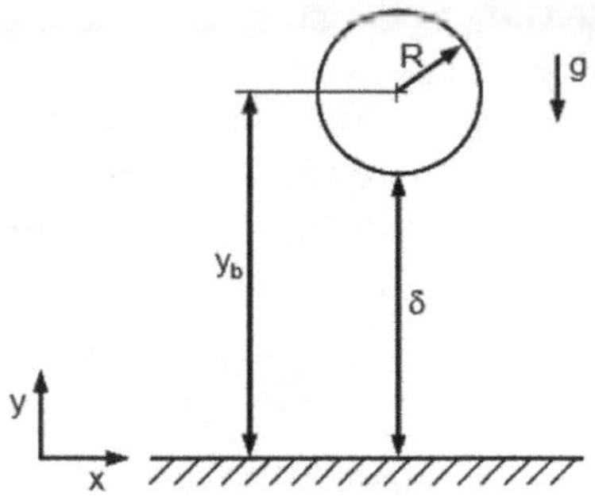

Fig. 5. Bouncing ball example [5]

MATLAB software is used to solve the equation of motion of the ball and software optiS-Lang 3.2.0, controls variation of input parameters k and b, respectively. MATLAB software version R2011b under MS Windows platform on single processor Core Duo T2400 computer with frequency of 1.83 GHz a 2 MB L2 cache, is used to numerically solve example of bouncing ball. MATLAB has implemented several numerical solvers for these stiff problems. However, it is not very straightforward to select a suitable one. In order to choose the best one for further applications, based on minimum calculation time, four stiff numerical solvers ODE are applied on the same system. Simulations of bouncing ball example with 1 s and 5 s duration time are presented.

3. Results

3.1. Arm gravity motion

Following Fig. 6a shows the motion of the elbow of the right arm. While solid curve represents 3D double pendulum simulation, dash-dot curve represents 2D simulation of arm model [13] and the points represent experimental results [13]. The second graph, see Fig. 6b, shows motion of the wrist, where curves are same with the Fig. 6a.

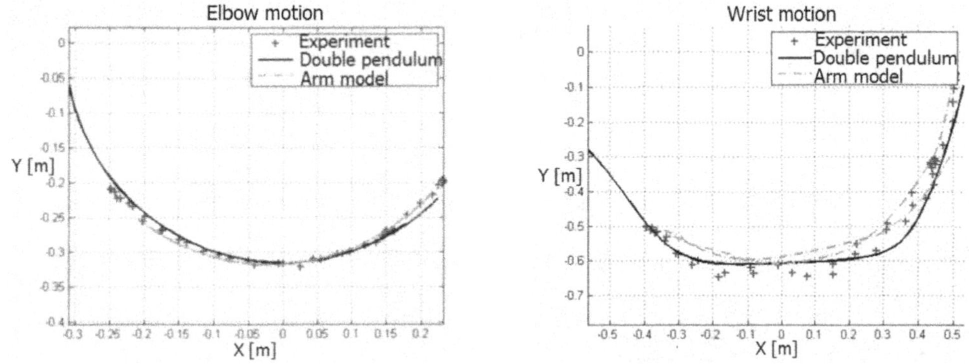

(a) Trajectories of an elbow (b) Trajectories of a wrist

Fig. 6. Comparison of human arm models and experiment

Previous figures show the results of simulations in comparison with experiment. Although the trajectory of the wrist slightly differs from the experiment and also from Valdmanova's simulation, the results refer to an equivalence of the systems.

3.2. Solver selection

The most suitable numerical solver for this particular problem are chosen based on computation time of four different stiff solvers. These are applied on the same system, namely the bouncing ball example. Simulation of 1 s and 5 s duration time are tested and computation times are compared.

Table 1 shows the computation times of particular simulations. Based on the results, solver ODE15s is used in MATLAB for the numerical integration in the further calculations.

Table 1. Calculation time of identical simulation with different solvers

Solver	Computation time [s] 1 s simulation	5 s simulation
ODE23t	169	1 820
ODE23tb	278	2 675
ODE15s	166	1 382
ODE23s	1 592	20 045

3.3. Numerical optimization of contact parameters

Solver ODE15s implemented in MATLAB software is used to solve the equation of motion of a bouncing ball. Software optiSLang controls variation of contact parameters k and b, respectively, to reach the most corresponding results of simulation to an original experiment. Mathematics optimization principle is applied on the three contact force models. Namely Hertz model, spring-dashpot model and non-linear damping model, respectively. Calculated position of ball centre of gravity together with the initial experiment are shown in following Fig. 7.

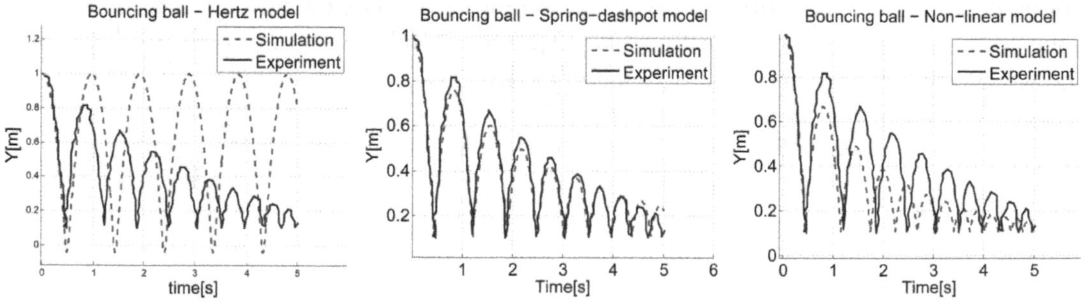

(a) Hertz model (b) Spring-dashpot model (c) Non-linear damping model

Fig. 7. Numerical optimizations results applied on bouncing ball example

Evaluated contact force parameters are displayed in Table 2.

Table 2. Contact parameters of particular force models

Model / Parameter	k	b
Hertz's	10 000	–
Spring-dashpot	3.303e+7	2.157e+4
Non-linear damping	3.009e+7	3.000e+4

3.3.1. Discussion

The Hertz model performs an elementary model suitable for the first approximation of impact. Since it does not take energy dissipation phenomena into account, it is not applicable for all configurations. In case of a fully elastic impact this model can provide satisfactory results. The spring-dashpot model takes energy dissipation effect into account, through damping coefficient that includes a coefficient of restitution. It refers to a more realistic situation, since it is not limited with an elastic impact. By varying with the coefficient of restitution between 0 and 1, phenomena between a fully plastic and a fully elastic impact can be captured. The non-linear damping force model also works with dissipation of energy, but the calculation states unstable. Compared to the spring-dashpot model, the curves of an experiment and a numerical simulation differ significantly. Based on the calculations, the spring-dashpot model provides results the most corresponding with the experiment. Due to this fact, it is used in further applications.

3.4. Double pendulum contacting a plain

The double pendulum system was described and validated to be a suitable approximation of a human arm. Purpose of this part is to evaluate results of the system including a contact with a plain. This can be applied in further applications such as the approximation of an arm or a leg undergoing into an impact with an infrastructure. Motion of the double pendulum that getting into a contact with plain is displayed in Fig. 8.

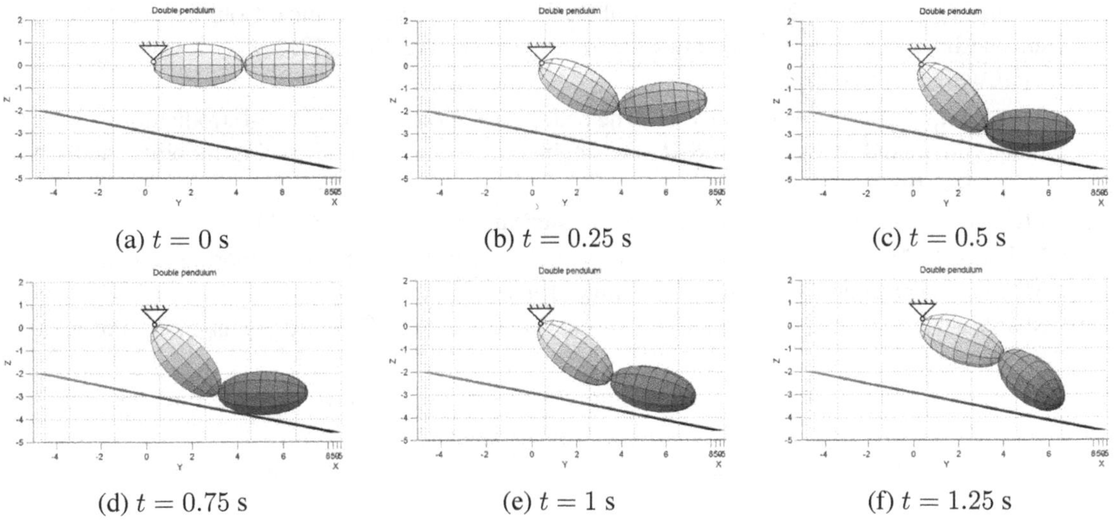

(a) $t = 0$ s (b) $t = 0.25$ s (c) $t = 0.5$ s

(d) $t = 0.75$ s (e) $t = 1$ s (f) $t = 1.25$ s

Fig. 8. Position of the double pendulum contacting a plain

4. Conclusion

Contact or impact scenario in virtual human body modelling plays significant role in biomechanics research. Various approaches in biomechanical modelling are currently developed The purpose of this work is to evaluate and test the algorithm for the double pendulum getting into a contact with a plain. The equations of motion of the double pendulum were derived using the Lagrange equation of second kind with multipliers. Evaluating of contact force parameters is performed using numerical optimization principle applied on bouncing ball example. Three contact force models are investigated, namely the Hertz model, the spring-dashpot model and the non-linear damping model. Optimized contact force parameters are used in the double pendulum impact scenario. Assuming a reference of biomechanics researches, the double pendulum system might approximate various segments of a human body. Application of a human arm problem was verified here. Impact scenario is demonstrated on the double pendulum getting into a contact with plain.

Acknowledgements

This work is supported by internal grant project SGS-2013-026.

References

[1] Boss, M., McPhee, J., Volumetric contact dynamics model and experimental validation of normal forces for simple geometries, Proceedings of the ASME 2011 International Design Engineering Technical Conferences & Computers and Information in Engineering Conference IDETC/CIE 2011, 2011.

[2] Drexel University. The Math forum @ Drexel 2010.

[3] Eberly, D., Intersection of ellipsoids, Geometric tools, LLC, 2010.

[4] Hajžman, M., Polach, P., Application of stabilization techniques in the dynamic analysis of multibody systems, Applied and Computational Mechanics 1 (2) (2007) 479–488.

[5] Machado, M. F., Flores, P., A novel continuous contact force model for multibody dynamics, Proceedings of the ASME 2011 International Design Engineering Technical Conferences & Computers and Information in Engineering Conference IDETC/CIE 2011, 2011.

[6] MathWorks. MATLAB R2010a.

[7] Moser, A., Steffan, H., Kasanický, G., The pedestrians model in PC-Crash-Introduction of a multibody system and its validation. SAE The engineering society for advancing mobility land sea air and space, 1999.

[8] Pfeiffer, F., Unilateral multibody dynamics, Meccanica 34 (1999) 437–451.

[9] Robbins, D. H., Anthropometry of motor vehicle occupants, Technical report, The University of Michigan, Transportation Research Institute, 1983.

[10] Shabana, A. A., Dynamics of multibody systems. 3rd edition, Cambridge University Press, Cambridge, 2005.

[11] Sohn, K. A., Juttler, B., Kim, M. S., Wang, W., Computing distance between surfaces using line geometry, IEEE Computer society, 2002.

[12] Stejskal, V., Valášek, M., Kinematics and dynamics in machinery, Marcel Dekker, New York, 1996.

[13] Valdmanová, L., Multibody model of upper extremity in 2D, Master thesis, University of West Bohemia, Pilsen, 2009. (in Czech)

[14] Wang, W., Choi, Y. K., Chan, B., Kim, M. S., Wang, J., Efficient collision detection for moving ellipsoids using separating planes. Computing 72 (2004) 235–246.

[15] Zhou, Q., Quade, M., Du, H., Concept design of a 4-DOF pedestrians legform. ESV Technical paper 07-0196, 2007.

7

A comparative study of 1D and 3D hemodynamics in patient-specific hepatic portal vein networks

A. Jonášová[a,*], O. Bublík[a], J. Vimmr[a]

[a]European Centre of Excellence NTIS – New Technologies for the Information Society, Faculty of Applied Sciences, University of West Bohemia, Univerzitní 8, 306 14 Pilsen, Czech Republic

---

Abstract

The development of software for use in clinical practice is often associated with many requirements and restrictions set not only by the medical doctors, but also by the hospital's budget. To meet the requirement of reliable software, which is able to provide results within a short time period and with minimal computational demand, a certain measure of modelling simplification is usually inevitable. In case of blood flow simulations carried out in large vascular networks such as the one created by the hepatic portal vein, simplifications are made by necessity.

The most often employed simplification includes the approach in the form of dimensional reduction, when the 3D model of a large vascular network is substituted with its 1D counterpart. In this context, a question naturally arises, how this reduction can affect the simulation accuracy and its outcome. In this paper, we try to answer this question by performing a quantitative comparison of 3D and 1D flow models in two patient-specific hepatic portal vein networks. The numerical simulations are carried out under average flow conditions and with the application of the three-element Windkessel model, which is able to approximate the downstream flow resistance of real hepatic tissue. The obtained results show that, although the 1D model can never truly substitute the 3D model, its easy implementation, time-saving model preparation and almost no demands on computer technology dominate as advantages over obvious but moderate modelling errors arising from the performed dimensional reduction.

Keywords: patient-specific model, blood flow, finite volume method, Windkessel model

---

1. Introduction

In the last two decades, human medicine has experienced a remarkable boom in the field of computer-aided imaging methods. With all the possibilities offered, for example, by the computed tomography (CT), it is not surprising that the latest efforts of the bioengineering community are directed toward the development of computational software that would help surgeons in their pre-operative planning and/or aid them during difficult and often life-threatening surgeries. However, compared to numerical simulations performed in industry, where a computation may take days or even weeks depending on the complexity of the solved problem, medicine and especially the clinical practice require results within a short time period and with minimal computational demand. With such strict requirements in mind, a development of reliable clinical software is, thus, not easy and it is only natural that a certain measure of simplification is inevitable and, in some cases, even necessary.

The impact of geometry and model simplifications on blood flow simulations is addressed in the present paper, which is one of the results of multidisciplinary research carried out at the University of West Bohemia in close co-operation with the medical doctors of the University

*Corresponding author. e-mail: jonasova@ntis.zcu.cz.

Hospital Pilsen and the Faculty of Medicine in Pilsen of the Charles University in Prague. The research, which is directed towards the development of clinical software for liver volumetry [1] and multiscale modelling of tissue perfusion [4], is primarily motivated by the growing need of vascular surgeons to improve the current pre-operative planning of liver surgeries. For example, a surgical removal of a tumour bearing part of the liver (liver resection) is usually performed on the basis of several on-site ultrasound measurements that help to identify the approximate boundaries of functionally independent hepatic segments (e.g., lobes). The final resection line is then chosen as an approximation of these boundaries and perceived as the 'optimal' surgical solution, although in most cases it is anything but. Thus, the need for a more accurate resection approach and with it associated development of a computer-aided pre-operative planning system arose.

Considering the complexity of the solved problem, which involves not only the simulation of vascular blood flow, but also the modelling of hepatic tissue perfusion, see, e.g., [3], several modelling simplifications have to be made. In terms of vascular blood flow, which involves the networks of the portal and hepatic veins (the hepatic arterial flow is neglected), the main simplification takes the form of dimensional reduction. In other words, the hepatic veins are modelled as a network of 1D segments instead of complex 3D structures. The use of this approach in a patient-specific model of human liver is apparent from Fig. 1, which shows the time evolution of contrast medium propagation within portal and hepatic vein systems computed with the help of the aforementioned 1D models of hepatic veins (visualised as black lines in Fig. 1).

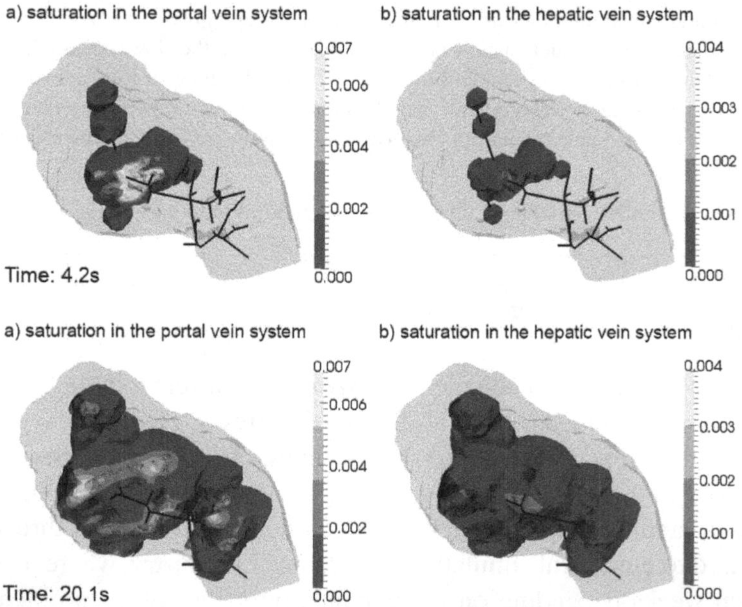

Fig. 1. Example of contrast medium propagation in a patient-specific model of human liver at two selected time instants: $t_1 = 4.2$ s (*top*) and $t_2 = 20.1$ s (*bottom*) (veins shown as black lines) [4]

As the length and distribution of inflow/outflow vessels can significantly affect the resulting tissue perfusion and the subsequent contrast medium propagation, it is crucial to understand the impact of model simplification on the overall simulation accuracy. With this in mind, the main objective of the present study is to compare and assess the flow fields computed by the 3D and 1D models representing two patient-specific hepatic portal vein systems of different complexity.

2. Models and methods

2.1. Hepatic portal vein networks

For the purpose of the present comparative hemodynamical study, we consider two patient-specific portal vein geometries with different levels of complexity and total number of outlets (9 and 39), Fig. 2. To both models, which originate from patient-specific data provided by the courtesy of the University Hospital Pilsen, we shall in the rest of this paper refer to as the simple and complex vascular networks. Their 3D representations shown in Fig. 2 are a result of a semi-automatic reconstruction process carried out on raw image segmentation data prepared by Miroslav Jiřík from the Department of Cybernetics at the University of West Bohemia. The final stage of the reconstruction process involves a smoothing of coarse surface meshes by the well-known Taubin smoothing algorithm [7] and their complete remeshing by the in-house software DICOM2FEM [2]. Finally, tetrahedral computational meshes for the two 3D network models are generated with the help of the software package HyperMesh v11.0 (Altair Engineering, Troy, USA). The number of cells contained in the simple and complex hepatic portal vein networks follows the results of a preliminary mesh sensitivity analysis, which revealed little flow changes in meshes refined near the walls and consisting of at least 816,547 and 2,042,156 cells, respectively. Both these meshes are used in the following blood flow simulations.

For the 1D analogue of the vascular networks mentioned above, the knowledge of lumens and their centres at planes of the CT scans is used, see Fig. 3 (left). To be more specific, each of the two 1D networks consist of simple segments connecting two points situated within the

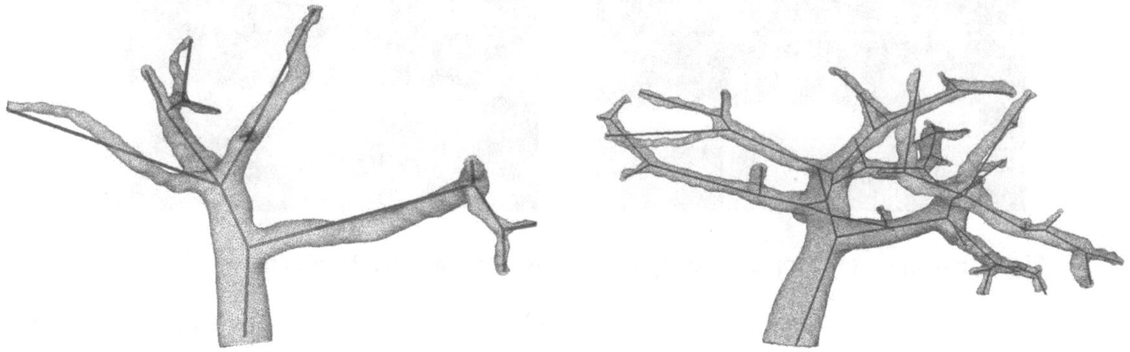

Fig. 2. 1D and 3D reconstructions of simple (*left*) and complex (*right*) hepatic portal vein networks consisting of 9 and 39 outlets, respectively

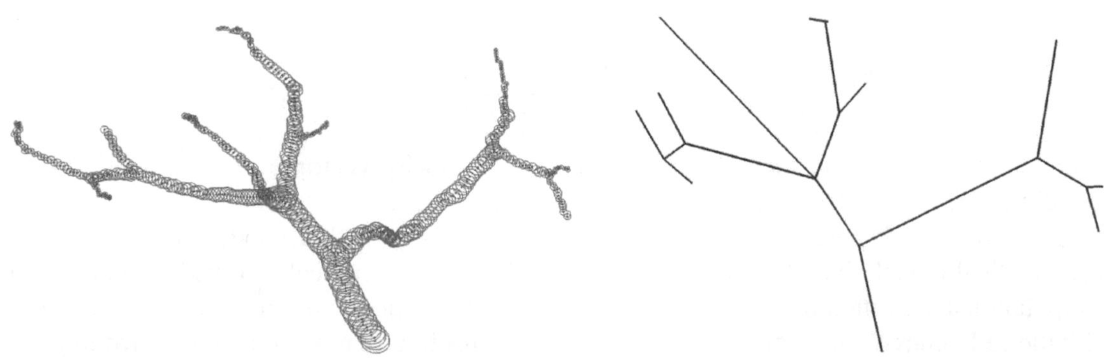

Fig. 3. Reconstruction of the non-segmented simple hepatic portal vein network with lumens and their centres (*left*) and with 1D segments shown as lines (*right*)

vascular network that either represent the inlet, outlet or bifurcation points, Fig. 3 (right). This way, the 1D analogue of the simple and complex hepatic portal vein networks shown in Fig. 2 is created consisting of 18 or 78 segments, respectively. For the purpose of blood flow modelling, which will be addressed later, the segments of the 1D networks also carry the information about their average inner diameter and overall length, both derived from the non-segmented vascular tree model, Fig. 3 (left).

2.2. *Three-dimensional blood flow*

For the 3D modelling of blood flow in this paper, the assumption of impermeable and inelastic vessel walls is used. The simplification in the form of neglected vascular elasticity is motivated by the fact that blood flow in the hepatic portal vein is known to have a relatively steady character, as illustrated by the examples of velocity waveforms in Fig. 4 measured by an ultrasound Doppler test. As for the impermeability assumption, the objective of the paper is only to consider the upper hierarchy of the perfusion tree, i.e., only the transport of blood into corresponding parts of the liver without the possibility of perfusion is considered. Because of the limitation of the 1D model introduced below, the blood in this study is taken as an incompressible Newtonian fluid with the density of $1\,050\,\mathrm{kg}\cdot\mathrm{m}^{-3}$ and dynamic viscosity of $0.003\,45\,\mathrm{Pa}\cdot\mathrm{s}$.

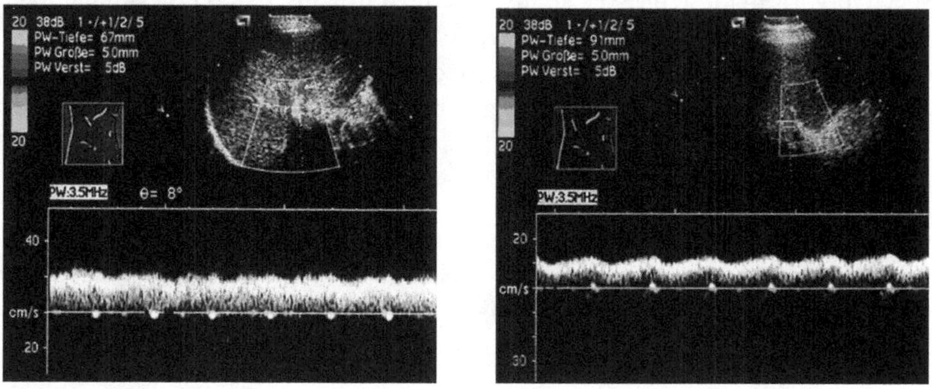

Fig. 4. Examples of Doppler velocity waveforms measured in a human hepatic portal vein

For the description of the blood flow in the reconstructed 3D venous networks, Fig. 2, the complete non-linear system of Navier-Stokes equations for an incompressible Newtonian fluid is used

$$\frac{\partial v_j}{\partial x_j} = 0, \tag{1}$$

$$\frac{\partial v_i}{\partial t} + \frac{\partial}{\partial x_j}(v_i v_j) + \frac{1}{\varrho}\frac{\partial p}{\partial x_i} = \frac{\eta}{\varrho}\frac{\partial^2 v_i}{\partial x_j \partial x_j}, \quad i, j = 1, 2, 3, \tag{2}$$

where t is the time, v_i is the i-th component of the velocity vector $\boldsymbol{v} = [v_1, v_2, v_3]^T$ corresponding to the Cartesian component x_i of the space variables vector $\boldsymbol{x} = [x_1, x_2, x_3]^T$, p is the pressure, ϱ and η are the density and dynamic viscosity of the blood, respectively. The mathematical model given by the system of Eqs. (1)–(2) is numerically solved using our own computational algorithm based on a stabilised variant of the projection method in combination with the cell-centred finite volume method formulated for hybrid unstructured tetrahedral grids. The principle of this algorithm, which we have verified and successfully implemented for the solution of various hemodynamical problems in the past, is described in detail in [8].

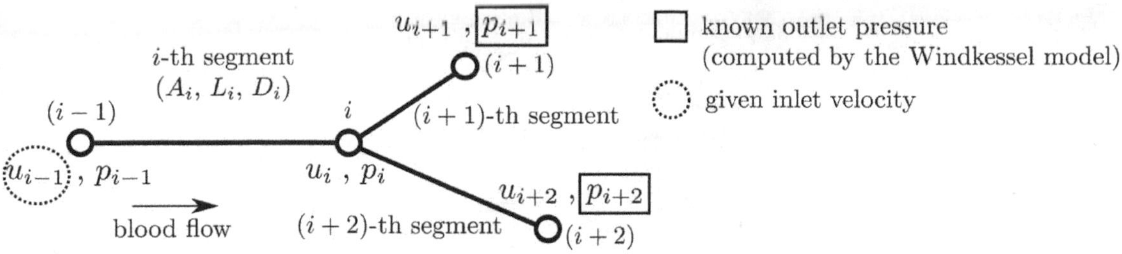

Fig. 5. Schematic drawing of a simple 1D vascular network — a bifurcating vessel

2.3. *One-dimensional blood flow*

Similarly to 3D flow problem addressed in the previous section, the blood flow in the 1D venous networks is assumed to be a steady flow of a Newtonian incompressible fluid in impermeable and inelastic 1D segments, see Fig. 5. For the modelling of pulsatile 1D blood flow in elastic vessels with variable mechanical properties, we refer the reader, for example, to [6].

Taking the aforementioned modelling simplifications into consideration, the blood flow in the inelastic segments of the 1D venous network is to be governed by the continuity equation and the Bernoulli equation completed with terms representing the friction loss in inelastic tubes. For illustration, let us consider an example of a simple bifurcating vessel, 1D analogue of which is shown in Fig. 5. Here, the motion of blood in the i-th segment of the vessel before the bifurcation can be mathematically described by the following two non-linear algebraic equations

$$A_{i-1}u_{i-1} = A_i u_i \,, \tag{3}$$

$$\frac{1}{2}\varrho u_{i-1}^2 + p_{i-1} = \frac{1}{2}\varrho u_i^2 + p_i + e_i^{\text{loss}} \,, \tag{4}$$

where A_{i-1} is the cross-sectional area of the inlet at the i-th segment, A_i is the average cross-sectional area of the i-th segment, u_i and p_i are the mean velocity and pressure computed at the end of the i-th segment (i.e., at the i-th point). For the approximation of losses originating from the viscous resistance, the term e_i^{loss} in Eq. (4) is computed proportional to the local velocity magnitude, i.e., as $e_i^{\text{loss}} = \frac{1}{2}\varrho u_i^2 \frac{L_i}{D_i}\frac{64}{\text{Re}_i}$, where D_i and L_i are the diameter and length of the i-th segment of the 1D venous network and $\text{Re}_i = u_i D_i \frac{\varrho}{\eta}$ is the corresponding Reynolds number. In the second part of the vessel, Fig. 5, the blood flow after the bifurcation point i is governed by the following three non-linear algebraic equations

$$A_i u_i = A_{i+1}u_{i+1} + A_{i+2}u_{i+2} \,, \tag{5}$$

$$\frac{1}{2}\varrho u_i^2 + p_i = \frac{1}{2}\varrho u_{i+1}^2 + p_{i+1} + e_{i+1}^{\text{loss}} \,, \tag{6}$$

$$\frac{1}{2}\varrho u_i^2 + p_i = \frac{1}{2}\varrho u_{i+2}^2 + p_{i+2} + e_{i+2}^{\text{loss}} \,, \tag{7}$$

where p_{i+1} and p_{i+2} denote known outlet pressures, each computed independently by one three-element Windkessel model, application of which is discussed below in the following section.

By generalising the principles demonstrated above to more complex 1D vascular networks such as the ones considered in this paper, we obtain a system of non-linear algebraic equations that is numerically solved using the well-known Newton method.

2.4. *Boundary conditions*

To be able to perform a quantitative comparison between the 3D and 1D flow models, we apply the same boundary conditions for both vessel representations. By taking into consideration

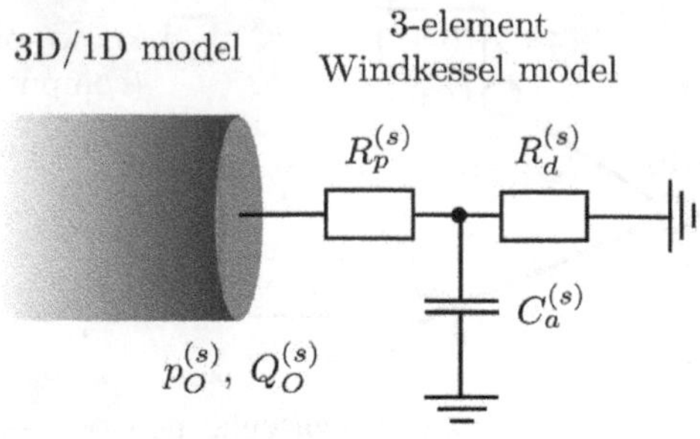

Fig. 6. Schematic drawing of the three-element Windkessel model used as an outflow boundary condition at the s-th outlet of the 3D/1D model of the hepatic portal vein

the relatively steady character of blood flow in real hepatic portal veins (as illustrated by the examples in Fig. 4), an average physiological velocity of $0.325\,\mathrm{m}\cdot\mathrm{s}^{-1}$ is prescribed at the inlets of the simple and complex vascular networks considered in this study, Fig. 2. Note that the inlet cross-sectional areas of the 3D/1D models are kept the same, thus, ensuring the prescription of the same inlet flow rate Q_0 in both models.

Because of the difficulties associated with clinical determination of physiological pressure in hepatic portal vein networks, each outlet of the 3D/1D models is coupled with a well-known lumped model — the three-element Windkessel model, schematic drawing of which is shown in Fig. 6. Compared to other modelling approaches such as the prescription of constant outlet pressure, which considering the complex geometry of the venous networks would be difficult to estimate, the Windkessel model is able to approximate the flow resistance of the downstream vascular bed and to provide a physiological value of pressure at all network outlets. For further details on the various types of the Windkessel model and their application, see, e.g., [9].

In general, the mathematical representation of the three-element Windkessel model coupled to the s-th network outlet is given by the following two equations for unknown pressures $p_d^{(s)}$ and $p_O^{(s)}$

$$\frac{\mathrm{d}}{\mathrm{d}t}p_d^{(s)} + \frac{1}{C_a^{(s)} R_d^{(s)}} p_d^{(s)} = \frac{1}{C_a^{(s)}} Q_O^{(s)}, \qquad p_O^{(s)} = p_d^{(s)} + R_p^{(s)}\, Q_O^{(s)}, \tag{8}$$

where $p_O^{(s)}$ and $Q_O^{(s)}$ are the pressure and flow rate determined at the s-th outlet of the 3D/1D models and $p_d^{(s)}$ is the distal pressure representing the pressure in arterioles and capillaries of the downstream vascular bed (here, the hepatic tissue). Note that the remaining parameters known as the lumped or Windkessel parameters of proximal $R_p^{(s)}$ and distal $R_d^{(s)}$ resistance and capacitance $C_a^{(s)}$ have to be calculated for each outlet prior to the numerical simulation. In this paper, these parameters are taken as a function of the outlet cross-sectional area $A_{\mathrm{out}}^{(s)}$, i.e.,

$$R_p^{(s)} = \frac{k_1}{A_{\mathrm{out}}^{(s)}}, \qquad R_d^{(s)} = \frac{k_2}{A_{\mathrm{out}}^{(s)}}, \qquad C_a^{(s)} = k_3\, A_{\mathrm{out}}^{(s)},$$

where $k_1 = 0.55 \times 10^4\,\mathrm{Pa}\cdot\mathrm{s}\cdot\mathrm{m}^{-1}$, $k_2 = 5.54 \times 10^4\,\mathrm{Pa}\cdot\mathrm{s}\cdot\mathrm{m}^{-1}$ and $k_3 = 324.6 \times 10^{-7}\,\mathrm{m}\cdot\mathrm{Pa}^{-1}$ are coefficients computed on the basis of data published in [5]. The outlet areas in the 3D and 1D network models are kept the same, thus, ensuring that the response of the relevant Windkessel model will be the same in case of identical outflows.

3. Numerical results

Considering the significance of portal veins in the perfusion tree hierarchy, where their main role is the transport of blood to relevant parts of the liver, the quantitative comparison between the introduced 3D and 1D flow models is aimed at the analysis of outlet flow rates. The flow rate values Q_{3D} and Q_{1D} computed for the simple and complex venous networks are listed in Tables 1 and 2, respectively, with corresponding outlet cross-sectional areas A_{out}. For the position of all the outlets listed in the aforementioned tables, we refer to Figs. 7 and 8, which also contain the information about the blood flow distribution within the two venous networks computed with the help of the three-element Windkessel models. For the sake of better analysis, let us introduce the absolute Δ and relative σ errors defined as

$$\Delta = |Q_{3D} - Q_{1D}|, \qquad \sigma = \frac{|Q_{3D} - Q_{1D}|}{Q_{3D}} \cdot 100\,\%. \tag{9}$$

Table 1. Overview of all outlet results for the simple portal vein network, as denoted in Fig. 7

outlet No.	A_{out} [mm^2]	flow rate Q [ml · s^{-1}] 3D model	1D model	Δ [ml · s^{-1}]	σ [%]	σ_A [%]
1	6.23	4.88	5.51	0.63	12.95	2.07
2	0.74	0.50	0.64	0.13	25.86	0.49
3	6.67	6.47	5.90	0.57	8.85	1.52
4	1.87	1.47	1.64	0.17	11.76	0.57
5	8.36	6.57	7.35	0.79	12.01	2.58
6	4.97	4.73	4.39	0.34	7.11	0.91
7	1.03	0.85	0.89	0.04	4.59	0.12
8	4.88	3.76	4.31	0.55	14.67	1.84
9	4.15	3.28	3.60	0.32	9.77	1.04

Table 2. Overview of selected outlet results for the complex portal vein network, as denoted in Fig. 8

outlet No.	A_{out} [mm^2]	flow rate Q [ml · s^{-1}] 3D model	1D model	Δ [ml · s^{-1}]	σ [%]	σ_A [%]
1	1.39	1.38	1.59	0.21	15.32	0.43
4	1.25	1.45	1.28	0.18	12.03	0.30
5	2.49	6.05	5.11	0.94	15.52	0.78
9	1.51	1.96	1.88	0.09	4.54	0.14
10	1.46	1.97	1.73	0.24	12.10	0.36
12	1.30	1.50	1.37	0.12	8.15	0.21
16	1.35	1.48	1.50	0.02	1.39	0.04
19	0.82	0.65	0.54	0.11	17.25	0.28
20	1.03	0.82	0.87	0.05	6.64	0.14
22	1.21	1.38	1.19	0.20	14.39	0.35
23	1.68	2.89	2.30	0.58	20.20	0.68
25	0.74	0.49	0.43	0.06	11.33	0.17
26	1.53	2.00	1.92	0.08	3.90	0.12
30	1.23	1.27	1.25	0.03	2.22	0.06
36	1.92	2.91	3.02	0.10	3.52	0.14
37	1.68	2.00	2.32	0.31	15.63	0.53

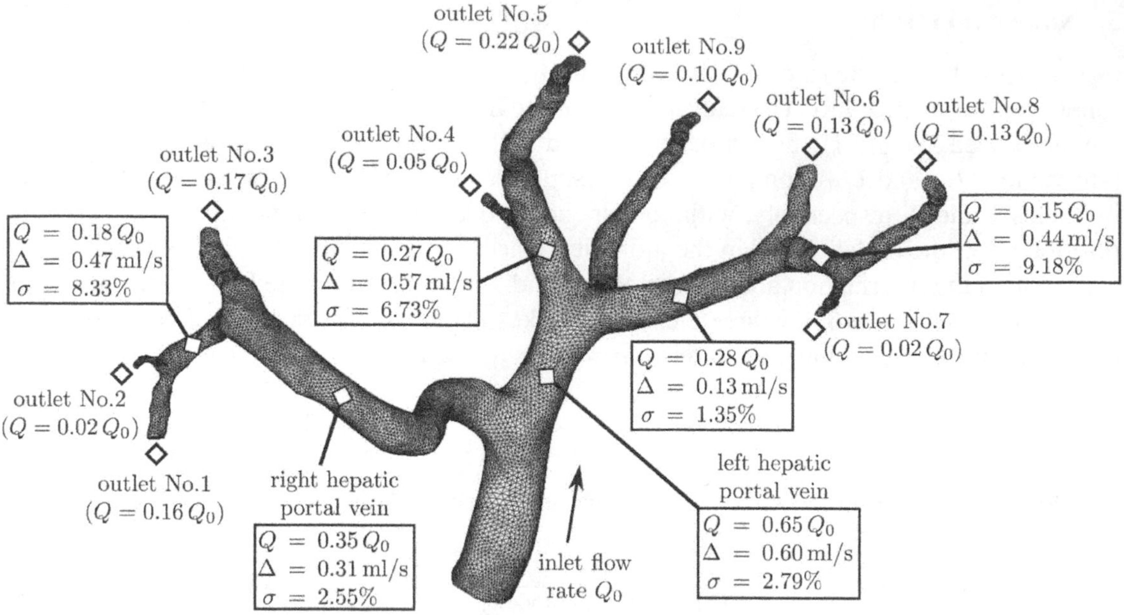

Fig. 7. Blood flow distribution in the simple hepatic portal vein network

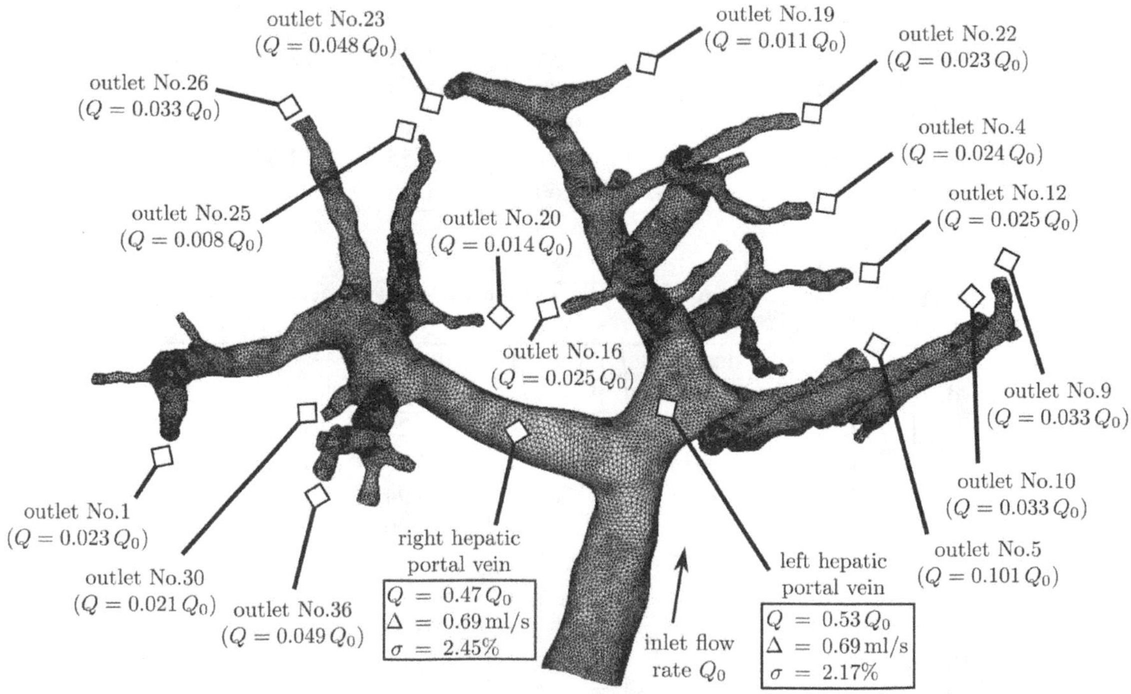

Fig. 8. Blood flow distribution in the complex hepatic portal vein network

On the basis of the errors, it can be observed that, in the case of the simple network (Table 1), the difference between the 3D and 1D flow models is mostly $\leq$ 0.8 ml/s or, in terms of the relative error, lies between 4 and 26 %. Although the error of about 26 % may seem quite high, it should be noted that it is associated with an outlet of small cross-sectional area, where the accuracy of the 3D flow model is probably not so reliable as at the large-sized outlets.

Therefore, to take into consideration the influence of A_{out} on the computed outflows, let us define the following area-weighted relative error

$$\sigma_A = \frac{A_{\text{out}}}{\sum A_{\text{out}}} \cdot \sigma, \tag{10}$$

where A_{out} is the cross-sectional area of the relevant outlet and $\sum A_{\text{out}}$ is the total outlet cross-sectional area of the network (simple model: $\sum A_{\text{out}} = 38.9\,\text{mm}^2$; complex model: $\sum A_{\text{out}} = 49.6\,\text{mm}^2$). Then from Table 1, it can be noted that the outflow at the outlet No. 2 loses its relevance in the overall context and the focus moves to the large-sized outlets No. 1 and 5 — each with the relative error (σ) around 12.5 %.

The same approach as in the simple network is also chosen for the complex one, with the exception that only 16 mostly large-sized outlets are selected for the analysis, Fig. 8. As apparent from the data listed in Table 2, the overall difference between the 3D and 1D flow models in the complex network is restricted to values ≤ 0.3 ml/s, except for the outlets No. 5 and 23, which are characterised by flow differences between 0.5 and 1 ml/s corresponding to $\sigma = 15.5 \div 20.2\,\%$.

4. Conclusions

In this paper, the portal blood flow was simulated in two patient-specific hepatic portal vein networks with the sole purpose to analyse the impact of dimensional reduction on the overall blood flow distribution. Compared to a friction-free 1D flow model, which during a preliminary analysis gave completely unrealistic flow fields with back flow appearing at some outlets, the model presented in this study demonstrated a considerable improvement in our effort to efficiently predict outflow velocities in simple as well as complex vascular networks.

By comparing not only the results obtained for both the 3D and 1D flow models, but also taking into account all the steps preceding any numerical simulation, several advantages (+) and disadvantages (−) of each modelling approach can be noted

- 3D blood flow described by the non-linear system of Navier-Stokes equations (1)–(2):
  - (−) time-consuming model preparation — includes the tasks such as the removal of non-anatomical branches and/or loops, which are usually caused by low-quality CT scans or uneven distribution of the contrast fluid within the blood, and generation of large tetrahedral computational meshes, which depending on the quality of the reconstructed network model, can take several days or even weeks,
  - (−) need for a stable and robust numerical method — requires at least a certain knowledge about special numerical methods used for the solution of the non-linear system of incompressible Navier-Stokes equations,
  - (−) numerical solution of the governing equations is computationally very demanding and strongly depends on the number of tetrahedral elements contained in the computational mesh (with normal computers, the process can take hours or even days to finish),
  - (+) detailed information about the portal hemodynamics at any network location;
- 1D blood flow described by the continuity (3) and Bernoulli equations (4):
  - (+) model preparation consists of only one step — the removal of non-anatomical branches and/or loops,

(+) need for a simple and reliable numerical method — does not require any extra knowledge in the field of computational fluid dynamics (CFD) or programming,
(+) numerical results available within several seconds even with normal (non-high performance) computers,
(−) no detailed information on the flow field is provided outside the network points.

On the basis of the observations made above, a clear conclusion can be drawn. Namely, that despite the existing differences in computed flow fields, which are a natural outcome considering the performed dimensional reduction, the benefits of the 1D approach clearly outweigh its slight inaccuracy when compared to its 3D counterpart. Although the outflow differences (σ mostly between 10 and 20 %) may seem high, it is important to recall the purpose of this study, which is to develop an efficient computational algorithm for the modelling of liver perfusion [3]. Finally, note that because of the efficiency requirement, the 1D flow model does not contain terms that would include 'bifurcation losses'. While these terms can improve the flow estimation (mostly in units of %), it is always at the cost of increased computational time.

Acknowledgements

This study was supported by the project NT13326 of the Ministry of Health of the Czech Republic and by the European Regional Development Fund (ERDF), project "NTIS — New Technologies for the Information Society", European Centre of Excellence, CZ.1.05/1.1.00/02.0090. The help of Miroslav Jiřík from the Department of Cybernetics, University of West Bohemia is also kindly acknowledged.

References

[1] Jiřík, M., Ryba, T., Svobodová, M., Mírka, H., Liška, V., LISA — Liver surgery analyzer software development, Proceedings of the 11th World Congress on Computational Mechanics (WCCM2014), Barcelona, 2014.

[2] Lukeš, V., DICOM2FEM — application for semi-automatic generation of finite element meshes, University of West Bohemia, http://sfepy.org/dicom2fem.

[3] Lukeš, V., Jiřík, M., Jonášová, A., Rohan, E., Bublík, O., Cimrman, R., Numerical simulation of liver perfusion: from CT scans to FE model, Proceedings of the 7th European Conference on Python in Science (EuroSciPy 2014), Cambridge, University of Cambridge, 2014, pp. 79–84.

[4] Jonášová, A., Rohan, E., Lukeš, V., Bublík, O., Complex hierarchical modeling of the dynamic perfusion test: Application to liver, Proceedings of the 11th World Congress on Computational Mechanics (WCCM2014), Barcelona, 2014, pp. 3 438–3 449.

[5] Sankaran, S., Moghadam, M. E., Kahn, A. M., Tseng, E. E., Guccione, J. M., Marsden, A. L., Patient-specific multiscale modeling of blood flow for coronary artery bypass graft surgery, Annals of Biomedical Engineering 40 (2012) 2 228–2 242.

[6] Sherwin, S. J., Franke, V., Peiró, J., Parker, K., One-dimensional modelling of a vascular network in space-time variables, Journal for Engineering Mathematics 47 (3–4) (2003) 217–250.

[7] Taubin, G., A signal processing approach to fair surface design, Proceedings of the 22nd annual conference on Computer graphics and interactive techniques — ACM SIGGRAPH 95, New York, 1995, pp. 351–358.

[8] Vimmr, J., Jonášová, A., Bublík, O., Numerical analysis of non-Newtonian blood flow and wall shear stress in realistic single, double and triple aorto-coronary bypasses, International Journal for Numerical Methods in Biomedical Engineering 29 (10) (2013) 1 057–1 081.

[9] Westerhof, N., Lankhaar, J.-W., Westerhof, B. E., The arterial Windkessel, Medical & Biological Engineering & Computing 34 (8) (2009) 1 049–1 064.

8

Modal parameters of a rotating multiple-disk-shaft system from simulated frequency response data

N. Khader[a,*]

[a]*Department of Mechanical Engineering, Jordan University of Science & Technology (JUST), P.O. Box 3030, Irbid 22110, Jordan*

Abstract

Modal parameters of a rotating multiple disk-shaft system are estimated in Multiple Input/Multiple Output (MIMO) scheme. The response at multiple output degrees of freedom (dofs) and excitations at multiple input (reference) dofs are related through the Frequency Response Function (FRF) matrix. The corresponding Impulse Response Function (IRF) matrix is obtained by Inverse Fast Fourier Transform (IFFT) of the FRF matrix. The resulting FRF matrix is not symmetric due to the gyroscopic effects introduced by rotation. The Eigensystem Realization Algorithm (ERA) and its equivalent low order time domain algorithm, based on the Unified Matrix Polynomial Approach (UMPA) are employed to estimate the desired modal parameters, i.e., system eigenvalues and the associated right hand and left hand eigenvectors. The right hand vectors are estimated from multiple columns of the FRF matrix with the structure rotating in one direction, and the left hand vectors are estimated from the multiple rows of the FRF matrix, which are calculated as the transpose of the same multiple columns of the FRF matrix, estimated with rotation in the opposite direction. The obtained results are found to be in excellent agreement with results obtained from Theoretical Modal Analysis (TMA).

Keywords: modal parameters, mechanical vibration, structural dynamics, rotor dynamics

1. Introduction

The multiple disk-shaft system is found in numerous mechanical and aerospace applications, such as compressors, turbines, and hard disk drives. Stringent requirements on such systems which operate at high rotational speeds, resulted in strong coupling between modes of constituent components, i.e., between modes of the shaft and individual disks. It is therefore of great importance to accurately predict their modal parameters to come up with a reliable design, free from resonance vibration during operation. This subject was examined by several researchers, who employed different theoretical, numerical and experimental approaches to address the problem [6–8,14,15,20,21,23].

Modal parameters of multiple disk shaft system were theoretically and experimentally estimated [11], where peaks in the FRF were used to identify the desired damped natural frequencies. However, it is known that the considered system has closely coupled or repeated frequencies, and it is essential to employ a MIMO estimation scheme to predict the modal parameters of such systems. This estimation technique was employed to estimate the modal parameters of the coupled vibration of a stationary flexible disk-flexible shaft system from theoretically generated FRF matrix [12,13]. To account for rotation, the present work simulates multiple reference testing of a flexible shaft carrying more than one flexible disk and rotating at a constant angular speed. Impulse forces are assumed to excite the system at a number of excitation points N_i, and

*Corresponding author. e-mail: nkhader@just.edu.jo.

the resulting response at a number of output points N_o is calculated. The assumed excitations and corresponding responses are used to estimate $N_o \times N_i$ FRF matrix, which relates multiple inputs and multiple outputs in the frequency domain.

The considered structure has isotropic rotating components and rotates at constant rotational speed. This results in Linear Time Invariant system, which depends only on the rotational speed [4] through the gyroscopic effects. The corresponding eigenvalue problem is non self-adjoint and the corresponding FRF matrix is non-symmetric. Along with system eigenvalues, both right hand and left hand eigenvectors are required to completely define the modal model of the system. According to Nordmann [16], the right hand and left hand eigenvectors can be estimated from a column and row of the FRF matrix, respectively. Since measuring a row of the FRF matrix is not practical, Gutiérrez and Ewins [5] suggested an alternative way to estimate the left hand eigenvector from a row of the FRF matrix. It was suggested to use a column of the FRF matrix to estimate right hand eigenvectors with rotation in one direction and to use the transpose of the same column of the FRF matrix, but with rotation in the opposite direction to estimate the left hand eigenvectors. This is true because all system matrices of the considered structure are symmetric, except for the gyroscopic matrix, which is skew-symmetric. This means that the FRF matrix of a structure spinning in one direction is the transpose of the FRF matrix, obtained with the structure spinning in the opposite direction.

The considered structure is known to have closely coupled and/or repeated modes due to its isotropy and circular symmetry, it is therefore essential to employ MIMO estimation algorithms, so as not to miss these modes. Therefore, multiple columns of the FRF matrix with rotation in one direction are used to estimate the right hand eigenvectors, and the transpose of the same multiple columns of the FRF matrix with rotation in opposite direction are used to estimate the left hand eigenvectors.

The IRF matrix is obtained by IFFT of an FRF matrix. ERA [9, 10] and its equivalent first order UMPA time domain algorithm [2,3,19] are employed to estimate the desired modal parameters of the considered system.

The considered simulation is based on the theoretical model described in [11], where Lagrange's equation was combined with the assumed modes method to derive equations of motion. The obtained results are found to be in excellent agreement with results from TMA.

2. Theoretical Analysis

2.1. *FRF Matrix of the multiple disk shaft system*

The examined rotor consists of two flexible disks, attached to a fixed-free flexible shaft. The shaft is modelled by a slender beam with circular cross section and uniformly distributed mass and stiffness. The flexible disk is modelled by an annular thin plate, with uniformly distributed mass and bending rigidity, and clamped at its inner radius to the outer radius of the shaft, as shown in Fig. 1.

Vibratory motion of the considered rotating multiple disk-shaft system is discretized by the application of the assumed modes method, where the flexible deformation of the disk or shaft is represented by summation of a number of time-dependent generalized coordinates, multiplied by assumed functions, as given bellow:

$$U_s(z,t) = \sum_m U_m(z) a_m(t), \tag{1}$$

$$V_s(z,t) = \sum_m V_m(z) b_m(t), \tag{2}$$

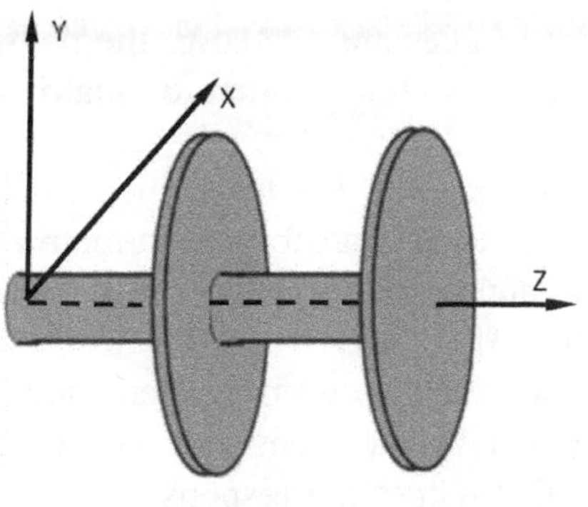

Fig. 1. Two flexible disk flexible shaft system

$$w(r,\theta,t) = \sum_m [w_m(r)\sin\theta q_{ms}(t) + w_m(r)\cos\theta q_{mc}(t)], \tag{3}$$

where U_s and V_s are flexible shaft deformations along X and Y directions in a fixed coordinate system, and w is the out-of-plane disk deformation in an attached to the disk coordinate system. Mode shapes of individual flexible disk or flexible shaft are taken as the assumed functions. In this study, where the coupled disk-shaft modes are examined, only one nodal diametral disk modes are accounted for, because they are the only flexible disk modes that couple with flexible shaft deformations. It is known that transverse flexible disk vibration takes place around the disk according to $\sin(n\theta)$ and $\cos(n\theta)$, where n denotes the number of nodal lines in a given mode. Modes shapes with $n > 1$ are known as reactionless modes because they don't result in net force or moment. This is not the case in modes with $n = 1$, where resulting forces and moments produce coupling between these flexible disk modes and its pitching and translation on the supporting flexible shaft.

The assumed modes method is combined with Lagrange's equation:

$$\frac{\mathrm{d}}{\mathrm{d}t}\left(\frac{\partial L}{\partial \dot{q}_j}\right) + \left(\frac{\partial D}{\partial \dot{q}_j}\right) - \left(\frac{\partial L}{\partial q_j}\right) = Q_j \tag{4}$$

to derive the governing equations of motion for the complete system, where $L = T - U$, D is the dissipation function, and q_j, Q_j, T and U are the j-th generalized coordinate, the j-th generalized force, total kinetic energy, and total strain energy, respectively. With assumed proportional damping, equations of motion of the considered system are expressed in the following matrix form:

$$[M]\{\ddot{q}\} + ([C] + [G])\{\dot{q}\} + [K]\{q\} = \{Q\}, \tag{5}$$

where, $[M]$, $[C]$, $[G]$, and $[K]$ are the mass, damping, gyroscopic, and stiffness matrices, and $\{q\}$ and $\{Q\}$ are vectors of the generalized coordinates and generalized forces, respectively.

When proportional damping is considered, the damping matrix $[C]$ is assumed to be a linear combination of the mass and stiffness matrices, i.e. $[C] = \alpha[M] + \beta[K]$, where α and β are constants. The FRF matrix, which relates the generalized coordinates to the generalized forces, can be expressed by:

$$[H]_{qq} = \left[[K] + j\omega\left([C] + [G]\right) - \omega^2[M]\right]^{-1}. \tag{6}$$

Knowing that the generalized forces are related to the physical forces by the transformation $\{Q\} = [\Psi_{\text{in}}]\{f\}$, where $[\Psi_{\text{in}}]$ is the transformation matrix constructed from the assumed functions of system components and the position of excitation points (references). Similarly, the displacements at response dofs can be expressed in terms of the generalized coordinates as: $\{x\} = [\Psi_{\text{out}}]\{q\}$, where $[\Psi_{\text{out}}]$ is the transformation matrix constructed from the assumed functions of system components and the position of response points. Details about these matrix transformations can be found in [13].

Using the given above relationships between the generalized coordinates and generalized forces on one side, and the physical displacements and physical excitation forces, on the other, one can write the FRF matrix that relates the response $\{X\}_{N_o \times 1}$ at N_o output dofs with the excitations $\{F\}_{N_i \times 1}$ at N_i input dofs in the form:

$$\{X(\omega)\}_{N_o \times 1} = [H(\omega)]_{N_o \times N_i} \{F(\omega)\}_{N_i \times 1} . \tag{7}$$

This FRF matrix (or the corresponding IRF matrix) can be subsequently employed in a MIMO estimation algorithm to extract the desired modal parameters of the considered system, as discussed in the following section.

2.2. *Modal parameters of the multiple disk shaft system*

In the formulated FRF matrix, the output dofs consist of shaft points with two orthogonal deformations for each point, and out-of-plane responses points for each disk. Similarly, the input dofs consist of transfer excitations at a number of shaft points and out-of-plane excitations at a number of points for each disk. Both input and output dofs will be discussed in details in the results section.

The Complex Mode Indicator Function (CMIF) of the estimated FRF matrix is examined first, which is a plot of the singular values obtained from Singular Value Decomposition (SVD) of the estimated FRF matrix for each spectral line [1,22] according to:

$$[H(\omega_k)]_{(N_o \times N_i)} = [U_k]_{(N_o \times N_i)} [\Sigma_k]_{(N_i \times N_i)} [V_k]^H_{(N_i \times N_i)}, \tag{8}$$

where $[U_k]_{(N_o \times N_i)}$ is the matrix of left singular vectors, $[\Sigma_k]_{(N_i \times N_i)}$ is a diagonal matrix of the singular values, and $[V_k]_{(N_o \times N_i)}$ is the matrix of right singular vectors at the k-th spectral line. Peaks in the CMIF curves occur at the damped natural frequencies of the considered structure, and the left and right singular vectors associated with these peaks give approximation to the corresponding mode shapes and modal participation factors, respectively.

ERA and its equivalent low order time-domain UMPA algorithm are used to extract the desired modal parameters from the estimated IRF matrix. The ERA algorithm was originally developed by NASA to construct a state space model from MIMO test data, and UMPA was developed as a general estimation approach, with the different estimation algorithms being special cases of this general approach in both time and frequency domains.

Formulation of the UMPA time domain is based on the following general equation:

$$\sum_{k=0}^{n} [\tilde{\alpha}_k]_{N_o \times N_o} [h_{t+k}]_{N_o \times N_i} = [0] \quad \text{for} \quad t = 0, \ldots, N, \tag{9}$$

where $[\tilde{\alpha}_k]$ is the k-th polynomial coefficient matrix and $[h_{t+k}]$ is the IRF at the k-th time shift. In the empolyed first order UMPA model, only $[\tilde{\alpha}_0]$ and $[\tilde{\alpha}_1]$ are used in the expansion. Additional time shifts are added to this basic form to make sure the size of the coefficient matrix is enough to determine the desired number of modes, which results in following expanded form of the equation:

$$[\alpha_0]\begin{bmatrix} h_0 & h_1 & h_2 & \cdots & \cdots & \cdots & h_N \\ h_1 & h_2 & \cdot^{\cdot^{\cdot}} & \cdots & \cdots & h_N & h_{N+1} \\ \vdots & \cdot^{\cdot^{\cdot}} & & & & & \vdots \\ h_{n-1} & & & & & & h_{N+n-1} \end{bmatrix} = -[\alpha_1]\begin{bmatrix} h_1 & h_2 & h_3 & \cdots & \cdots & \cdots & h_{N+1} \\ h_2 & h_3 & \cdot^{\cdot^{\cdot}} & \cdots & \cdots & h_{N+1} & h_{N+2} \\ \vdots & \cdot^{\cdot^{\cdot}} & & & & & \vdots \\ h_n & & & & & & h_{N+n} \end{bmatrix} \tag{10}$$

or in compact form:

$$[\alpha_0][H_0] = -[\alpha_1][H_1], \tag{11}$$

$[\alpha_0]$ and $[\alpha_1]$ are rectangular matrices of order $n \times N_o$, and $[H_0]$ and $[H_1]$ are the Hankel matrices, constructed from the IRF matrix $[h(t)]_{N_o \times N_i}$. The corresponding matrix coeffiecient characteristic polynomial is given by:

$$[\alpha_1]z^{(1)} + [\alpha_0]z^{(0)} = 0, \tag{12}$$

where $z = \mathrm{e}^{\lambda \Delta t}$ and Δt is the time between consecutive samples.

For high order normalization, i.e., $[\alpha_1] = [I]$, the companion matrix is simply $[\alpha_0]$ and the resulting eigenvalue problem is:

$$[\alpha_0]\{\phi\} = \lambda_z\{\phi\}, \tag{13}$$

where $\{\phi\} = \begin{Bmatrix} \lambda_z^{n-1}\psi \\ \lambda_z^{n-2}\psi \\ \vdots \\ \lambda_z\psi \\ \psi \end{Bmatrix}$.

On the other hand, when low order normalization is used, the resulting eigenvalue problem is:

$$[I]\{\phi\} = \lambda_z[\alpha_1]\{\phi\}. \tag{14}$$

The desired eigenvalues $\lambda_r = \sigma_r + j\omega_r$ of considered sytem are obtained from eigenvalues $(\lambda_z)_r$ of the formulated eigevalue problems above, with $\sigma_r = \mathrm{Re}\left(\frac{\ln(\lambda_z)_r}{\Delta t}\right)$, and $\omega_r = \mathrm{Im}\left(\frac{\ln(\lambda_z)_r}{\Delta t}\right)$.

3. Results and Discussion

The geometric and material properties of the shaft-disk system are:
$E_{\text{disk}} = E_{\text{shaft}} = 200$ GPa, $\nu_{\text{disk}} = \nu_{\text{shaft}} = 0.3$, $\rho_{\text{disk}} = \rho_{\text{shaft}} = 7\,800$ kg/m^3,
$\left(\frac{R_{\text{in}}}{R_{\text{out}}}\right)_{\text{disk}} = 0.2$, $(R_{\text{out}})_{\text{disk}} = 0.25$ m, $h_{d_1} = 0.002$ m, $h_{d_2} = 0.002\,5$ m,
$Z_{d_1} = 0.5$ m, $Z_{d_2} = 0.75$ m, $L_{\text{shaft}} = 0.75$ m, $(R_{\text{out}})_{\text{shaft}} = 0.05$ m, $(R_{\text{in}})_{\text{shaft}} = 0.048$ m,
where h_{d_k} and $Z_{d_k}$are the thickness and spanwise position of the k-th disk. An FRF matrix, which relates 86 response and 7 input dofs is estimated. The response dofs consist of two transverse orthogonal deformations at 7 shaft points, and 36 out-of-plane response dofs for each disk, located at the intersections of 6 radial and 6 circular lines, uniformly distributed over the disk. The excitation is assumed to take place at input (reference) points, which consist of a shaft excitation along the transverse X-direction (dof # 7), a shaft excitation along the transverse

Y-direction (dof # 11), shown in Fig. 2. The first disk is excited by two out of plane excitations at dofs # 17 and # 34, and the second disk is excited by three out-of-plane excitations at dofs # 51, # 68 and # 85, shown in Fig. 3. The driving point FRF, associated with dof # 34, located at the first disk, is presented in Fig. 4 for angular speed $\Omega = 300$ Hz. The cross point FRFs $H(34, 85)$ and $H(85, 34)$, which relate dof # 34 (located on the first disk) and dof # 85 (located on the second disk) are shown in Fig. 5. These FRFs are estimated at angular speed $\Omega = 300$ Hz with the same sense of rotation. The non-symmetry property of the FRF matrix is noticed from the shown phase plot. As discussed earlier, the skew symmetric nature of the gyroscopic matrix results in an FRF matrix for a rotating structure in one direction to be the transpose of the FRF matrix for the same structure when it is rotating in the opposite direction. This is demonstrated in Fig. 6, which shows the cross point FRF $H(34, 85)$ at $\Omega = 300$ Hz in one direction, and the cross point FRF $H(85, 34)$ at $\Omega = 300$ Hz in the opposite direction. The agreement between the presented cross point FRFs demonstrates the fact stated earlier, i. e., the FRF matrix with rotation in one direction is equivalent to the transpose of the FRF matrix with rotation in the opposite direction.

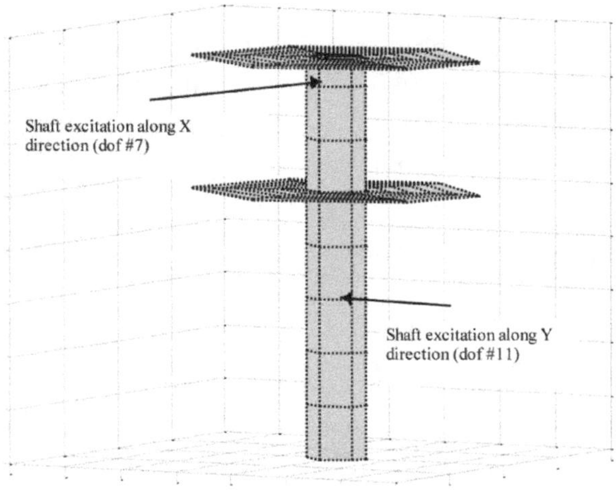

Fig. 2. Transverse shaft excitation dofs

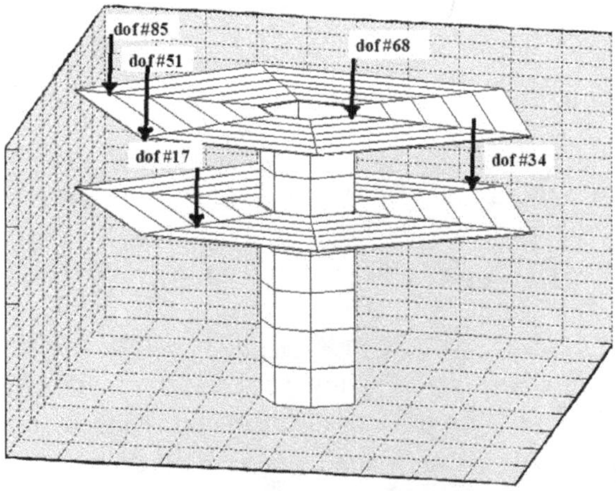

Fig. 3. Out of plane disk excitation dofs

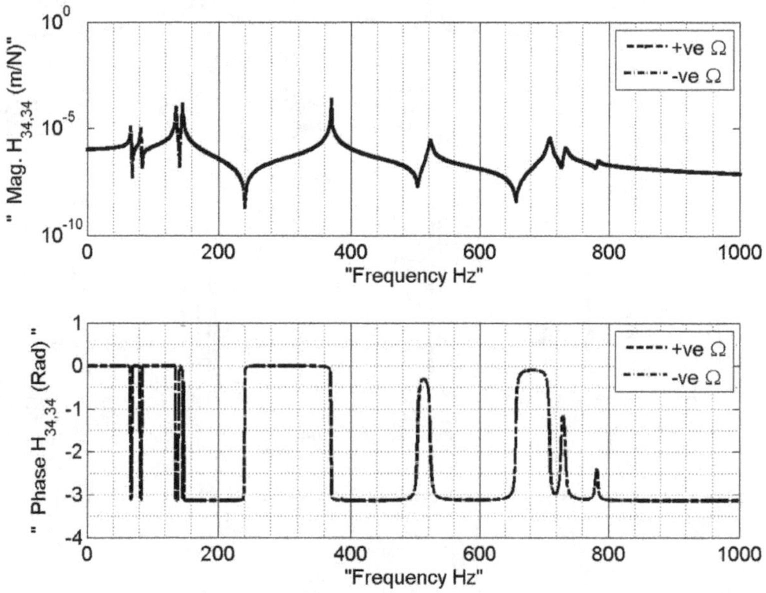

Fig. 4. Driving point FR $H(34, 34)$ with positive and negative Ω

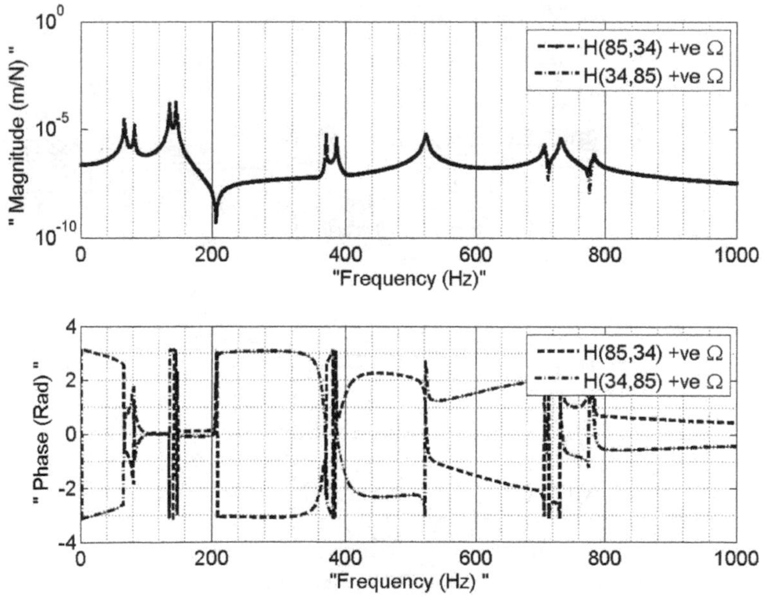

Fig. 5. Cross point FRFs $H(85, 34)$ and $H(34, 85)$ with positive Ω

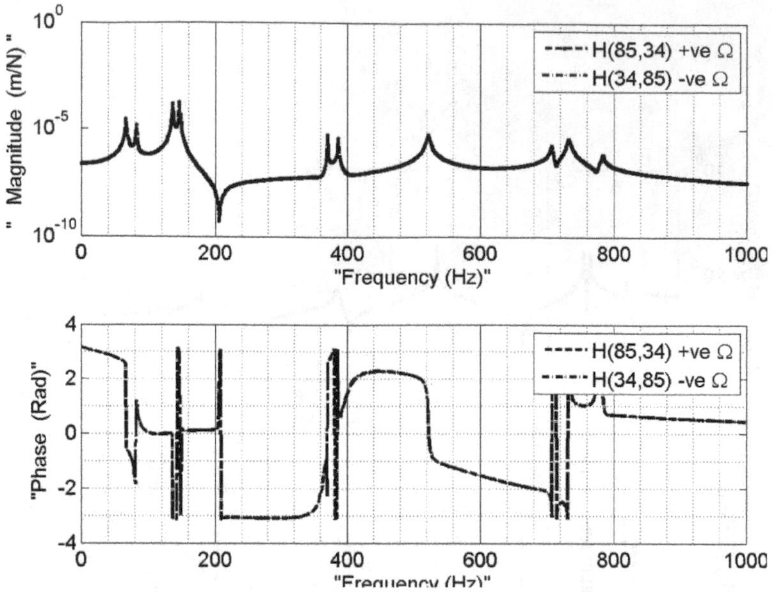

Fig. 6. Cross point FRFs $H(85, 34)$ with positive Ω and $H(34, 85)$ with negative Ω

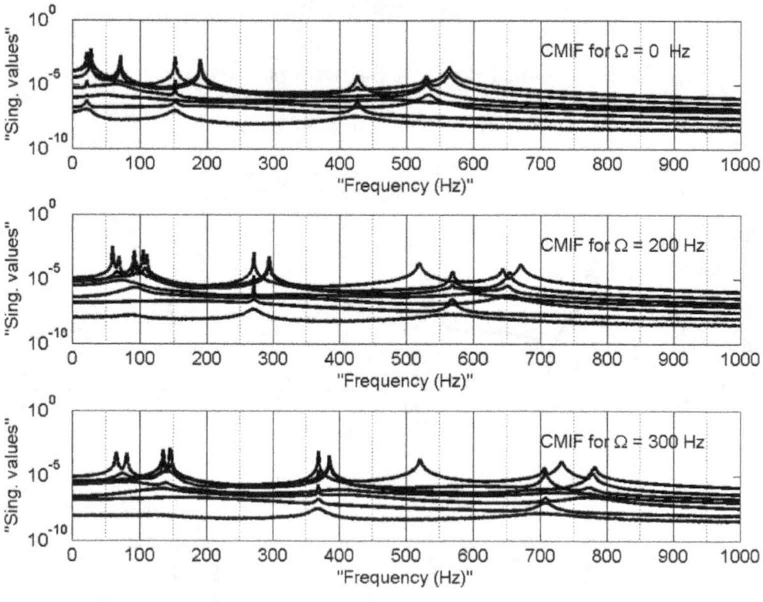

Fig. 7. CMIF of the FRF matrix at $\Omega = 0, 200$ and 300 Hz

Fig. 7 presents the CMIF of the estimated FRF matrix at rotor speeds $\Omega = 0, 200$ and 300 Hz, where split in the resulting natural frequencies due to the gyroscopic effect is demonstrated.

Since the eigenvalue problem associated with rotating structure is non self-adjoint and its FRF matrix is not symmetric, it is necessary to estimate both right hand and left hand eigenvectors, associated with the same set of eigenvalues. Multiple columns of the FRF matrix with rotation in a given direction are used to estimate eigenvalues and the corresponding right hand eigenvectors. Based on the discussion above, the multiple rows, required to estimate left hand eigenvectors, are obtained as the transpose of the same multiple columns of the FRF matrix, estimated with rotation in the opposite direction.

Thus, the same reference and same excitation points are used to obtain the required FRF matrix data, necessary to estimate the system eigenvalues and both right hand and left hand eigenvectors. The natural frequencies and damping ratios are estimated using both ERA and the first order UMPA time domain algorithms, and the results are shown in Table 1 along with the corresponding results from TMA, where an excellent agreement between the results is noticed. The available FRF data along with the estimated eigenvalues are used to estimate residues, which can be used to obtain normalized sets of the right hand and left hand eigenvectors.

Table 1. Estimated natural frequencies (Hz) by different estimation methods

Mode #	ERA	UMPA	TMA
1	65.391	65.391	65.391
2	80.726	80.728	80.726
3	134.755	134.756	134.755
4	134.775	134.775	134.775
5	145.140	145.140	145.140
6	147.113	147.113	147.113
7	369.011	369.010	369.011
8	369.057	369.058	369.058
9	385.694	385.696	385.694
10	386.039	386.039	386.039
11	521.193	521.192	521.193
12	706.340	706.336	706.341
13	708.175	708.171	708.175
14	731.825	731.825	731.825
15	777.624	777.624	777.624
16	781.672	781.670	781.672

In modal analysis, the FRF matrix at a given frequency ω_k is expressed as a superposition of the contribution of individual modes as:

$$[H(\omega_k)] = \sum_{r=1}^{N} \left(\frac{[A_r]}{j\omega_k - \lambda_r} + \frac{[A_r^*]}{j\omega_k - \lambda_r^*} \right), \tag{15}$$

where $[A_r]$ is the residue matrix and λ_r is the r-th eigenvalue associated with the r-th mode, and $(\ )^*$ denotes complex conjugate. The k-th vector of the residue matrix, associated with the r-th eigenvalue, can be expressed in terms of the normalized r-th right hand eigenvector and the k-th element of the normalized r-th left hand eigenvector by:

$$\{A_r\}_k = \{\phi_R\}_r (\phi_L)_{kr}, \tag{16}$$

where $\{\phi_R\}_r$ and $\{\phi_L\}_r$ are r-th right hand and left hand eigenvectors. If a unit value is assigned to the k-th element of the r-th left hand eigenvector, i.e., $(\phi_L)_{kr} = 1$, then the r-th right hand eigenvector $\{\phi_R\}_r$ is determined from the k-th vector of the residue matrix i.e., $\{\phi_R\}_r = \{A_r\}_k$ where the residue matrix $[A_r]$ can be estimated from the available FRF matrix data. Other elements of the left hand eigenvector can be estimated from the corresponding rows of the residue matrix.

The estimated modal parameters are used to synthesize the FRF matrix, which is then compared to the original FRF matrix as a way to check the accuracy of the estimated results. One driving point and two cross point original and synthesized FRFs are shown in Figs. 8–10. The presented data clearly demonstrate the agreement between these FRFs, which confirms the validity of the estimated eigenvalues and corresponding right hand and left hand eigenvectors.

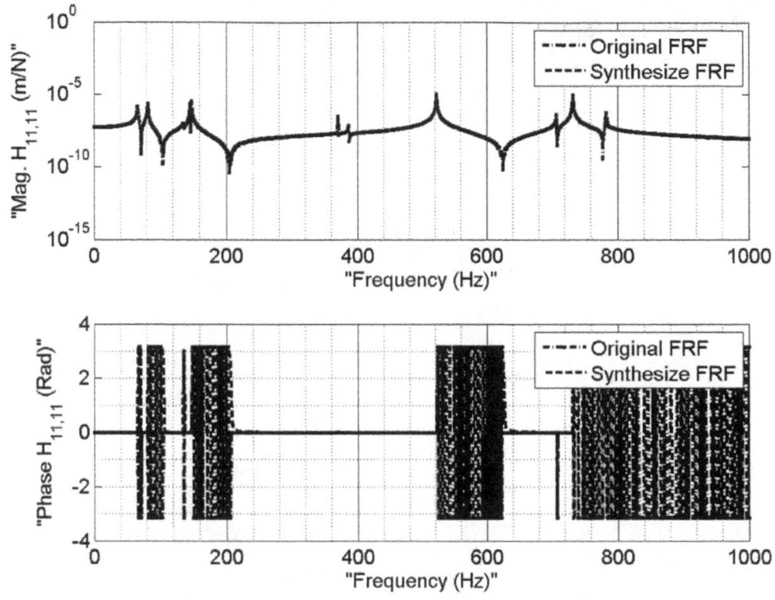

Fig. 8. Original and synthesized driving point FRF $H(11, 11)$

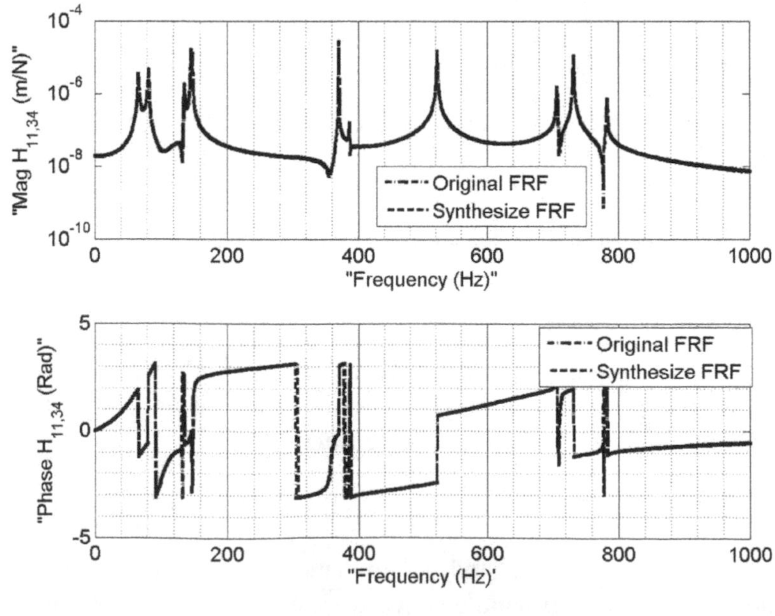

Fig. 9. Original and synthesized cross point FRF $H(11, 34)$

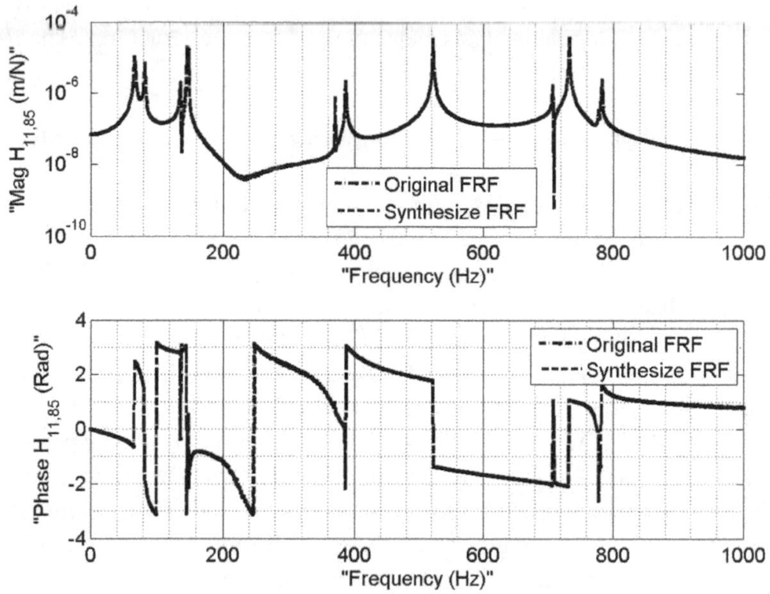

Fig. 10. Original and synthesized cross point FRF $H(11, 85)$

4. Conclusions

MIMO estimation algorithm is employed to determine modal parameters of a rotating multiple-disk-shaft system from theoretically generated FRF and IRF matrices. A first order time domain estimation algorithm is used to estimate modal frequencies. Residue matrix, obtained from an FRF matrix with rotation in one direction is used to estimate right hand eigenvectors, and residue matrix, obtained from an FRF matrix with rotation in the opposite direction is used to estimate left hand eigenvectors. With the adopted approach, the same excitation and response dofs can be used to estimate both right hand and left hand eigenvectors. The obtained eigenvalues and eigenvectors are in excellent agreement with the results from TMA, which confirms the ability of the presented method to handle experimental data.

Acknowledgements

This paper presents part of the work the author carried out during his stay at University of Cincinnati Structural Dynamics Research Lab. (UC-SDRL) while on a sabbatical leave from Jordan University of Science & Technology (JUST). The author acknowledges the financial support provided by JUST, as well as the valuable discussions with Dr. Brown and Dr. Allemang from UC-SDRL.

References

[1] Allemang, R. J., Brown, D. L., A complete review of the complex mode indicator function (CMIF) with applications, Proceedings of International Conference on Noise and Vibration Engineering (ISMA), Katholieke Universiteit Leuven, Belgium, 2006.

[2] Allemang, R. J., Brown, D. L., A unified matrix polynomial approach to modal identification, Journal of Sound and Vibration 211 (1998) 301–322.

[3] Brown, D. L., Phillips, A. W., Allemang, R. J., A first order extended state vector expansion approach to experimental modal parameter estimation, Proceedings of International Modal Analysis Conference, Orlando, 2005.

[4] Bucher, I., Ewins, D. J., Modal analysis and testing of rotating structures, The Royal Society 359 (2001) 61–96.

[5] Gutiérrez-Wing, E. S., Ewins, D. J., Modal characterisation of rotating machines, Proceedings of 19th International Model Analysis Conference, Soc Experimental Mechanics Inc., 2001, pp. 1 249–1 256.

[6] Hili, M. A., Fakhfakh, T., Haddar, M., Vibration analysis of a rotating flexible shaft-disk system, Journal of Engineering Mathematics 57 (2007) 351–363.

[7] Jang, G. H., Lee, S. H., Jung, M. S., Free vibration analysis of a spinning flexible disk-spindle system supported by ball bearing and flexible shaft using the finite element method and substructure synthesis, Journal of Sound and Vibration 251 (2002) 59–78.

[8] Jia, H. S., On the bending coupled natural frequencies of a spinning, multispan Timoshenko shaft carrying elastic disks, Journal of Sound and Vibration 221 (1999) 623–649.

[9] Juang, J. N., Pappa, R. S., An eigensystem realization algorithm for modal parameter identification and model reduction, AIAA Journal of Guidance, Control, and Dynamics 8 (1985) 620–627.

[10] Juang, J. N., Mathematical correlation of modal parameter identification methods via system realization theory, Journal of Analytical and Experimental Modal Analysis 2 (1987) 1–18.

[11] Khader, N., Atoum, A., Al-Qaisia, A., Theoretical and experimental modal analysis of multiple flexible disk-flexible shaft system, Presented at the SEM annual Conference, Springfield, Massachusetts, 2007.

[12] Khader, N., Modal parameters of a flexible disk-flexible shaft system from simulated data, International Journal of Vehicle Noise and Vibration 8 (2012) 60–73.

[13] Khader, N., Modal parameter estimation of a two-disk-shaft system by the unified matrix polynomial approach, Presented at XXXII IMAC Orlando, 2014.

[14] Lee, C. W., Chun, S. B., Vibration analysis of a rotor with multiple flexible disks using assumed modes method, Journal of Vibration and Acoustics 120 (1998) 87–94.

[15] Lee, C. W., Jia, H. S., Kim, C. S., Chun, S. B., Tuning of simulated natural frequencies for a flexible shaft-multiple flexible disk system, Journal of Sound and Vibration 207 (1997) 435–451.

[16] Nordmann, R., Identification of the modal parameters of an elastic rotor with oil film bearings, Transactions of the ASME Journal of Vibration, Acoustics and Reliability in Design 106 (1984) 107–112.

[17] Ouisse, M., Foltete, E., Reduced models identification from experimental modal analysis of non-self adjoint systems-rotor dynamics, active control and vibroacoustics applications, Proceedings of Conference on Noise and Vibration Engineering, Katholieke Universiteit Leuven, Belgium, 2010.

[18] Ouisse, M., Foltete, E., On the properness condition for modal analysis of non-symmetric second order systems, Mechanical Systems and Signal Processing, 25 (2011) 601–620.

[19] Phillips, A. W., Allemang, R. J., The unified matrix polynomial approach to understanding modal parameter estimation: An update, Proceedings of International Conference on Noise and Vibration Engineering, Katholieke Universiteit Leuven, Belgium, 2004.

[20] Shahab, A. S., Thomas, J., Coupling effect of disk flexibility on the dynamic behavior of multi disk-shaft system, Journal of Sound and Vibration 114 (1987) 435–446.

[21] Shen, Jr-Yi, Tseng, C. W., Shen, I. Y., Vibration of rotating disk/spindle system with flexible housing/stator assemblies, Journal of Sound and Vibration 271 (2004) 725–756.

[22] Shih, C. Y., Tsuei, Y. G., Allemang, R. J., Brown, D. L., Complex mode indication function and its application to spatial domain parameter estimation, Journal of Mechanical Systems and Signal Processing 2 (1988) 367–372.

[23] Wu, F., Flowers, G. T., A transfer matrix technique for evaluating the natural frequencies and critical speeds of a rotor with multiple flexible disks, Journal of Vibration and Acoustics 114 (1992) 242–248.

Evaluation of human thorax FE model in various impact scenarios

M. Jansová[a,*], L. Hynčík[a], H. Čechová[a], J. Toczyski[b], D. Gierczycka-Zbrozek[b], P. Baudrit[c]

[a]New Technologies – Research Centre, University of West Bohemia, Univerzitní 8, 306 14 Plzeň, Czech Republic
[b]Institute of Aeronautics and Applied Mechanics, Warsaw University of Technology, ul. Nowowiejska 29, 00-665 Warsaw, Poland
[c]CEESAR – European Centre of Studies on Safety and Risk Analysis, 132, Rue des Suisses, 92000, Nanterre, France

Abstract

The study focused on the validation of the 50th percentile male model — a detailed FE model of the thoracic segment of the human body developed within project *Development of a Finite Element Model of the Human Thorax and Upper Extremities* (THOMO) co-funded by the European Commission (7th Framework Programme). The model response was tested in three impact scenarios: frontal, lateral and oblique. The resulting impactor contact force vs. time and chest deflection vs. time responses were compared with experimental results. The strain profile of the 5th rib was checked with lateral and oblique strain profiles from post-mortem human subject (PMHS) experiments. The influence of heart and lungs on the mechanical response of the model was assessed and the material data configuration, giving the most biofidelic thorax behaviour, was identified.

Keywords: human body, thorax, validation, impact, finite element model

1. Introduction

In the last decade, the legalisation of computer simulated crash tests with virtual human body models is one of the goals of automotive industry [5]. This testing requires a biofidelic model of the human body. There is a wide range of existing virtual human body models with various level of accuracy — ranging from simplified multibody models, allowing relatively fast simulations and approximated injury prediction, up to models respecting anatomical details that serve for detailed analysis of injury assessment under impact loading. Some models even allow scaling to different anthropometry.

Detailed review of development of biomechanical numerical models of the human body was performed by Yang [20]. The Human Model for Safety (HUMOS) programme running from 1997 to 2001 was the first step toward the development of commonly accepted models and computer methods [13]. It was followed by the HUMOS2 project from 2001 to 2006 [17].

One of the more detailed full human body models is the Total Human Model for Safety (THUMS^{TM}) developed by Toyota Motor Corporation and Toyota Central R&D Labs to predict internal organ injury [6,16].

Currently the Global Human Body Model Consortium (GHBMC) — an international consortium of nine automakers and two suppliers working with research institutes and government agencies creates the world's most detailed midsize (50th percentile) male human body model [3].

*Corresponding author. e-mail: jansovam@ntc.zcu.cz.

The paper describes the validation process of the thorax model developed within the THOMO project in cooperation with GHBMC. The goal of THOMO project was the development of numerical models, focusing on the validation of rib strain fields, in order to exhibit the rib fracture mechanisms.

2. Methodology

2.1. Reference model

In order to facilitate the thorax model development, the complete GHBMC model (mid-size male) was simplified by the Thorax Center of Expertise at the University of Virginia and forwarded to the THOMO consortium members for refinement. Material properties of particular tissues were provided by Centers of Expertise (COE). The ribs were defined as breakable with a threshold of plastic failure strain of 1.8 %.

Further development of the model focused on resolving geometrical and numerical instabilities, re-meshing, updating the musculature and expanding the range of available body positions of the model. Initial configuration (seated with upper arms aligned with the thorax) was modified in order to obtain a model with lifted arms, necessary for lateral and oblique pendulum impact setups (Fig. 1). The PMHS tests, carried out within the THOMO project were performed with arms lifted [10].

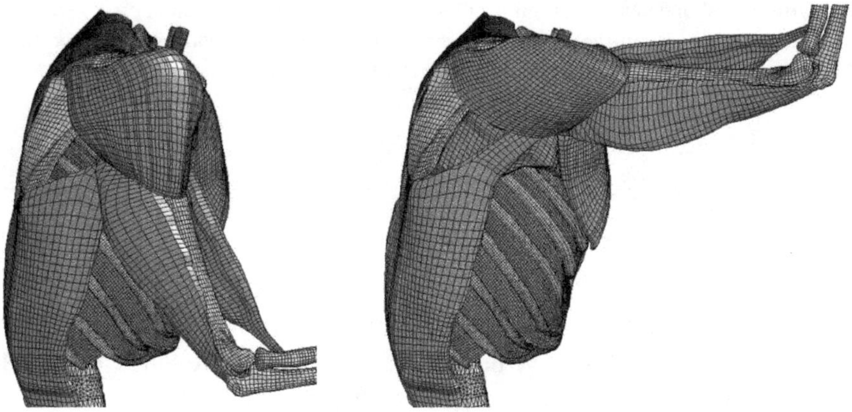

Fig. 1. Detail of the muscles of right upper arm of the THOMO model for frontal impact (left) and lateral and oblique impact (right)

2.2. Validation tests

The extensive review of available validation test data was performed within the THOMO project. These data are in the form of validation corridors formed by the response of cadavers to given test setup. Based on the review results, suitable tests were chosen in cooperation with the Thorax Center of Expertise at the University of Virginia. The THOMO thorax model was validated in three validation scenarios, namely frontal impact, lateral impact and oblique impact, using the pendulum impactor setup.

The impact test scenarios have been prepared in LS-Dyna software. According to the experimental test setups, the occupant model was seated on a rigid plate and a sliding contact was defined between the plate, pelvis and lower extremities. The rigid plate was fixed at all degrees of freedom (DOF). The occupant model had a full range of motion at all DOF. The impactor was modelled as a rigid body with initial velocity that corresponded to a reference experimental test. The impactor had only one DOF in direction of the impact for frontal and lateral tests. For the oblique test it had two DOF in the transverse plane at the level of impact direction. The influence of gravity was not considered in the study.

The strain profiles of the 5th rib were drawn for the lateral and oblique setups. The strain distribution was plotted as a function of the curvilinear abscissa with costo-transverse joint at $s = 0\,\%$ and costo-chondral joint at $s = 100\,\%$ (Fig. 2). These strain profiles can be used as a tool for rib fracture prediction [10].

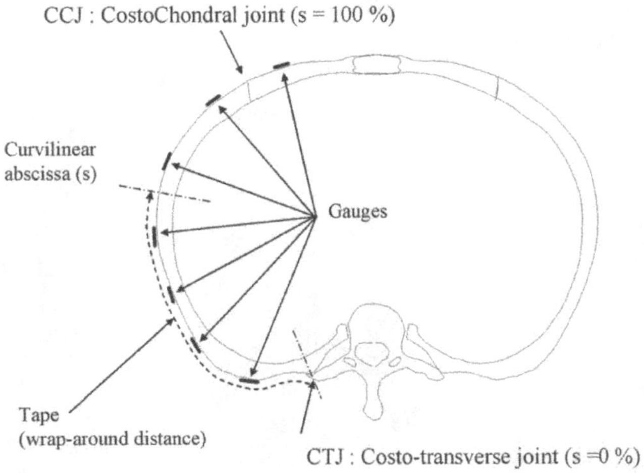

Fig. 2. Definition of the curvilinear abscissa for rib strain profile [10]

Initial series of simulations was performed using the reference model, and then a parametric study was performed. The range of changes covered material properties of internal organs, rib and sternum in order to obtain more biofidelic response in the validation scenarios. The lungs are represented by 10 639 solid elements and the heart by 3 380 solid elements (Fig. 3, left). The ribcage has significantly higher number of solid elements — 49 000 for ribs, 10 400 for costal cartilage and 1 720 for sternum (Fig. 3, right).

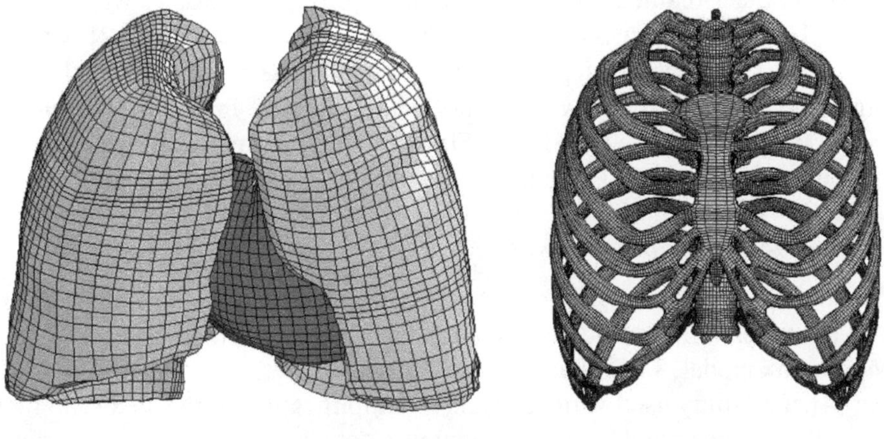

Fig. 3. FE model of heart and lungs (left) and ribcage (right)

Frontal impact

For the frontal impact (according to Kroell [8,9]), the impactor with mass of 23.4 kg and outer diameter of 152.4 mm was used at two energy levels, defined by the impact velocity of 4.3 m/s and 6.7 m/s. The impactor was aligned with the sternum centreline, at the level of 4th sternocostal joint (Fig. 4, left). The reference test responses, external thorax deflection and impactor contact force were monitored.

Oblique impact

For the oblique impact, a test scenario developed at the Ohio State University (OSU) was used as a reference [15]. The impactor of mass of 23.97 kg and outer diameter of 152.4 mm was centred on the middle point of the 6th rib and rotated by 30° of pure lateral, Fig. 4, middle. The impact velocity was 2.5 m/s. The impactor contact force, thorax deflection and strain along the ribs were monitored.

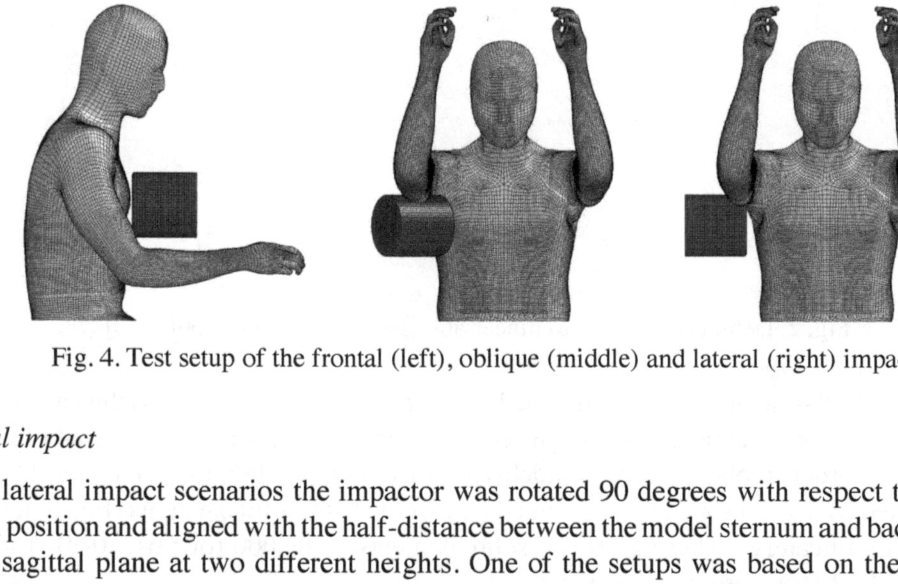

Fig. 4. Test setup of the frontal (left), oblique (middle) and lateral (right) impact

Lateral impact

In the lateral impact scenarios the impactor was rotated 90 degrees with respect to the initial frontal position and aligned with the half-distance between the model sternum and back measured in the sagittal plane at two different heights. One of the setups was based on the Ohio State University (OSU) configuration with the impactor at the level of 6th rib [15] — the impactor mass of 23.97 kg, outer diameter of 152.4 mm and the impact velocity was 2.5 m/s. Second setup based on HSRI (*Highway Safety Research Institute*) test scenario [12] with impactor at the level of the 4th sternocostal joint. The impactor had a mass of 23.4 kg and outer diameter of 152.4 mm. The impact velocity was 4.3 m/s. The difference in impactor position is 9 mm, therefore only OSU lateral setup is shown in Fig. 4, right. The impactor contact force and strain along the ribs were the monitored outputs.

2.3. Material models for heart and lungs

The THOMO thorax model is too stiff in the frontal impact in comparison to validation corridors. Therefore an alteration of heart and lungs material models is considered as one modification of the THOMO thorax model. The influence of material parameters for heart and lungs has been presented in [4]. The study used various levels of simplification of the THOMO thorax model and has shown that the heart and lungs behaviour was not physical and some material definitions caused numerical instability of the model. A similar parametric study of the material properties was performed in current study. The material properties used for the analysis are listed in Table 1

Table 1. Material models and parameters of heart materials

	Parameters	Material type and id in LS-Dyna	Reference
Reference material	Mass density: $\rho = 1\,000$ kg/m^3, Diastolic material coefficients: $C = 1.085$ kPa, $B_1 = 24.26, B_2 = 40.52, B_3 = 1.63,$ Pressure in the muscle tissue: $P = 2.482\,5$ GPa	Hyperelastic MAT_128	[2]
Material 1	Mass density: $\rho = 1\,000$ kg/m^3 Bulk modulus: $K = 1.33$ MPa Short-time shear modulus: $G_0 = 0.22$ MPa Long-time shear modulus: $G_{\text{inf}} = 0.075$ MPa Decay constant: $\beta = 1$ 1/ms	Viscoelastic MAT_006	[7]
Material 2	Nominal strain [%] vs. Nominal stress [kPa] 10 5.4 20 12.4 30 28.0 50 374.9	Low density foam MAT_057	[6]
Material 3	Mass density: $\rho = 1\,000$ kg/m^3 Bulk modulus: $K = 0.066$ MPa Short-time shear modulus: $G_0 = 0.014$ MPa Long-time shear modulus: $G_{\text{inf}} = 0.010\,7$ MPa Decay constant: $\beta = 1$ 1/ms	Viscoelastic MAT_006	[11]
Material 4	Mass density: $\rho = 1\,000$ kg/m^3 Bulk modulus: $K = 2.6$ MPa Short-time shear modulus: $G_0 = 0.44$ MPa Long-time shear modulus: $G_{\text{inf}} = 0.15$ MPa Decay constant: $\beta = 0.25$ 1/ms	Viscoelastic MAT_006	[14] quoting [18,19]
Material 5	Data from THOMO consortium partners		

(heart) and Table 2 (lungs). Because heart and lungs in the model show hourglassing effect, the influence of two types of hourglass control (parameter IHQ) was tested as well by changing the stiffness form of type 2 Flanagan-Belytschko (IHQ $= 4$) to Flanagan-Belytschko with exact volume integration (IHQ $= 3$).

2.4. *Material models for ribs and sternum*

Assuming that the mechanical response of the thorax is affected significantly by the ribcage stiffness [1], the influence of ribs and sternum material parameters on the model behaviour was studied as well. The ribs and the sternum cortical bone in the reference model were modelled by an elasto-plastic rate dependent material with parameters listed in Table 3. The lowest value of Young's modulus for the rib cortical bone found in the literature was 2.32 GPa [21]. This value was used for the "altered ribs" model, while other material parameters were kept same as for the reference model. For sternum, parameters of an elastic material with density of 1 800 kg/m^3 and Young's modulus of 2.5 GPa were found in [11]. Therefore in the "altered sternum" model these parameters were used with an elastic definition for the sternum material.

Table 2. Material models and parameters of lungs materials

	Parameters	Material type and id in LS-Dyna	Reference
Reference material	Mass density: $\rho = 288$ kg/m^3, Bulk modulus: $K = 2.66$ MPa, Material coefficients: $\Delta = 0.1$ mm, $C = 1.115\text{e}{-3}$ MPa, $\alpha = 0.213, \beta = -0.343$, $C_1 = 1.002\text{e}{-3}$ MPa, $C_2 = 2.04$	Transversely anisotropic MAT_129	[22]
Material 1	Mass density: $\rho = 600$ kg/m^3 Young's modulus: $E = 0.01$ MPa	Low density foam MAT_057	[7] quoting [18]
Material 2	Nominal strain [%] vs. Nominal stress [kPa] 10 7.9 20 14.1 30 20.1 50 31.7	Low density foam MAT_057	[6]
Material 3	Mass density: $\rho = 1\,000$ kg/m^3 Bulk modulus: $K = 0.066$ MPa Short-time shear modulus: $G_0 = 0.014$ MPa Long-time shear modulus: $G_{\text{inf}} = 0.010\,7$ MPa Decay constant: $\beta = 1$ 1/ms	Viscoelastic MAT_006	[11]
Material 4	Mass density: $\rho = 1\,000$ kg/m^3 (taken from Mat3) Bulk modulus: $K = 0.066$ MPa (taken from Mat3) Short-time shear modulus: $G_0 = 22.4$ kPa Long-time shear modulus: $G_{\text{inf}} = 7.5$ kPa Decay constant: $\beta = 0.25$ 1/ms	Viscoelastic MAT_006	[14]
Material 5	Mass density: $\rho = 600$ kg/m^3 Bulk modulus: $K = 0.22$ MPa Short-time shear modulus: $G_0 = 0.02$ MPa Long-time shear modulus: $G_{\text{inf}} = 0.075$ MPa Decay constant: $\beta = 1$ 1/ms (based on other publications)	Viscoelastic MAT_006	[14] quoting [18,19]
Material 6	Mass density: $\rho = 600$ kg/m^3 Bulk modulus: $K = 0.22$ MPa Short-time shear modulus: $G_0 = 0.02$ MPa Long-time shear modulus: $G_{\text{inf}} = 0.075$ MPa Decay constant: $\beta = 0.25$ 1/ms	Viscoelastic MAT_006	[14] quoting [18,19]
Material 7	Mass density: $\rho = 600$ kg/m^3 (taken from Mat5) Bulk modulus: $K = 0.22$ MPa (taken from Mat5) Short-time shear modulus: $G_0 = 22.4$ kPa Long-time shear modulus: $G_{\text{inf}} = 7.5$ kPa Decay constant: $\beta = 0.25$ 1/ms	Viscoelastic MAT_006	[14]

Table 3. Material parameters of rib and sternum cortical bone

	Mass density ρ [kg/m^3]	Young's modulus E [GPa]	Poisson's ratio	Yield stress [GPa]	Tangent modulus [GPa]	Reference
Reference material (ribs and sternum)	2 000	10.18	0.3	0.065 3	2.3	[23]
Altered ribs	2 000	2.32	0.3	0.065 3	2.3	[21]
Altered sternum	1 800	2.5	0.3	–	–	[11]

3. Results

3.1. Mechanical response

The validation tests presented in paragraph 2.2 were performed with reference material parameters for heart (Table 1), lungs (Table 2) and rib and sternum (Table 3).

Frontal impact — high speed (6.7 m/s)

The contact impactor force results of the frontal test are shown in Fig. 5, left. The force was slightly lower than the corridor in the initial phase, the first peak agreed well with the corridor but was delayed in time, and the second peak exceeded the upper corridor boundary. The resulting thorax deflection (Fig. 5, right) fit the corridor until 13ms and then decreased at a higher rate than the experimental reference. The contact impactor force versus thorax deflection results are shown in Fig. 6. Only the initial part fit the corridor. The occupant model thorax was stiffer compared to the experimental test.

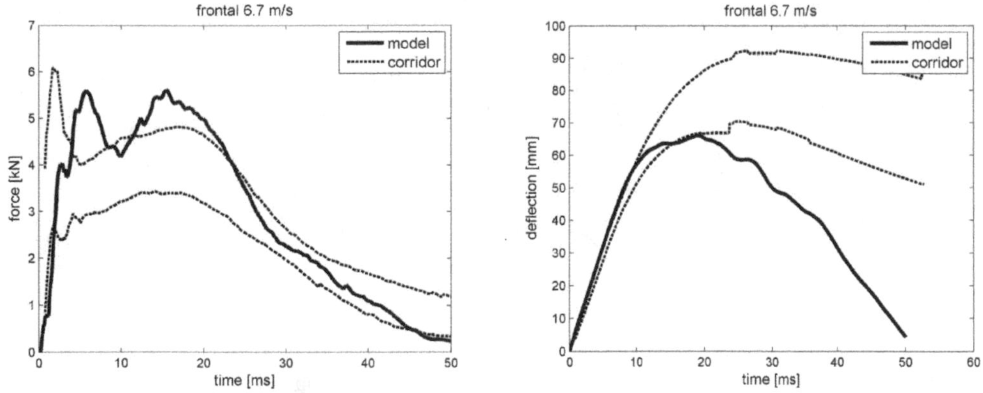

Fig. 5. The impactor contact force (left) and the thorax deflection (right) results of high speed frontal test

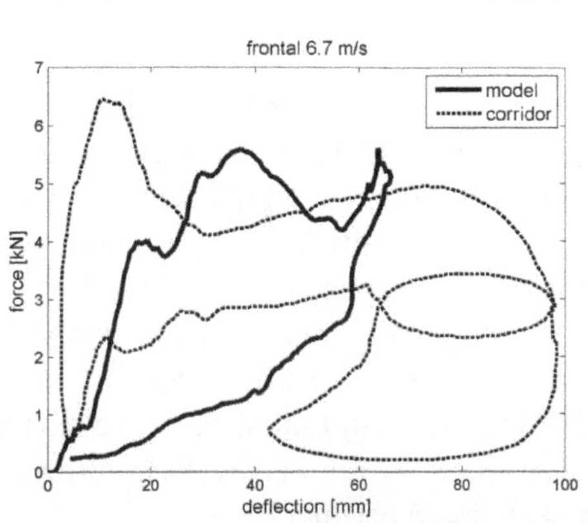

Fig. 6. The impactor contact force versus the thorax deflection for high speed frontal test

Frontal impact — low speed (4.3 m/s)

The contact impactor force results of the frontal test are shown in Fig. 7, left. The force fit the corridor for the first few milliseconds and then it increased above the corridor. There was a second force peak which did not fit in the corridor trend and range. The resulting thorax deflection (Fig. 7, right) fit the corridor until 15 ms and then it decreased below the corridor range. The contact impactor force versus thorax deflection results are shown in Fig. 8. Only the initial part fit the corridor. The occupant model thorax was stiffer compared to the experimental test.

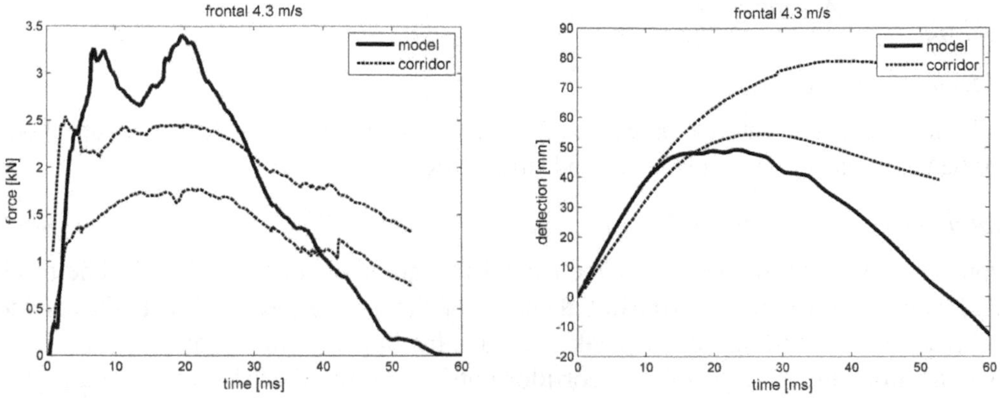

Fig. 7. The impactor contact force (left) and the thorax deflection (right) results of low speed frontal test

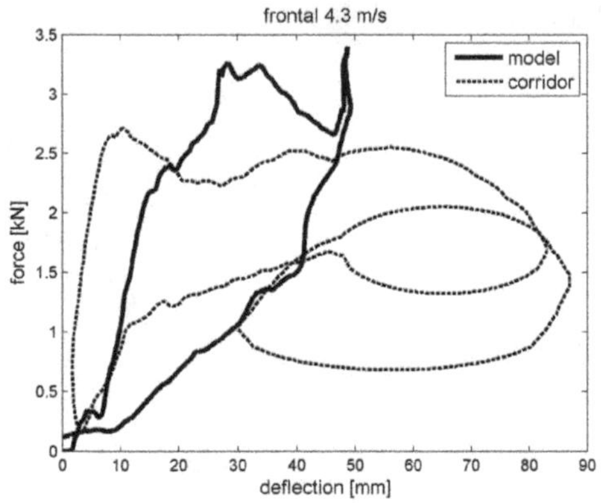

Fig. 8. The impactor contact force versus the thorax deflection for low speed frontal test

Oblique impact — OSU

The impactor contact force of the OSU oblique test is shown in Fig. 9, left. The simulation force response had a higher slope and its peak value was about 50 % higher than the corridor upper boundary. The thorax deflection (Fig. 9, right) was below the corridor range. The model response indicated that the model is too stiff in oblique impact scenario. The strain along the 5th ribs at the time of the highest deflection is shown in Fig. 10, right. There are negative strain values on the external side of the rib on the impacted side, positive strain values at the rear part of the rib on the impacted side, and lower strain values on the rib on the opposite side. The peak compressive strain on the right rib is around 80 % of the rib length, corresponding to strain profiles from the PMHS tests in oblique setup [10]. As well as for the frontal test, the thorax response was stiffer comparing to the experiment.

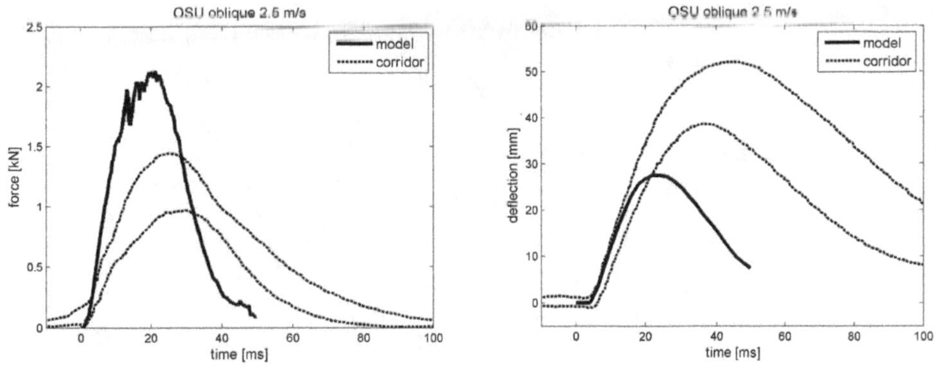

Fig. 9. The impactor contact force (left) and the thorax deflection (right) results of the oblique test

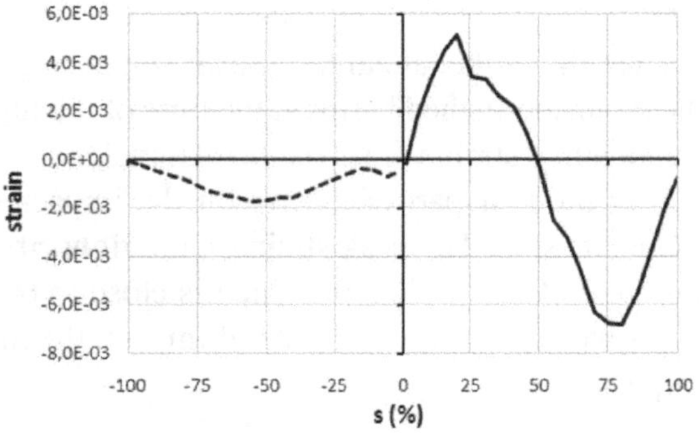

Fig. 10. Strain along the 5th ribs for the oblique test

Lateral impact — OSU

The impactor contact force of OSU oblique test is shown in Fig. 11, left. The force had slower progress and its peak exceeded the upper corridor boundary slightly. The thorax deflection (Fig. 11, right) fit the corridor well almost to the end of the unloading phase. The strain along the 5th ribs at the time of the highest deflection is shown in Fig. 12, right. There are negative strain values on the external side of the rib on the impacted side, positive strain values at the rear part of the rib on the impacted side, and lower strain values on the rib on the opposite side. The peak strain on the right rib is around 65 % of the rib length, which is close to 60 % reported with the PMHS tests in lateral setup [10].

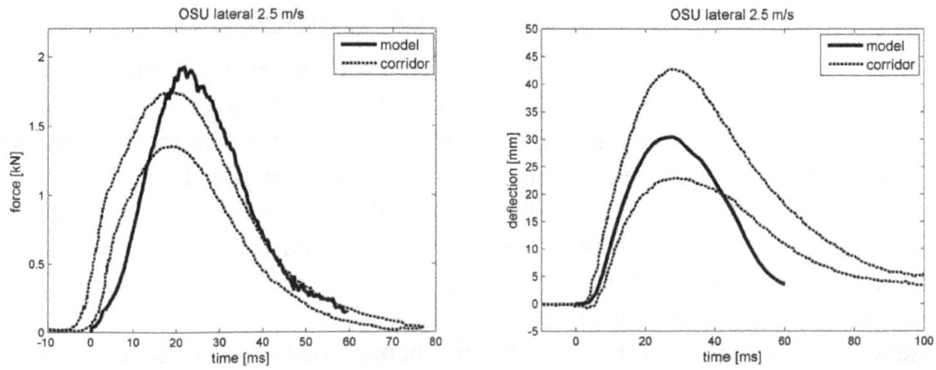

Fig. 11. The impactor contact force (left) and the thorax deflection (right) results of OSU lateral test

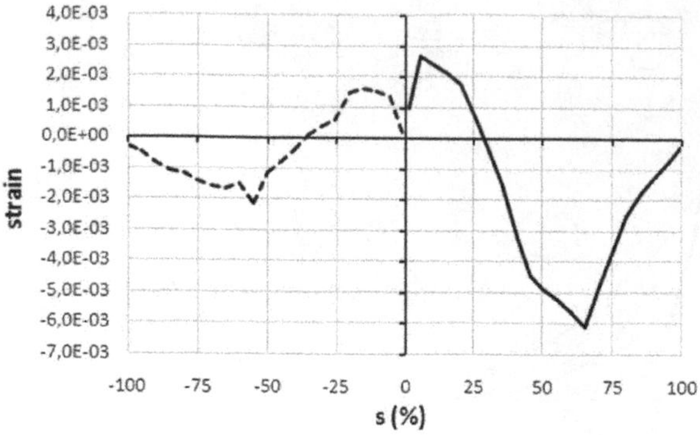

Fig. 12. Strain along the 5th ribs for the lateral test

Lateral impact — HSRI

The contact impactor force results of the lateral test are shown in Fig. 13, left. The force fit the experimental corridor. The strain along the 5th ribs at the time of the highest deflection is shown in Fig. 13, right. There are negative strain values on the external side of the rib on the impacted side, positive strain values at the rear part of the rib on the impacted side, and lower strain values on the rib on the opposite side. The peak strain on the right rib is at same location as in OSU lateral setup, around 65 % of the rib length, which is close to 60 % reported with PMHS tests in lateral setup [10]. As well as for the OSU lateral impact, the model exhibits reasonable behaviour in the lateral impact.

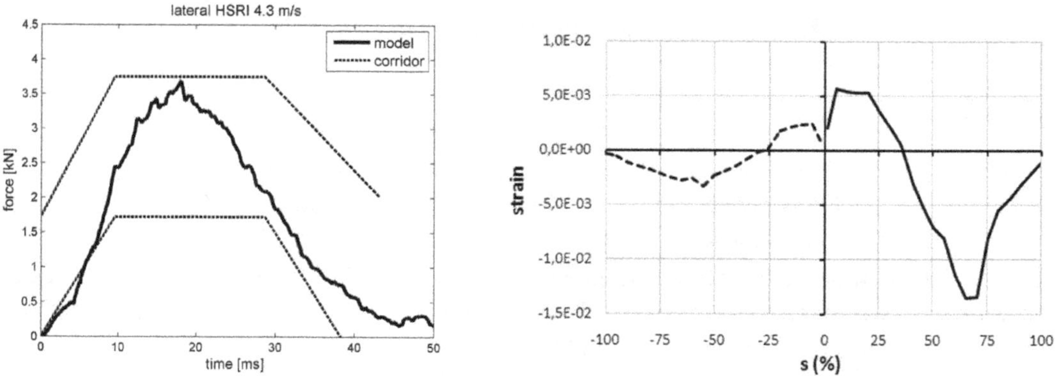

Fig. 13. The impactor contact force (left) and strain along the 5th ribs for HSRI lateral test

3.2. *Influence of heart and lungs model on the global response*

The frontal and oblique validation tests indicated that the model was stiffer than the PMHS during the experimental tests. In order to verify the influence of the thorax internal organs on the mechanical response of the model, the heart or lungs were removed from the model. Three tests were run with the altered model without heart — frontal low speed, OSU lateral and OSU oblique. In the frontal low speed test, the first peak of the contact impactor force decreased by 0.7 kN (Fig. 14, left) and the peak thorax deflection increased by 8 mm (Fig. 14, right) with respect to the reference model. There was no difference in the mechanical response of the OSU lateral and OSU oblique tests for this test configuration.

To assess the influence of the lungs on the mechanical response of the model, the left and the right lung were removed. Three tests were run with the altered model — frontal low speed, OSU lateral and OSU oblique. In the contact impactor force of frontal low speed test, there was

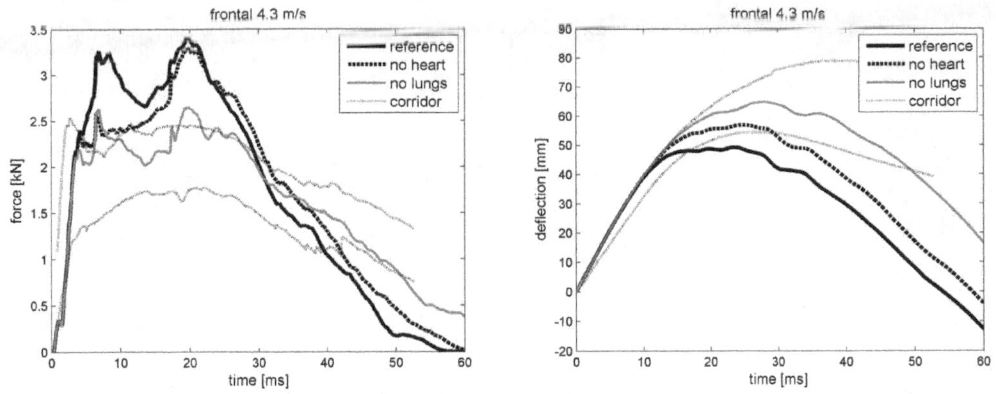

Fig. 14. The impactor contact force (left) and the thorax deflection (right) results of low speed frontal test of models with heart or lungs removed compared to the reference model

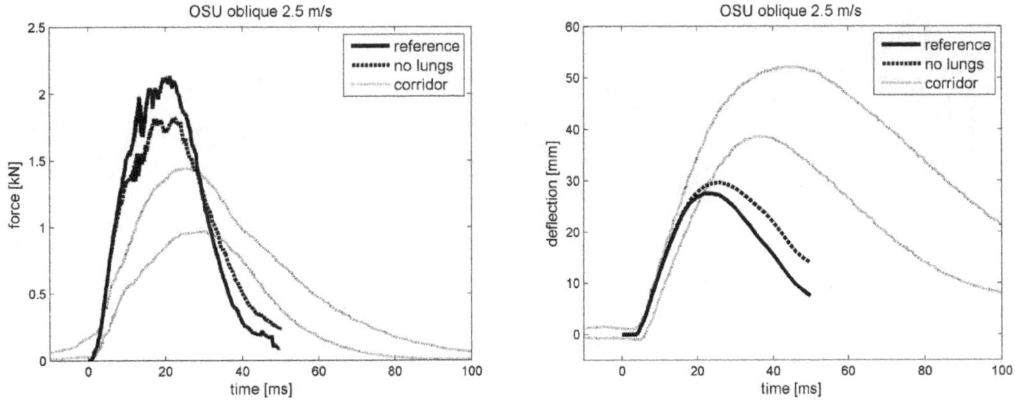

Fig. 15. The impactor contact force (left) and the thorax deflection (right) results of OSU oblique test of models with heart or lungs removed compared to the reference model

a significant decrease of 0.64 kN of the first peak and 0.75 kN of the second peak (Fig. 14, left) and the peak thorax deflection increased by 16 mm (Fig. 14, right).

There was no difference in the mechanical response of the OSU lateral test. In the OSU oblique test, the peak impactor contact force decreased by 0.32 kN (Fig. 15, left), the peak thorax deflection increased by 2 mm (Fig. 15, right).

The first peak of the impactor contact force and the thorax deflection in the frontal test were sensitive to the absence of the heart part. In the frontal test, the absence of the lungs significantly influenced both peaks of the contact impactor force and the thorax deflection. The lungs part also influenced the impactor contact force and thorax deflection in the OSU oblique test.

3.3. Influence of material parameters of heart and lungs on the global response

Modifications of the heart material slightly influenced the first peak of the impactor contact force (Fig. 16, left). The lowest value was found for Material 3, however this material model caused early termination of the simulation — at time 16.9 ms. Influence of the material choice on the second peak of the impactor force and the thorax deflection could be neglected (Fig. 16).

The simulations with most of the selected material models of lungs from Table 2 resulted with numerical instabilities and premature simulation termination. In terms of numerical stability, Material 7 was the best choice. The alteration of the lungs material model affected both the first

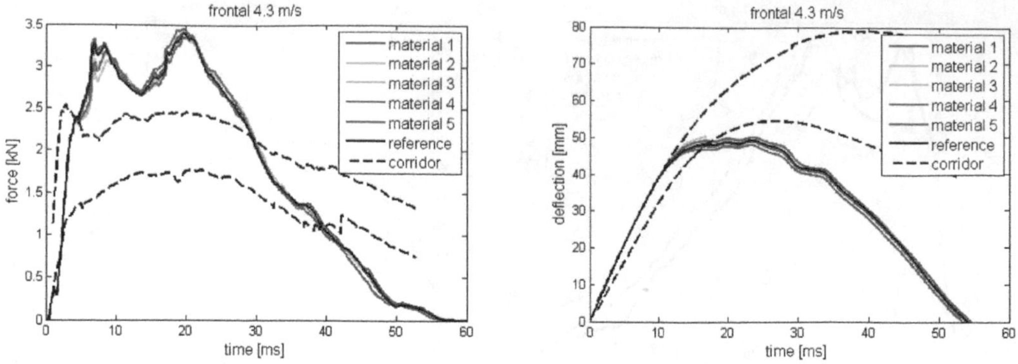

Fig. 16. The impactor contact force (left) and the thorax deflection (right) results of low speed frontal test with models with various material models of heart

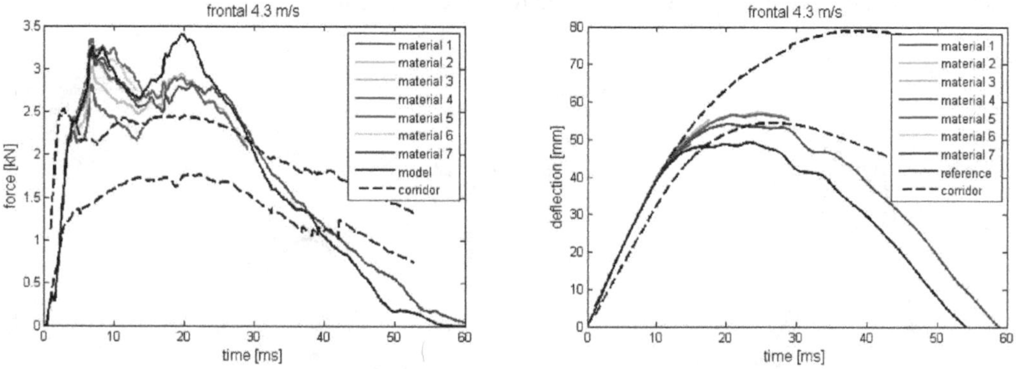

Fig. 17. The impactor contact force (left) and the thorax deflection (right) results of low speed frontal test with models with various material models of lungs

and the second peak of the impactor contact force, depending on the material choice (Fig. 17, left). The first peak was the lowest in case of Material 1. The second peak decreased for all material modifications. Thoracic deflection increased in all simulations (Fig. 17, right).

3.4. Influence of rib and sternum cortical bone material on the global response

For each model with altered Young's modulus of the cortical bone, the first peak of the impactor force decreased and the thorax deflection slightly increased compared to the reference material model (Fig. 18). The most significant change was observed when both rib and sternum cortical bone material were altered at once (Fig. 18).

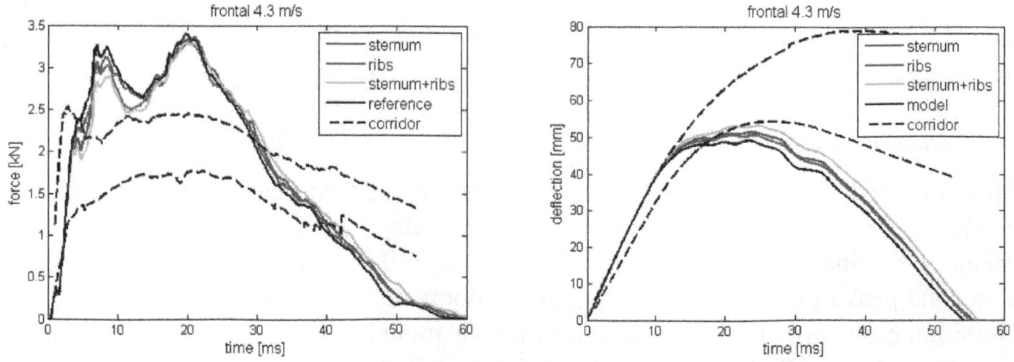

Fig. 18. The impactor contact force (left) and the thorax deflection (right) results of low speed frontal test with models with altered rib and sternum cortical bone material parameters

3.5. Influence of heart and lungs material combined with rib and sternum cortical bone material

As the next step, both rib and sternum cortical bone materials were altered. For the heart model, Material 2, 3 and 5 properties were chosen. For the lungs model, Material 4 and 7 were applied. The material models were adjusted based on the mechanical response that was the closest to the corridors. Although the simulation with lungs Material 4 ended with errors when lungs material properties only were altered, in combination with heart Material 2 and 5 the simulations ended normally (Fig. 20). The simulation with heart Material 3 and lungs Material 4 was unstable; however, combination of heart Material 3 with lungs Material 7 terminated normally (Fig. 19). The simulations with lungs Material 7 had the second peak of impactor contact force slightly lower and delayed by approximately 5 ms compared to the lungs Material 4 (Fig. 19, left). The deflection in the simulations with lungs Material 7 was slightly lower than the one in the simulations with lungs Material 4 (Fig. 19, right).

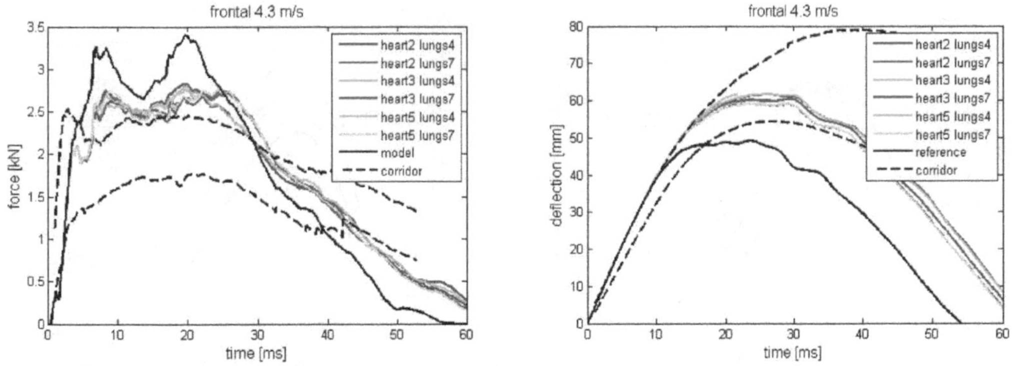

Fig. 19. The impactor contact force (left) and the thorax deflection (right) results of low speed frontal test with models with combination of altered heart and lungs material parameters

The parameter IHQ affecting the hourglassing was changed from a stiffness form of type 2 Flanagan-Belytschko (IHQ = 4) to Flanagan-Belytschko with exact volume integration (IHQ = 3) and tested in three material combinations in frontal scenario — first with combination of heart Material 2 and lungs Material 4, second with heart Material 5 and lungs Material 7 and third with heart Material 2 and lungs Material 7. These material combinations were selected because the simulations ended normally in the low speed frontal test (Fig. 19). No significant effect of the parameter IHQ variation on the model response was observed.

To verify the effect of heart and lungs material modifications on the global response in the oblique and lateral scenarios, three models were tested — first with combination of heart Material 2 and lungs Material 4, second with heart Material 5 and lungs Material 7 and third with heart Material 2 and lungs Material 7. These material combinations were selected because the simulations ended normally in the low speed frontal test (Fig. 19). All three models had altered rib and sternum cortical bone material (Table 3) because it provided model response closer to the corridors. As the results showed that the heart Material 2 and lungs Material 7 had almost no hourglassing effect, their combination was tested as well.

In the oblique scenario, the peak impactor contact force decreased by 7 % and the thorax deflection increased by 12 %, both getting closer to the corridors (Fig. 20). In lateral scenario, the peak impactor contact force decreased by 5 %, with the peak values getting almost into the corridors (Fig. 21, left). The thorax deflection increased approximately by 20 % and the unloading phase remained within the corridor for additional 11 ms of the simulation (Fig. 21, right).

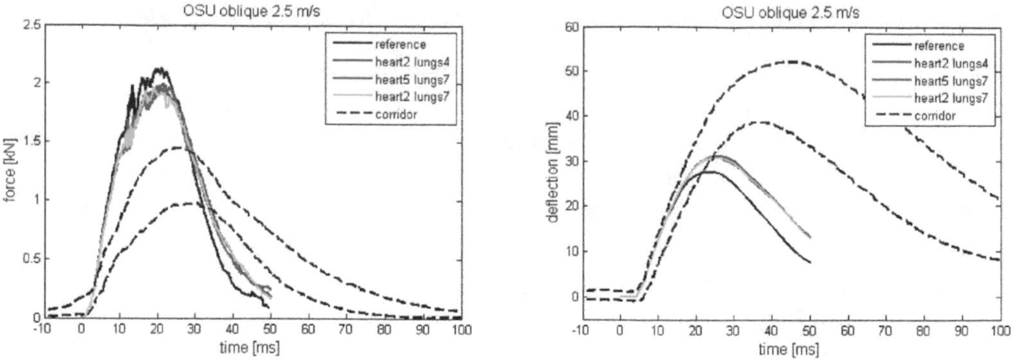

Fig. 20. The impactor contact force (left) and the thorax deflection (right) results of OSU oblique test with models with combination of altered heart and lungs material parameters

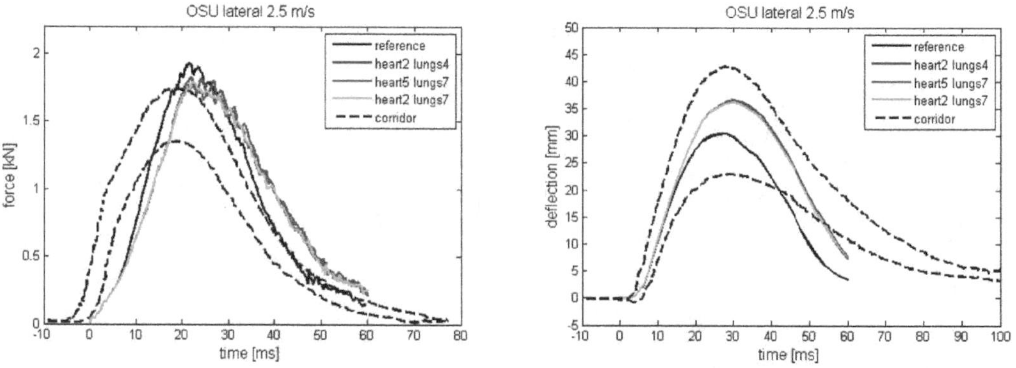

Fig. 21. The impactor contact force (left) and the thorax deflection (right) results of OSU lateral test with models with combination of altered heart and lungs material parameters

4. Discussion

A series of pendulum impact simulations using the reference material models has been performed. The model response was tested in three impact scenarios: frontal, lateral and oblique.

The frontal impact with the high speed shows good performance in the model during the loading phase (Fig. 5). The force dependent on deflection copies the experimental response with just a small time shift. Although the unloading phase is faster, concerning the injury prediction, the global response is well validated taking into account the maximum values of force and deflection and the shape of curves. Fig. 6 is just a consequence of the combination of curves on Fig. 5.

The low speed frontal impact shows worse behaviour (Fig. 7). The problem might be caused by the higher overall stiffness of the model in order to fit the high speed response. Taking into account the potential injury predictability and good high speed response, the model is tuned to be used in safety applications.

The oblique impact shows higher stiffness and faster unloading phase. The problem is probably caused by the simplicity of the model, because in the biological reality, the human thorax is softer in the oblique direction. Taking into account the potential injury predictability, the model should be tuned for injury risk assessment also in the oblique impact.

The OSU lateral impact shows almost perfect behaviour of the model, where the signal obtained from the calculations fits the experimental signal corridor quite well. The same behaviour can be observed in the HSRI lateral impact, where the model response fits the corridor.

Additionally, the analysis concerning the model design was also performed. It focused on variation of material properties of the internal organs, which were assumed to constitute the differences between model behaviour and experimental reference, to verify the effect of material modifications on the global response and tune the model with the experimental corridors. The analysis shows significant improvement of the model response in frontal impact. It is caused mainly by different models of lungs and by altering rib and sternum parameters.

Taking the results into account, the model can be perfectly used for injury analysis in the lateral direction. The frontal and oblique impacts shows higher stiffness of the model, however, concerning the injury risk predictability the model is safe to be used for impact analysis.

5. Conclusion

The parametric study and validation of the 50th percentile THOMO model were performed. The model response was tested in three impact scenarios: frontal, lateral and oblique.

In the frontal test scenarios, the initial mechanical response of the model was too stiff, compared to the reference experimental tests for high speed and low speed impact. For oblique test, the initial mechanical response was too stiff as well. For lateral tests, the results correspond reasonably to the corridors from PMHS tests.

The material properties of the major thoracic internal organs (heart and lungs) and the rib and sternum cortical bone were found to have a major influence on the thoracic response mainly in the frontal impact. Adjusting the material properties could improve the global response of the model.

Shift in the strain profile between lateral and oblique tests setup is around 15 %. The value of 20 % is observed in PMHS tests [10].

Acknowledgements

This paper is a part of the THOMO project (http://www.thomo.eu) co-funded by the FP7 of the European Commission. The authors would like to gratefully thank the project partners, the GHBMC and the COE. The contents of this publication are sole the responsibility of the authors and do not necessarily represent the views of the EC and of the GHBMC.

Reference

[1] Brolin, K., Mendoza-Vazquez, M., Song, E., Lecuyer, E., Davidsson, J., Design implication for improving an anthropometric test device based on human body simulations, Proceedings of the International IRCOBI Conference, Dublin, Ireland, 2012, pp. 843–855.

[2] Deng, Y. C., Kong, W., Ho, H., Development of a finite element human thorax model for impact injury studies, SAE International Congress and Exposition, Detroit, Michigan, SAE Paper 1999-01-0715, 1999.

[3] GHBMC website: http://www.ghbmc.com.

[4] Gierczycka-Zbrożek, D., Rzymkowski, C., Toczyski, J., Influence of the material definition on the biomechanical response of a simplified human torso FE model, Proceedings of the 9th International Forum of Automotive Traffic Safety, Changsha, China, 2011, pp. 231–238.

[5] Haug, E., Beaugonin, M., Montmayeur, N., Marca, C., Choi, H., Towards legal virtual crash tests for vehicle occupant safety design using human models, International Crashworthiness and Design Symposium, 2003.

[6] Iwamoto, M., Nakahira, Y., Tamura, A., Kompars, H., Watanabe, I., Miki, K., Development of advanced human models in THUMS, Proceedings of the 6th European LS-DYNA Users' Conference, Gothenburg, Sweden, 2007, pp. 47–56.

[7] Kimpara, H., Lee, J. B., Yang, K. H., King, A. I., Iwamoto, M., Watanabe, I., Miki, K., Development of a three-dimensional finite element chest model for the 5th percentile female, Stapp Car Crash Journal 49 (2005) 251–269.

[8] Kroell, C., Schneider, D., Nahum, A., Impact tolerance and response of the human thorax, SAE Technical Paper 710851, 1971, doi:10.4271/710851.

[9] Kroell, C., Schneider, D., Nahum, A., Impact tolerance and response of the human thorax II, SAE Technical Paper 741187, 1974, doi:10.4271/741187.

[10] Leport, T., Baudrit, P., Potier, P., Trosseille, X., Lecuyer, E., Vallancien, G., Study of rib fracture mechanisms based on the rib strain profiles in side and forward oblique impact, Stapp Car Crash Journal 55 (2011) 99–250.

[11] Lizee, E., Robin, S., Song, E., Bertholon, N., Le Coz, J.-Y., Besnault, B., Lavaste, F., Development of a 3D finite element model of the human body, SAE Technical Paper 983152, 1998, doi:10.4271/983152.

[12] Robin, S., HUMOS: Human model for safety — A joint effort towards the development of refined human-like car occupant models, Proceedings of the 17th International Technical Conference on the Enhanced Safety of Vehicles, Paper No. 297, Amsterdam, 2001.

[13] Robbins, D. H., Lehman, R. J., Nusholtz, G. S., Melvin, J. W., Benson, J. B., Stalnaker, R. L., Culver, R., Quantification of thoracic response and injury: the gathering of data; Final Report, Contract DOT-HS-4-00921, 1982.

[14] Ruan, J., El-Jawahri, R., Chai, L., Barbat, S., Prasad, P., Prediction and analysis of human thoracic impact responses and injuries in cadaver impact using a full human body finite element model, Stapp Car Crash Journal 47 (2003) 299–321.

[15] Shaw, J. M., Herriott, R. G., McFadden, J. D., Donnelly, B. R., Bolte, J. H., Oblique and lateral impact response of the PMHS thorax, Stapp Car Crash Journal 50 (2006) 147–167.

[16] Shigeta, K., Kitagawa, Y., Yasuki, T., Development of next generation human FE model capable of organ injury prediction, Proceedings of the 21st International Technical Conference on the Enhanced Safety of Vehicles, Paper No. 09-0111, Stuttgart, 2009.

[17] Vezin, P., Verriest, J., Development of a set of numerical human models for safety, Proceedings of the 19th International Technical Conference on the Enhanced Safety Vehicles Conference, Paper No. 05-0163, Washington D.C., 2005.

[18] Wang, H.-C. K., Development of a side impact finite element thorax model, Ph.D. Thesis, Wayne State University, Detroit, 1995.

[19] Yamada, H., Strength of Biological Materials, Williams & Wilkins Company, Baltimore, 1970.

[20] Yang, K. H., Hu, J. W., White, N. A., King, A. I., Chou, C. C., Prasad, P., Development of numerical models for injury biomechanics research: A review of 50 years of publications in the Stapp Car Crash Conference, Stapp Car Crash Journal 50 (2006) 429–490.

[21] Yoganandan, N., Pintar, F. A., Biomechanics of human thoracic ribs, Journal of Biomechanical Engineering 120 (1998) 100–104.

[22] Yuen, K. F., The development of numerical human body model for the analysis of automotive side impact lung trauma, Master Thesis, University of Waterloo, Canada, 2009.

[23] Zhao, J., Narwani, G., Development of a human body finite element model for restraint system R&D applications, Proceedings of the 19th International Technical Conference on the Enhanced Safety of Vehicles, Paper No. 05-0399, Washington D.C.; 2005.

10

Thermal analysis of both ventilated and full disc brake rotors with frictional heat generation

A. Belhocine[a,*], C.-D. Cho[b], M. Nouby[c], Y. B. Yi[d], A. R. Abu Bakar[e]

[a] *Faculty of Mechanical Engineering, USTO Oran University 31000 Oran, Algeria*

[b] *Department of Mechanical Engineering, Inha University, Incheon, 402-751, Republic of Korea*

[c] *Department of Mechanical Engineering, Faculty of Engineering, South Valley University, Qena-83523, Egypt*

[d] *Department of Mechanical and Materials Engineering, University of Denver, 2390 S York St Denver, CO 80208, USA*

[e] *Department of Automotive Engineering, Universiti Teknologi Malaysia, 81310 UTM Skudai, Malaysia*

Abstract

In automotive engineering, the safety aspect has been considered as a number one priority in development of a new vehicle. Each single system has been studied and developed in order to meet safety requirements. Instead of having air bags, good suspension systems, good handling and safe cornering, one of the most critical systems in a vehicle is the brake system. The objective of this work is to investigate and analyze the temperature distribution of rotor disc during braking operation using ANSYS Multiphysics. The work uses the finite element analysis techniques to predict the temperature distribution on the full and ventilated brake discs and to identify the critical temperature of the rotor. The analysis also gives us the heat flux distribution for the two discs.

Keywords: brake disc, convection, heat-transfer coefficient, thermal analysis, stress

---

1. Introduction

Braking system is one of the important safety components of a vehicle. It is mainly used to decelerate vehicles from an initial speed to a given speed. A friction based braking system is a common device to convert kinetic energy into thermal energy through a friction between the brake pads and the rotor faces. Because high temperatures can lead to overheating of the brake fluid, seals and other components, the stopping capability of a brake increases with the rate at which heat is dissipated due to forced convection and thermal capacity of the system [24].

Brake disc convective cooling has been historically studied by means of experimental and theoretical methods [17, 18] and the optimization was only boosted with the advent of modern computational resources in the late 1980s [6]. Currently, although of not simple usage and requiring previous understanding of the basics of fluid mechanics and heat transfer coupled with the knowledge of numerical flow modeling, computational fluid dynamics (CFD) has significantly gained preference in the automotive industry design process as a tool for the prediction of complex flow and heat transfer behavior in regions, where otherwise very laborious and time consuming experimental set up work would be needed. As a result, brake disc convective cooling analysis and optimization is nowadays mostly carried out using CFD commercial codes, see, e.g., [23].

*Corresponding author. e-mail: al.belhocine@yahoo.fr.

Many investigations of heat flow through ventilated disc brakes are reported in the literature. Michael and Roland [8] discussed the airflow patterns in the disc rotors. Wallis et al. [28] carried out a numerical study using the software Fluent on disc rotor blades to examine the effects of local heat and mass transfer of the axial gap distances for a single co-rotating disc. The study of the single rotating disc showed that heat and mass transfer coefficients are enhanced considerably by decreasing the hub height.

The friction heat generated between two sliding bodies causes thermoelastic deformation, which alters the contact pressure distribution. This coupled thermo-mechanical process is referred to as the frictionally-excited thermoelastic instability (TEI) [14]. Other works have studied the transient brake analysis [15,26,30,32]. Zagrodzki [30] analyzed sliding systems with frictional heating, which exhibit thermoelastic instability (TEI) in friction clutches and brakes when the sliding speed exceeds a critical value. Zhu et al. [32] established the theoretical model of a three-dimensional (3D) transient temperature field to predict the change of brake shoe's temperature field during hoist's emergency braking. Voldřich [26] postulated that the exceeding of the critical sliding velocity in brake discs causes formation of hot spots, non-uniform contact pressure distribution, vibration, and permanent damage of the disc. The analytical model of TEI development was published by Lee and Barber [15].

Many researchers investigated the heat generation phenomenon between contact surfaces in automotive clutches and brakes to predict the temperature distribution and especially the maximum temperature during the clutch engagement and braking to avoid failure before an estimated lifecycle. This process is very complex because of the following characteristics pressure, coefficient of friction and sliding speed. The researchers used different numerical techniques such as the finite element and finite difference methods to compute the sliding surface temperature [13,27].

Abu Bakar and Ouyang [2] adjusted the surface profiles using measured data of the surface height and produced a more realistic model for brake pads. Direct contact interaction between the disc brake components is represented by a combination of node-to-surface and surface-to-surface contact elements [1].

Gao and Lin [7] stated that there was considerable evidence to show that the contact temperature is an integral factor reflecting the specific power friction influence of combined effect of load, speed, friction coefficient, and the thermo physical and durability properties of the materials of a frictional couple. Experiments showed that the friction coefficient in general decreased with increasing sliding speed and applied load, but increased with increasing disc temperature up to 300 °C and then decreased above this temperature. The specific wear rate was found to increase with increasing sliding speed and disc temperature [22].

Nouby et al. [20] introduced a nontraditional evaluation tool to examine the effects of different materials that are used in fabricating disc brake components commonly used or special by manufactured for heavy-duty performance and racing cars. An extension of the FE models discussed earlier is a three-dimensional FE model of the disc brake corner that incorporates a wheel hub and steering knuckle developed and validated at both components and assembly levels to predict disc brake squeal. In addition, the real pad surface topography, negative fiction-velocity slope, and friction damping were considered to increase the prediction accuracy of the squeal. Brake squeal is a high frequency noise produced when driver decelerates and/or stops the vehicle moving at low speed. It is the noise caused by a self-excited vibration generated by the friction force variation between the friction material and the rotor. It is a phenomenon of dynamic instability that occurs at one or more natural frequencies of the brake system.

In work of Mosleh et al. [19], a brake dust generated in vehicle brakes causes discoloration of wheels and, more importantly, the emission of particles suspected of health hazard in the environment. Laboratory testing of brake pad materials against cast iron discs revealed that the majority of wear particles are submicrometer in size. Wear particles with a size of 350 nm had the highest percentage in the particle size distribution plots, regardless of the magnitude of the nominal contact pressure and the sliding speed. Due to their predominantly submicrometer size, a significant amount of brake dust particles may be inhalable in environmental and occupational exposure situations.

According to Altuzarra et al. [3], if the sliding speed is high, the resulting thermo-mechanical feedback is unstable, leading to the development of non-uniform contact pressure and local high temperature with important gradients called 'hot spots'. The formation of such localized hot spots is accompanied by high local stresses that can lead to material degradation and eventual failure [10]. Also, the hot spots can be a source of undesirable frictional vibrations, known in the automotive disc brake community as 'hot roughness' or 'hot judder' [29].

A ventilated disc is lighter than a solid one, and with additional convective heat transfer occurring on the surface of the vent hall. Thus, the ventilated disc can control its temperature rise and minimize the effects of thermal problems such as the variation of the pad friction coefficient, brake fade and vapor lock [5,9]. The ventilated disc, however, may increase judder problems by inducing an uneven temperature field around the disc. Also, the thermal capacity of the ventilated disc is less than that of the solid disc, and the temperature of the ventilated disc can rise relatively faster than that of the solid disc during repetitive braking [11]. Therefore, thermal capacity and thermal deformation should be carefully considered when modifying the shape of the ventilated disc.

In this study, we will model of the thermal behaviour of a dry contact between the discs of brake pads during the braking phase; the strategy of calculation is based on the software ANSYS 11. This last is comprehensive mainly for the resolution of the complex physical problems. As a current study of the problem, ANSYS simulations with less assumption and less program restrictions have been performed for the thermo-mechanical case. A temperature distribution obtained by the transient thermal analysis is used in the calculations of stresses on the disc surface.

2. Numerical procedure

During the braking process, the temperature field changes input heat flux and heat exchange conditions. The input heat flux is mainly dependent on the friction coefficient and angular velocity of the brake disc, while the heat exchange is connected with the friction pair materials and external environmental factors.

Based on the first law of thermodynamics, the kinetic energy during braking is converted to thermal energy. The conversion of energy takes place because of the friction between the main components of the brake disc (pads and disc). Initially, the generated thermal energy is transferred by conduction to the components in contact and further by convection to surrounding. The initial heat flux q_0 into the rotor face is directly calculated using the following formula [25]:

$$q_0 = \frac{1-\phi}{2}\frac{mgv_0z}{2A_d\varepsilon_p}, \tag{1}$$

where $z = a/g$ is the braking effectiveness, a is the deceleration of the vehicle [m/s^2], ϕ is the rate distribution of the braking forces between the front and rear axle, A_d is the disc surface swept by a brake pad [m^2], v_0 is the initial speed of the vehicle [m/s], ε_p is the factor load distributed on the disc surface, m is the mass of the vehicle [kg] and $g = 9.81$ is the acceleration of gravity [m/s^2].

The disc rotor vane passage and the sector chosen for the numerical analysis are shown in Fig. 1. The dimensions of the ventilated disc for the basic study are outer diameter of 262 mm, inner diameter of 66 mm and 36 vanes. The geometric model was created using the ANSYS Workbench 11.

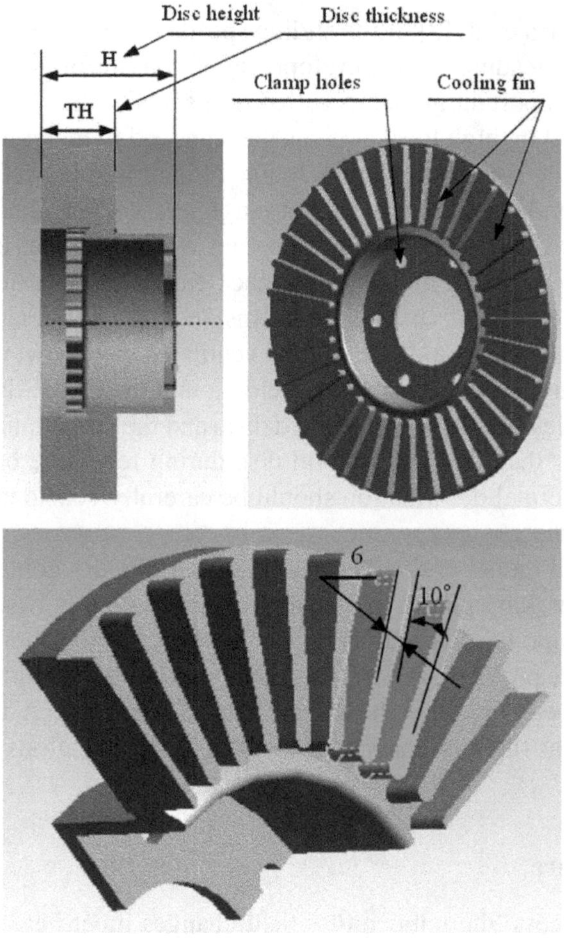

Fig. 1. Geometrical characteristics of the ventilated disc

The loading corresponds to the heat flux on the disc surface. The dimensions and the parameters used in the thermal calculation are summarised in Table 1.

The material of the brake disc is gray cast iron (FG 15) with high carbon content and good thermophysical characteristics. The thermoelastic characteristics of the material adopted in the rotor simulation are listed in Table 2. There are three reasons why rotors are made of the cast iron [4]:

- It is relatively hard and resists wear.
- It is cheaper than steel or aluminum.
- It absorbs and dissipates heat well and helps in cooling of the brakes.

Table 1. Input parameters

Input parameter	Values
Inner disc diameter [mm]	66
Outer disc diameter [mm]	262
Disc thickness (TH) [mm]	29
Disc height (H) [mm]	51
Vehicle mass m [kg]	1 385
Initial speed v_0 [m/s]	28
Deceleration a [m/s^2]	8
Effective rotor radius R_{rotor} [mm]	100.5
Rate distribution of the braking forces ϕ [%]	20
Tire radius R_{tire} [mm]	380
Factor of charge distribution of the disc ε_p	0.5
Surface disc swept by the pad A_d [mm^2]	35 993

Table 2. Material properties of the brake disc

Material properties	Disc
Thermal conductivity k [W/m °C]	57
Density ϱ [kg/m^3]	7 250
Specific heat C [J/Kg °C]	460
Poisson's ratio v	0.28
Thermal expansion $\alpha \cdot 10^{-6}$ [1/K]	10.85
Elastic modulus E [GPa]	138

It is very difficult to exactly model the brake disc, as there are still investigations going on to find out transient thermal behaviour of disc brakes during braking. In this case, to model a complex geometry, some simplifications are always necessary. These simplifications are made, keeping in mind the difficulties involved in the theoretical calculation and the importance of the parameters that are taken and those that are ignored. In modelling, we usually ignore the things of less importance and with little impact on the analysis. The assumptions are always made depending upon the details and accuracy required in the modelling.

By applying brakes on the car disc brake rotor, heat is generated by friction and the thermal flux has to be conducted and dispersed across the disc rotor cross section. The condition of braking is very severe and thus the thermal analysis has to be carried out. The thermal loading as well as structure is axisymmetric. Although, axisymmetric analysis can be performed, in this study we performed a 3D analysis, which is an exact representation of the thermal analysis. With the above load structural analysis, the thermal analysis is also carried out for analyzing the stability of the structure.

To simplify the analysis, several assumptions were also been made as follows [12]:

- All kinetic energy at disc brake rotor surface is converted into frictional heat or heat flux.
- The heat transfer involved in this analysis takes place only by conduction and convection. Heat transfer by radiation can be neglected as it amounts only to 5 % to 10 % [16].

- The disc material is considered as homogeneous and isotropic.
- The domain is considered as axisymmetric.
- Inertia and body force effects are negligible during the analysis.
- The disc is stress free before the brake application.
- In this analysis, the ambient temperature and initial temperature is set to 20 °C.
- All other possible disc brake loads are neglected.
- Only certain parts of disc brake rotor experience convection heat transfer such as the cooling vanes area, the outer ring diameter area and the disc brake surface.
- Uniform pressure distribution generated by the brake pad onto the disc brake surface is considered.

The thermal conductivity and specific heat are a function of temperature, as illustrated in Figs. 2 and 3.

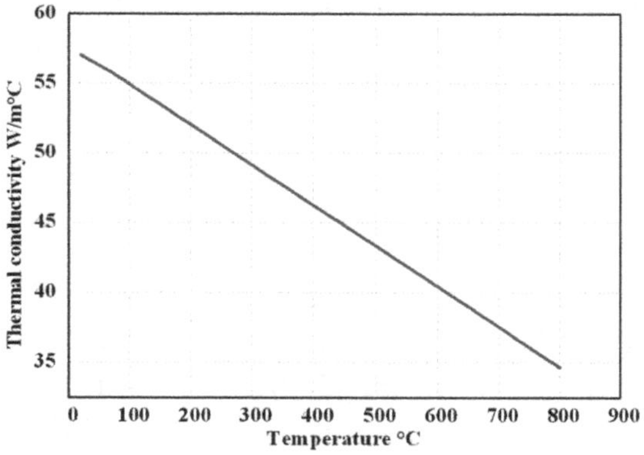

Fig. 2. Thermal conductivity versus temperature

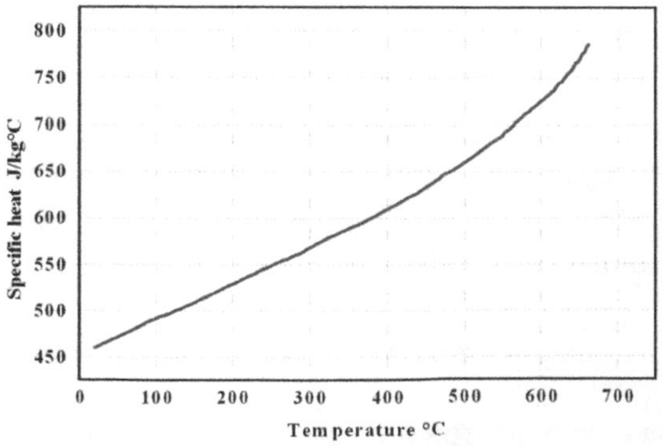

Fig. 3. Specific heat versus temperature

3. Finite element formulation for heat conduction

The unsteady heat conduction equation of each body for an axisymmetric problem described in the cylindrical coordinate system is given as follows:

$$\varrho c \frac{\partial T}{\partial t} = \frac{1}{r} \partial \partial r \left(r k_r \frac{\partial T}{\partial r} \right) + \frac{\partial}{\partial z} \left(k_z \frac{\partial T}{\partial z} \right) \tag{2}$$

with the boundary conditions and initial condition

$$T = T^* \quad \text{on} \quad \Gamma_0, \tag{3}$$

$$q_n = h(T - T_\infty) \quad \text{on} \quad \Gamma_1, \tag{4}$$

$$q_n = q_n^* \quad \text{on} \quad \Gamma_2, \tag{5}$$

$$T = T_0 \quad \text{at time} = 0, \tag{6}$$

where ϱ, c, k_r, and k_z are the density, specific heat and thermal conductivities in the r and z directions of the material, respectively. Also, T^* is the prescribed temperature, h is the heat transfer coefficient, q_n^* is the heat flux at each contact interface caused by friction, T_∞ is the ambient temperature, T_0 is the initial temperature and Γ_0, Γ_1 and Γ_2 are the boundaries on which temperature, convection and heat flux are imposed, respectively.

Using the Galerkin's approach, a finite element formulation of unsteady heat equation (2) can be written in the following matrix form as

$$C_T \dot{T} + KH_T T = R, \tag{7}$$

where C_T is the capacity matrix, KH_T is the conductivity matrix, T and R are the nodal temperature and heat source vector, respectively.

The most commonly used method for solving Eq. (7) is the direct integration method based on the assumption that the temperature T_t at time t and the temperature $T_{t+\Delta t}$ at time $t + \Delta T$ have the following relation:

$$T_{t+\Delta t} = T_t + [(1-\beta)\dot{T}_t + \beta \dot{T}_{t+\Delta t}]\Delta t. \tag{8}$$

Eq. (8) can be used to reduce the ordinary differential equation (7) to the following implicit algebraic equation:

$$(C_T + b_1 KH_T)T_{t+\Delta t} = (C_T - b_2 KH_T)T_t + b_2 R_t + b_1 R_{t+\Delta t}, \tag{9}$$

where the variable b_1 and b_2 are given by

$$b_1 = \beta \Delta t, \qquad b_2 = (1-\beta)\Delta t. \tag{10}$$

For different values of β, a well-known numerical integration scheme can be obtained [21]. In this study, $0.5 \leq \beta \leq 1.0$ was used, which results in an unconditionally stable scheme.

4. Model geometry and mesh

4.1. Fluid field

Considering symmetry in the disc, we took only a quarter of the geometry of the fluid field (Fig. 4) by using the mesh generation software ANSYS ICEM CFD.

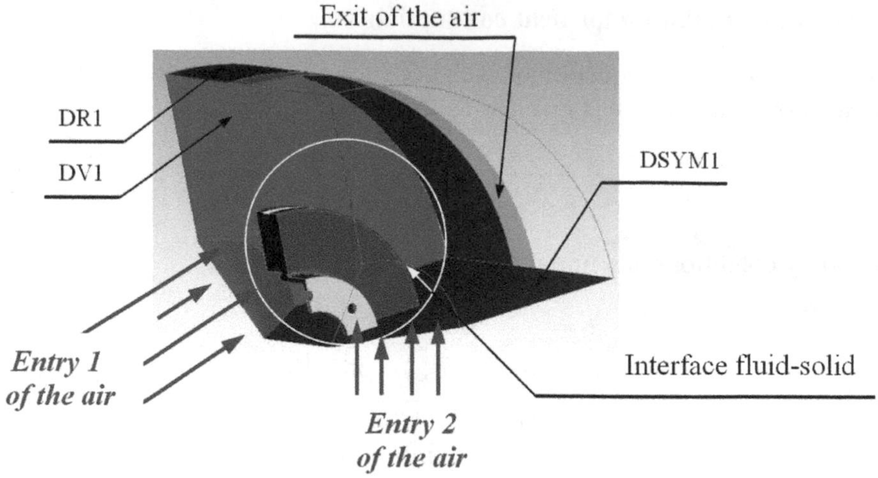

Fig. 4. Definition of surfaces of the fluid field

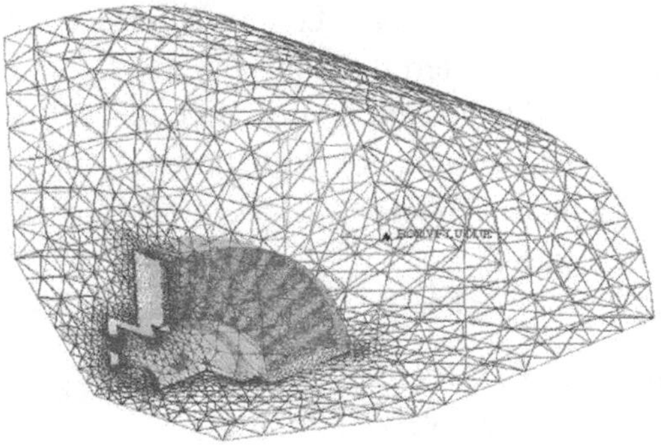

Fig. 5. Mesh of the fluid field

4.2. Mesh

This stage includes the mesh preparation in the fluid field. In our case, we used a linear tetrahedral element with 30 717 nodes and 179 798 elements (Fig. 5).

During the braking process, a part of the frictional heat escapes into the ambient air by convection and radiation. Consequently, the determination of the heat transfer coefficients is essential. Their exact calculation is, however, rather difficult, because these coefficients depend on the location and the construction of the braking system, the speed of the vehicle and consequently, on the air circulation or the air flow conditions (laminar, turbulent, or transition flow). Since the process of heat transfer by radiation is not too significant, we will determine only the convection coefficient h of the disc using the ANSYS CFX 11.0 code. The parameter will be utilized to determine the 3D temperature distribution of the disc.

Because of disc symmetry, we took only a quarter of the geometry in the case of the full and ventilated discs; one kept the tetrahedral form to generate the mesh of the discs (Figs. 6 and 7).

4.3. Modelling in ANSYS CFX

For the mesh generation of the CFD model, it is necessary to define various surfaces of the disc in the integrated computer aided engineering and manufacturing (ICEM) CFD as shown

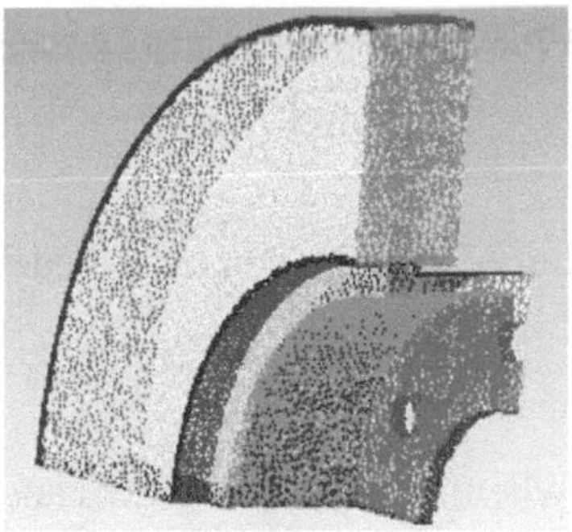

Fig. 6. Mesh of the full disc (Number of elements 272 392)

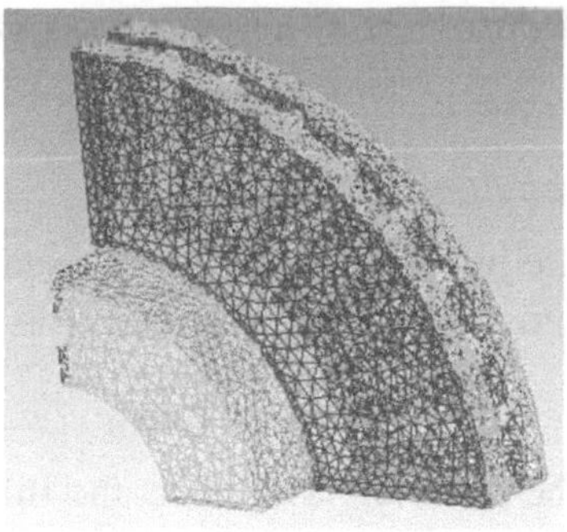

Fig. 7. Mesh of the ventilated disc (Number of elements 27 691)

in Fig. 8. In our case, we used a linear tetrahedral element with 30 717 nodes and 179 798 elements. In order to facilitate the calculation, a non-uniform mesh is used with refinement in locations with gradients.

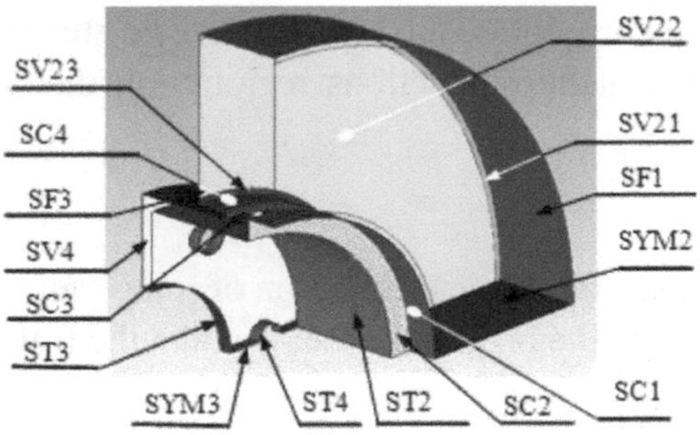

Fig. 8. Definition of surfaces of the full disc

Table 3. Boundary conditions

Boundary	Boundary condition	Parameters
Inlet	Pressure inlet	Atmospheric pressure and temperature
Outlet	Pressure outlet	Atmospheric pressure and temperature
Domain edges	Symmetry	Symmetry
Disc surface	Wall	800 K temperature, thermal properties of grey cast iron

a) *Physical model*

In this step, all physical characteristics of the fluid and the solid are declared. After the meshing, all the parameters of the different models are defined in order to start the analysis.

b) *Definition of the domains*

Initially, the elaborated models are validated and are activated in the option "Thermal Energy — the calculation of heat transfer — Heat Transfer".

Fluid domain — Speed entry: $V_{\text{ent non.st}} = V_{\text{ent}} - Va \cdot t$,
Disc domain — Entering flux: $\text{FLUX}_{\text{non.st}} = (CF)\,(V_{\text{ent non.st}})$,

$CF = 149\,893.838$,

$V_{\text{ent non.st}} = V_{\text{ent}} - Va \cdot t$.

$\text{FLOW}_{\text{non.st}}$: Non stationary flux entering.
$V_{\text{ent non.st}}$: Non stationary speed entering of the air.

c) *Definition of materials*

The physical properties of used materials are introduced into the computer code. In this study, we selected the cast iron material (FG 15).

d) *Definition of the boundary conditions*

The first step is to select the Inlet and Outlet faces where the heat flux enters and leaves. These options are found in the insertion menu "Boundary Conditions" in the CFX Pre. Interface.

In the next step, boundary conditions concerning the pads are defined by selecting the options "Wall" and "Symmetry", because there will be the possibility of adjusting a certain number of parameters in the boundary conditions such as the flux entering the disc.

e) *Application of the interfaces domains*

The areas of interfaces are commonly used to create the connection or linkage areas. Surfaces located between the interactions regions (air-disc) are reported as solid-fluid interface.

f) *Initial conditions*

Since the study is focused on the determination of temperature field in a disc brake of average class vehicle during the braking phase, we take the following temporal conditions:

- Braking time = 3.5 [s],
- Increment time = 0.01 [s],
- Initial time = 0 [s].

The airflow through and around the brake disc is analyzed using the ANSYS CFX software package. The ANSYS CFX solver automatically calculates the heat transfer coefficient at the wall boundary. Afterwards, the heat transfer coefficients considering convection are calculated and organized in such a way that they are used as a boundary condition in the thermal analysis. Averaged heat transfer coefficient had to be calculated for both discs using the software ANSYS CFX Post as indicated in Fig. 9.

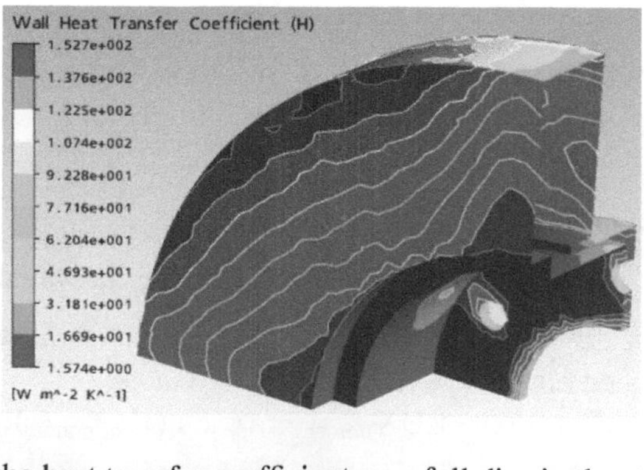

Fig. 9. Distribution of the heat transfer coefficient on a full disc in the steady-state case (FG 15)

g) *Calculation of the heat transfer coefficient h*

The heat transfer coefficient is a parameter related to the air velocity, the brake disc shape, and many other factors. For different air velocity, the heat transfer coefficient in different parts of the brake disc changes with time [31]. It depends on air flow in the region of brake rotor and on the vehicle speed, but it does not depend on the material. In our simulation, this coefficient is determined as an average value of the coefficient h "Wall heat Transfer Coefficient" and variable with time.

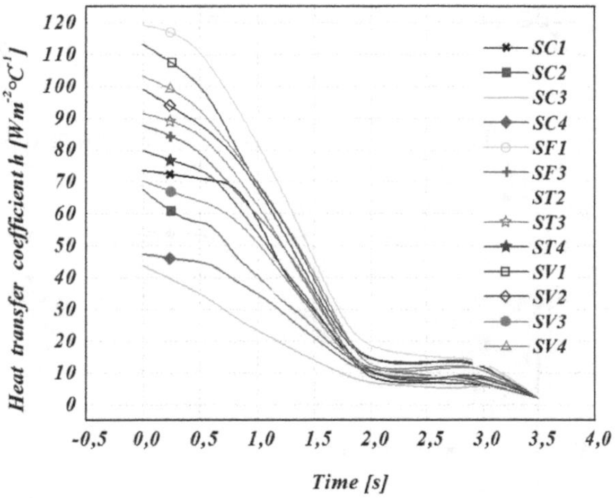

Fig. 10. Variation of the heat transfer coefficient h for various surfaces of the full disc in a transient case (FG 15)

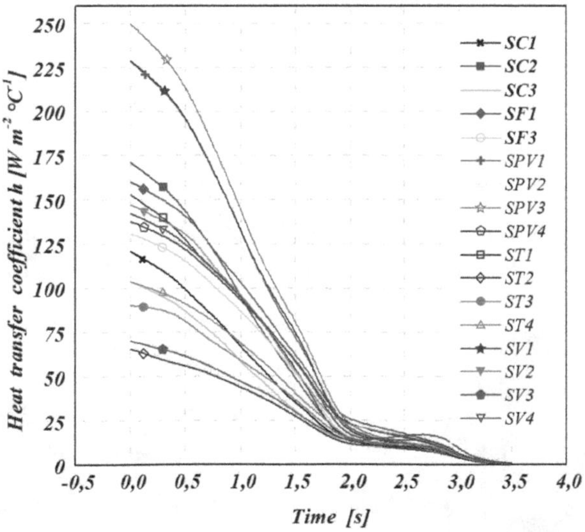

Fig. 11. Variation of the heat transfer coefficient h for various surfaces of the ventilated disc in a transient case (FG 15)

4.4. Determination of disc temperature

The modelling of disc temperature is carried out by simulating stop braking of a middle class car (braking of type 0). The characteristics of the vehicle and of the disc brake are listed in Table 1. The vehicle speed decreases linearly with time to the value 0 as shown in Fig. 12. The variation of the heat flux during the simulation time is depicted in Fig. 13.

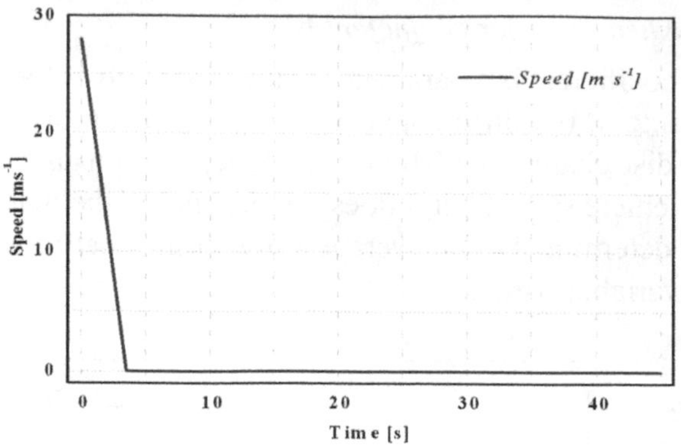

Fig. 12. Speed of the vehicle versus time (braking of type 0)

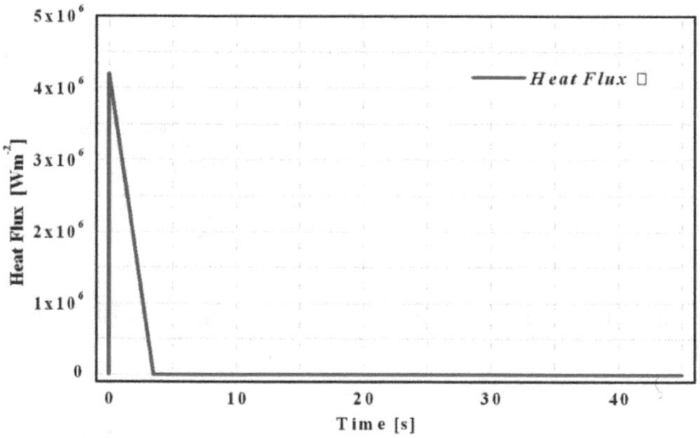

Fig. 13. Heat Flux versus time

4.5. Finite element mesh for heat transfer analysis

The mesh generation of the full and ventilated disc is carried out using ANSYS Multiphysics. The basic element used for the meshing is of tetrahedral shape (Figs. 14 and 15).

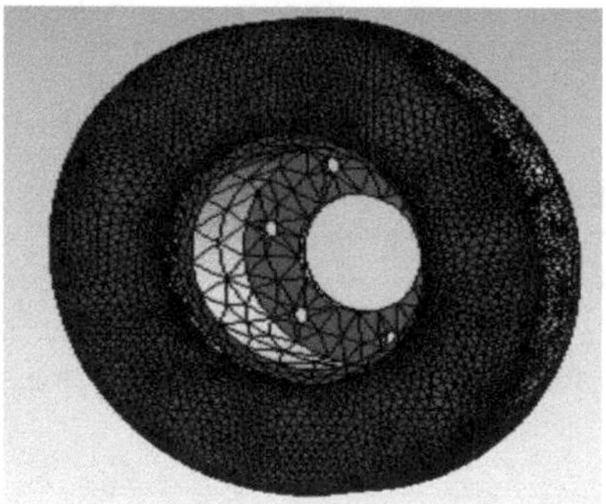

Fig. 14. Full type disc mesh model (total number of nodes 172 103 — total number of elements 114 421)

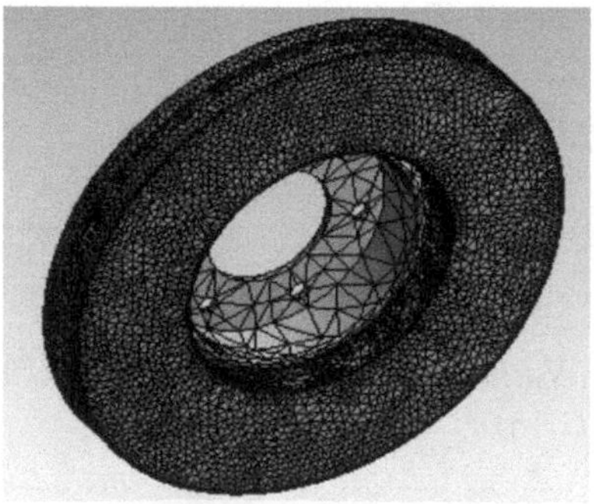

Fig. 15. Ventilated type disc mesh model (total number of nodes 154 679 — total number of elements 94 117)

4.6. *Boundary conditions*

The boundary conditions are introduced into within the module ANSYS Workbench [Multiphysics] by choosing the mode of the first simulation (permanent or transitory), and by defining the physical properties of the materials. These conditions constitute the initial conditions of our simulation. After having fixed these parameters, we introduce appropriate boundary conditions associated with each surface and specify the following computational parameters:

- Total time of simulation = 45 s.
- Increment of initial time = 0.25 s.
- Increment of minimal initial time = 0.125 s.
- Increment of maximal initial time = 0.5 s.
- Initial temperature of the disc = 20 °C.
- Materials: Grey Cast iron FG 5.
- Convection: We introduce the values of the heat transfer coefficient h obtained for each surface in the shape of a curve (Figs. 10 and 11).
- Heat flux: We apply the values obtained by means of the CFX code.

5. Results and discussions

The temperature in the disc brake is modeled with regard to the variation of a certain number of parameters such as the type of braking, the cooling mode of the disc and the choice of disc material. The brake discs are made of cast iron with high carbon content. The contact surface of the disc receives an entering heat flux calculated by Eq. (1).

The model presents a 3D solid disc squeezed between two finite-width friction materials. The entire surface S of the disc has three different regions including S_1 and S_2. On S_1, the heat flux is specified with respect to the frictional heating between the pads and the disc, and S_2 is defined as a convection boundary. The rest of the surfaces without S_1 and S_2 is either temperature specified or assumed to be insulated: the inner and outer rim areas of the disc. Since an axisymmetric model is considered, all the nodes on the hub radius are fixed so that the nodal displacements in the hub become zero, i.e., in radial, axial and angular directions.

5.1. *Influence of construction of the disc*

Fig. 16 shows the variation of temperature with time during the total time simulation of braking for both the full and ventilated discs. The highest temperatures are reached on the contact surface of the disc pads. The initial sudden increase in temperature is due to the short duration of the braking phase and to the speed of the physical phenomenon. For the two types of discs, one disc has an immediate, fast temperature increase followed by a decrease in temperature after a certain time of braking.

On the basic of these observations, we can conclude that the geometric design of the disc is an essential factor in the improvement of the cooling process of the discs.

Figs. 17 and 18 show the temperature variation according to the radius and thickness, respectively. It is noted that there is an appreciable variation of temperature between the two types of discs. The influence of ventilation on the temperature field appears clearly at the end of the braking ($t = 3.5$ s).

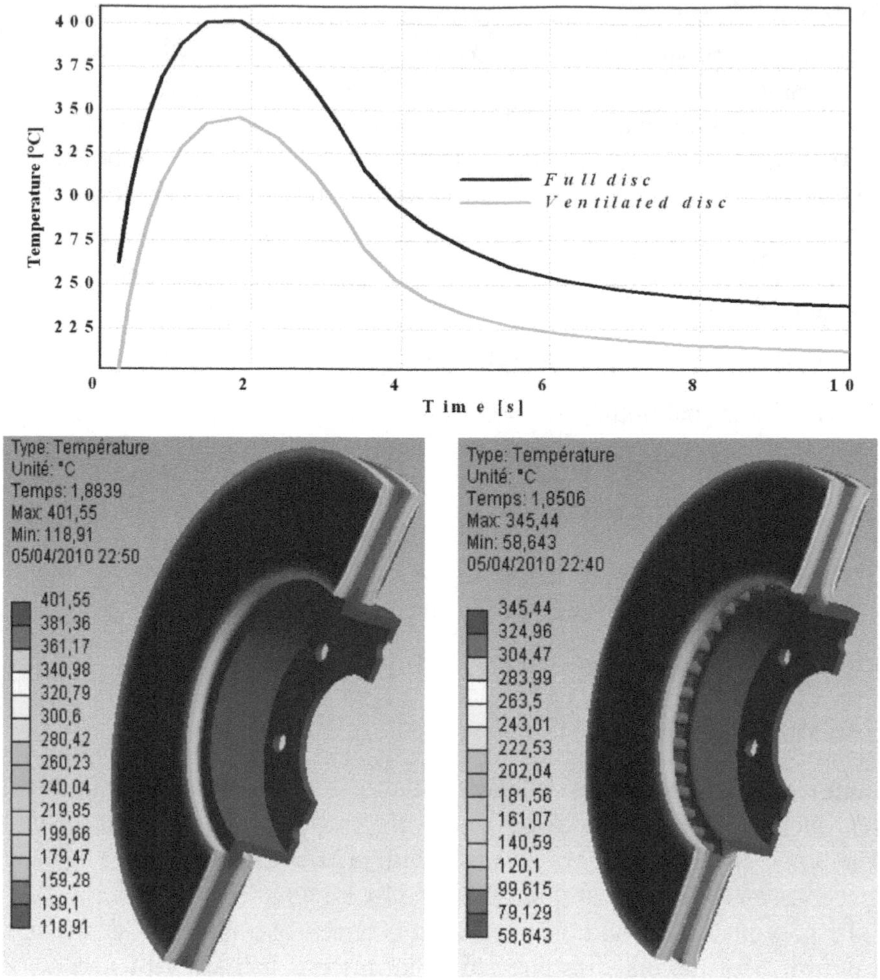

(a) (b)

Fig. 16. Temperature distribution of a full (a) and ventilated disc (b) of cast iron (FG 15)

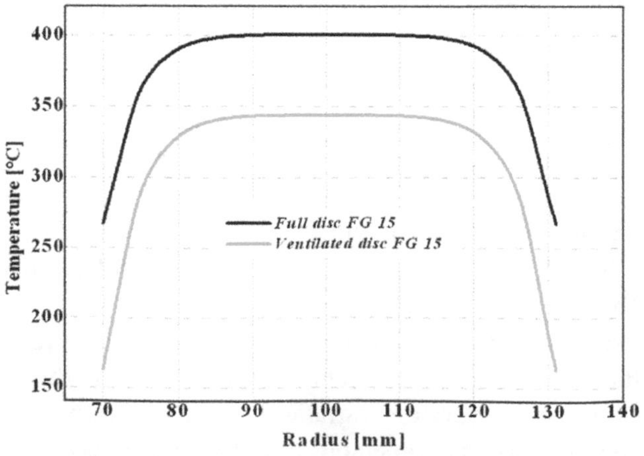

Fig. 17. Temperature variation through a radius for both designs of the same material (FG 15)

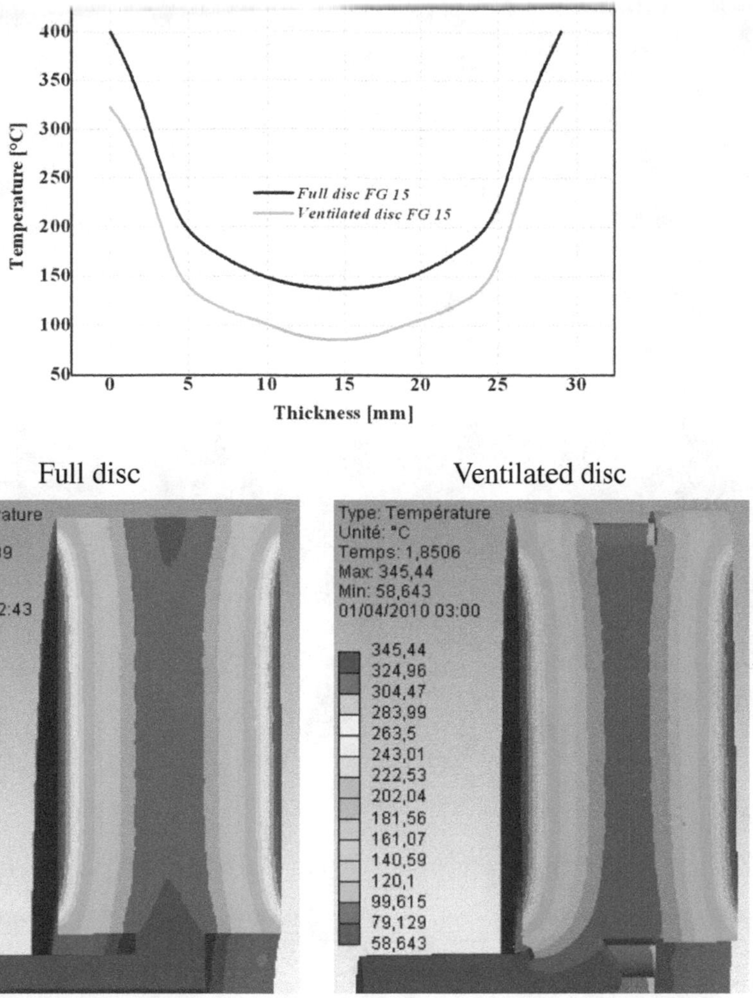

Fig. 18. Temperature variation through the thickness for both designs of the same material (FG 15)

5.2. Comparison between the full and ventilated discs

In this part, the maps of total and directional heat flux as well as the temperature distribution in the full and ventilated discs of cast iron FG 15 for each braking phase are presented. The temperature distribution of the disc at the beginning of the braking ($t = 0.25$ s) is inhomogenous (see Fig. 19). According to experimental tests, the braking often begins with the formation from hot circles relatively on the uniform surfaces of the disc in the circumferential direction, moving radially and then transforming into hot points (hot spot). The appearance of the phenomenon of the hot points is due to the non-uniform dissipation of the heat flux.

According to Figs. 20 and 23, the maximum value of the total heat flux is located on the level of the calorific throat at the end of braking ($t = 3.5$ s); this is explained by the increase in the gradients and the thermal concentrations in this zone. The calorific throat is manufactured so as to limit the heat flux coming from the friction tracks and moving towards the bowl of the disc brake, avoiding an excessive heating of the rim and the tire. During the heating, the disc is tightened to dilate in the hot zones from where creating of compressive stresses with plasticization. On the other hand, residual stresses of traction appear during the cooling. The rotating disc is, thus, subjected to constraints traction/compression.

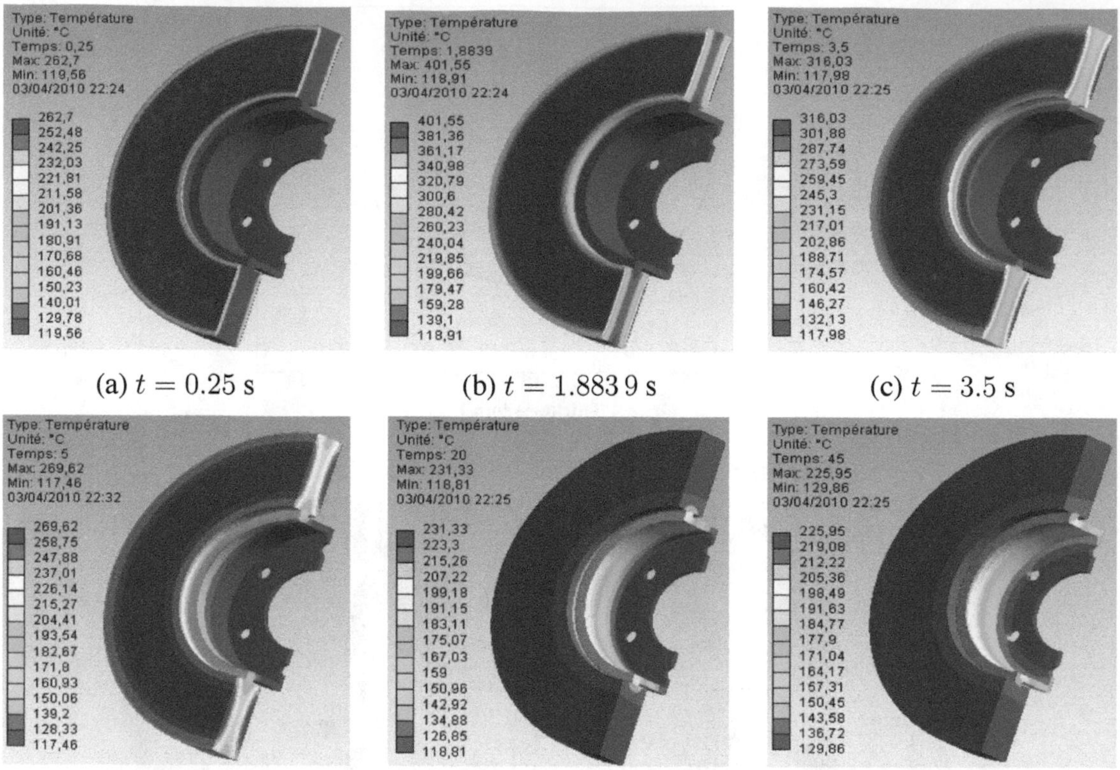

(a) $t = 0.25$ s (b) $t = 1.883\,9$ s (c) $t = 3.5$ s

(d) $t = 5$ s (e) $t = 20$ s (f) $t = 45$ s

Fig. 19. Temperature distribution for the full disc of material FG 15

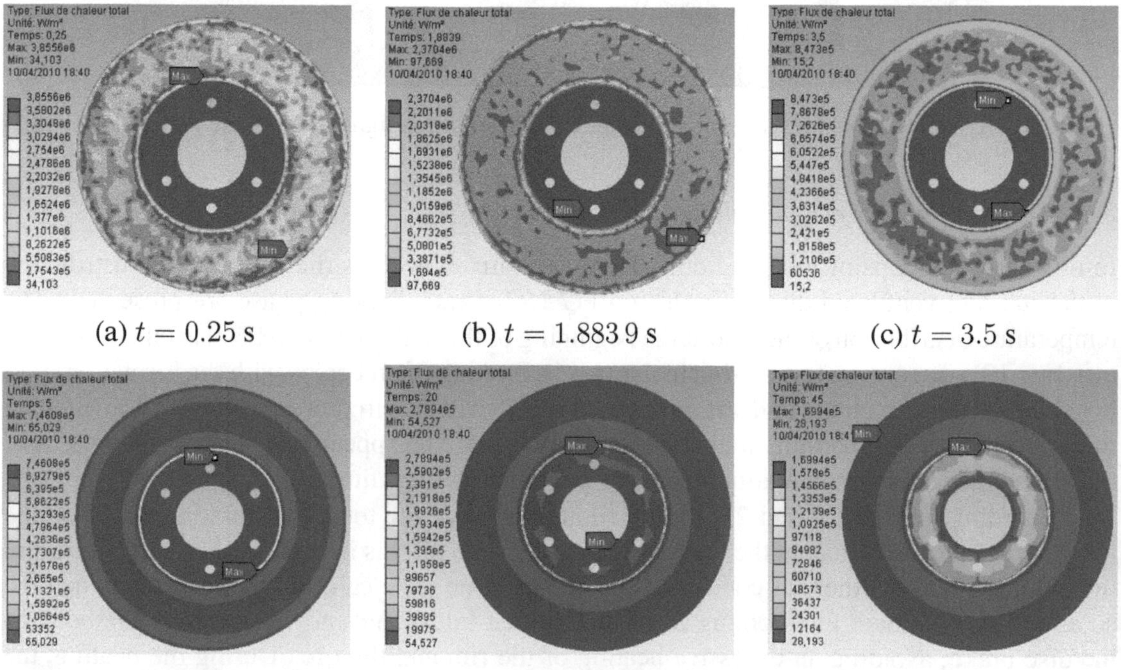

(a) $t = 0.25$ s (b) $t = 1.883\,9$ s (c) $t = 3.5$ s

(d) $t = 5$ s (e) $t = 20$ s (f) $t = 45$ s

Fig. 20. Distribution of total heat flux for the full disc of material FG 15

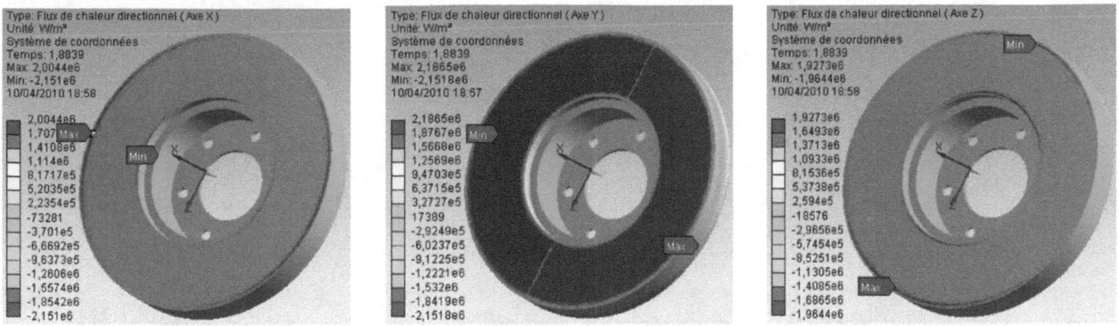

(a) Along the X axis (b) Along the Y axis (c) Along the Z axis

Fig. 21. Distribution of directional heat flux at the time $t = 1.883\,9$ s along the X, Y, Z axes for the full disc of material FG 15

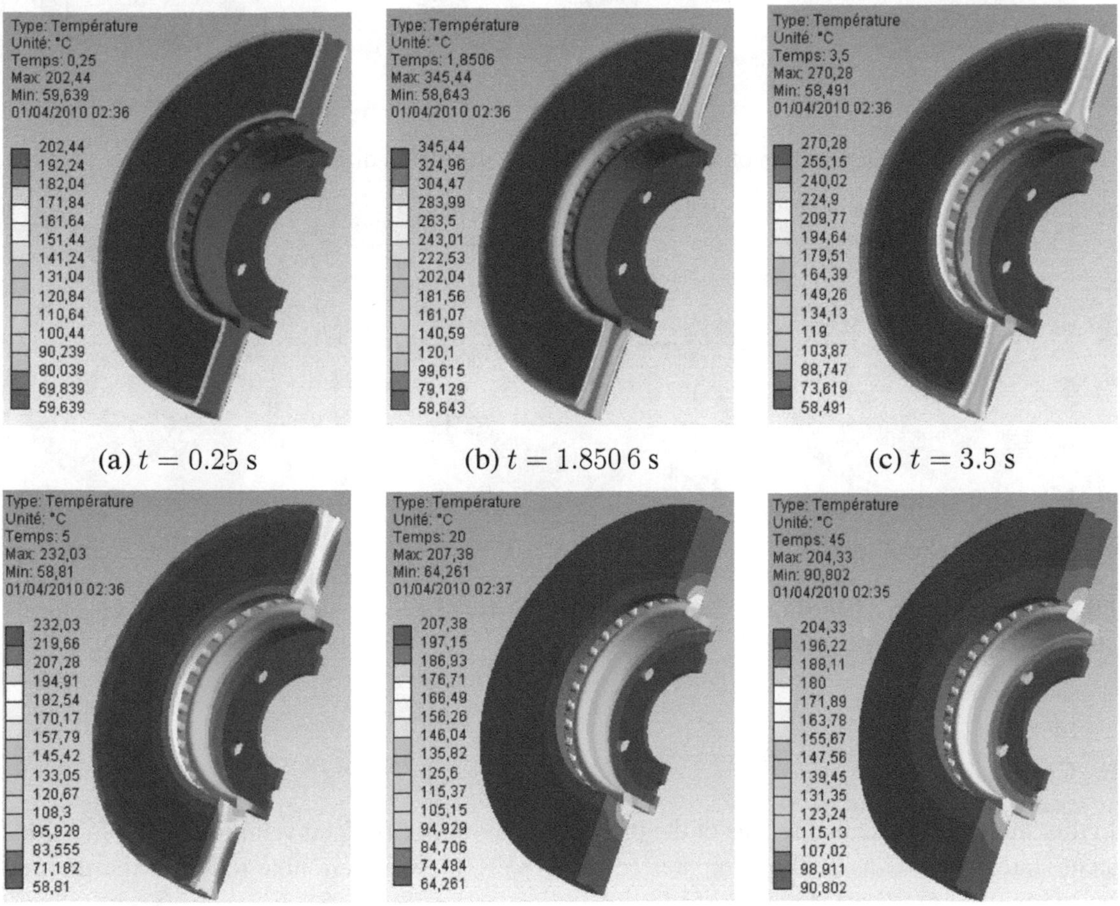

(a) $t = 0.25$ s (b) $t = 1.850\,6$ s (c) $t = 3.5$ s

(d) $t = 5$ s (e) $t = 20$ s (f) $t = 45$ s

Fig. 22. Temperature distribution for the ventilated disc of material FG 15

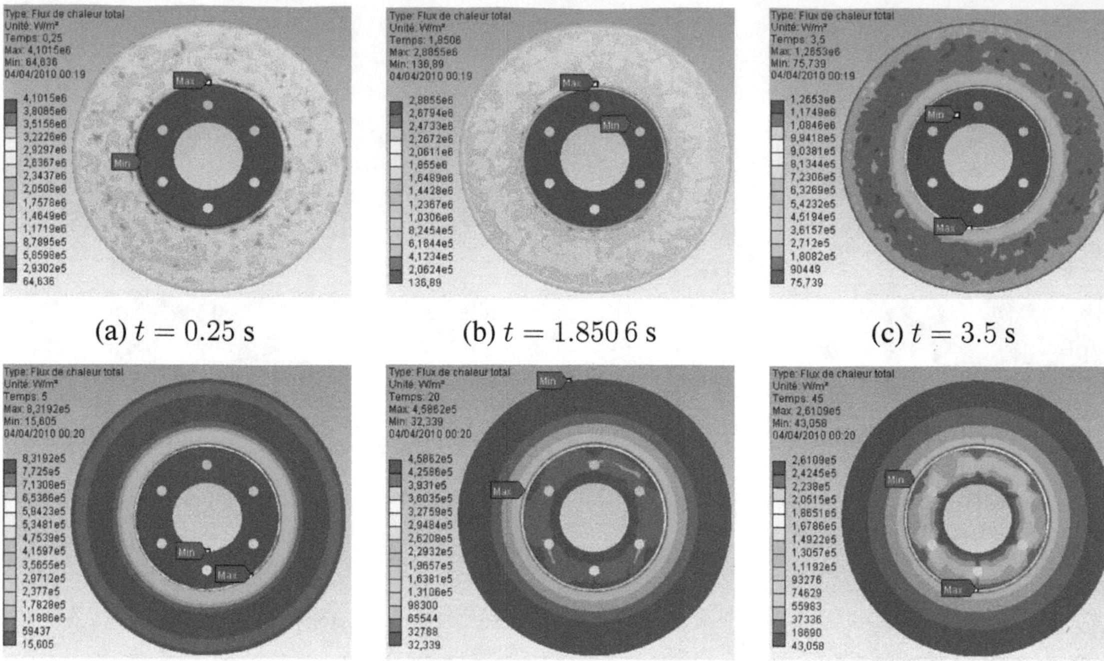

(a) $t = 0.25$ s (b) $t = 1.850\,6$ s (c) $t = 3.5$ s

(d) $t = 5$ s (e) $t = 20$ s (f) $t = 45$ s

Fig. 23. Distribution of total heat flux for the ventilated disc of material FG 15

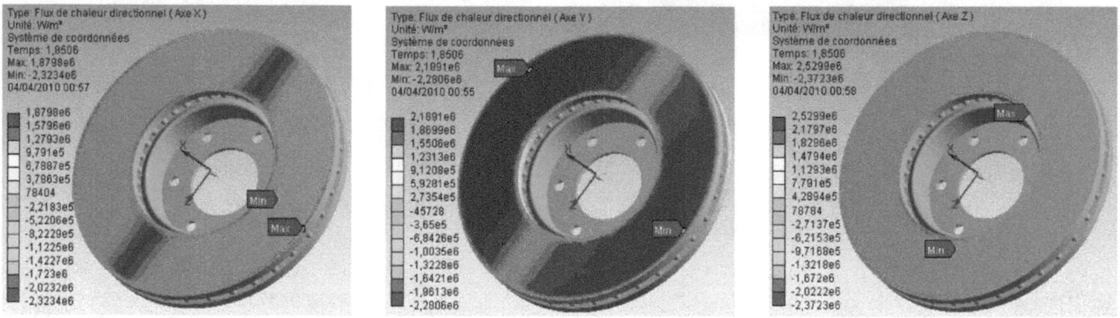

(a) Along the X axis (b) Along the Y axis (c) Along the Z axis

Fig. 24. Distribution of directional heat flux at the time $t = 1.850\,6$ s along the X, Y, Z axes for the ventilated disc of material FG 15

6. Conclusion

In this study, we presented results of the thermal behaviour of full and ventilated discs in a transient state. By means of the computer code ANSYS 11, we were able to study the thermal behaviour of a gray cast iron (FG 15).

In addition, the influence of disc ventilation on the thermal behaviour of the discs brake was analysed. The numerical simulation shows that radial ventilation plays a very significant role in cooling of the disc during the braking phase. Through the numerical simulation, we could note that the quality of the results concerning the temperature field is influenced by several parameters such as:

(i) Technological parameters illustrated by the design.

(ii) Numerical parameters represented by the number of elements and the step of time.

(iii) Physical parameters expressed by the types of materials.

Regarding the calculation results, we can say that they are satisfactorily in agreement with those commonly found in literature investigations. It would be interesting to solve the problem in thermomechanical disc brakes with an experimental study to validate the numerical results in order to demonstrate a good agreement between the model and reality.

Regarding the outlook, there are three possible improvements related to disc brakes that can be done to further understand the effects of thermomechanical contact between the disc and the pads. They are as follows:

(i) Experimental study to verify the accuracy of the numerical model developed.

(ii) Tribology and vibrations study of the contact disc — pads.

(iii) Study of dry contact sliding under the macroscopic aspect (macroscopic state of the disc and pad surfaces).

References

[1] Abdo, J., Experimental technique to study tangential to normal contact load ratio, Tribology Transactions 48 (2005) 389–403.

[2] Abu Bakar, A. R., Ouyang, H., Wear prediction of friction material and brake squeal using the finite element method, Wear 264 (11–12) (2008) 1 069–1 076.

[3] Altuzarra, O., Amezua, E., Aviles, R., Hernandez, A., Judder vibration in disc brakes excited by thermoelastic instability, Engineering Computations 19 (4) (2002) 411–430.

[4] Basha Shaik, A. F., Srinivas, Ch. L., Structural and thermal analysis of disc brake with and without crossdrilled rotar of race car, International Journal of Advanced Engineering Research and Studies 1 (4) (2012) 39–43.

[5] Choi, B. K., Park, J. H., Kim, M. R., Simulation of the braking condition of vehicle for evaluating thermal performance of disc brake, Proceedings of KSAE Autumn Conference, 2008, pp. 1 265–1 274.

[6] Dhaubhadel, M. N., Review: CFD applications in the automotive industry, Journal of Fluids Engineering 118 (4) (1996) 647–653.

[7] Gao, C. H., Lin, X. Z., Transient temperature field analysis of a brake in a non-axisymmetric threedimensional model, Journal of Materials Processing Technology 129 (2002) 513–517.

[8] Hudson, M., Ruhl, R., Ventilated brake rotor air flow investigation, SAE Technical Paper 971033, 1997.

[9] Jacobsson, H., Aspects of disc brake judder, Proceedings of the Institution of Mechanical Engineers, Part D: Journal of Automobile Engineering 217 (6) (2003) 419–430.

[10] Jang, Y. H., Ahn, S. H., Frictionally-excited thermoelastic instability in functionally graded material, Wear 262 (2007) 1 102–1 112.

[11] Jung, S. P., Park, T. W., Kim, Y. G., A study on thermal characteristic analysis and shape optimization of a ventilated disc, International Journal of Precision Engineering and Manufacturing 13 (1) (2012) 57–63.

[12] Khalid, M. K., Mansor, M. R., Abdul Kudus, S. I., Tahir, M. M., Hassan, M. Z., Performance investigation of the UTeM Eco-Car disc brake system, International Journal of Engineering & Technology 11 (6) (2011) 1–6.

[13] Komanduri, R., Hou, Z. B., Analysis of heat partition and temperature distribution in sliding systems, Wear 251 (1–12) (2001) 925–938.

[14] Lee, K. J., Barber, J. R., An experimental investigation of frictionally-excited thermoelastic instability in automotive disk brakes under a drag brake application, Journal of Tribology 116 (1994) 409–414.

[15] Lee, K., Barber, J. R., Frictionally excited thermoelastic instability in automotive disk brakes, ASME Journal of Tribology 115 (1993) 607–614.

[16] Limpert, R., Brake design and safety: 2nd Edition, Warrendale, Pennsylvania: Society of Automotive Engineering Inc., 1999, pp. 137–144.

[17] Limpert, R., The thermal performance of automotive disc brakes, SAE Technical Paper 750873, 1975.

[18] Morgan, S., Dennis, R., A theoretical prediction of disc brake temperatures and a comparison with experimental data, SAE Technical Paper 720090, 1972.

[19] Mosleh, M., Khemet, B. A., A surface texturing approach for cleaner disc brakes, Tribology Transactions 49 (2) (2006) 279–283.

[20] Nouby, M., Abdo, J., Mathivanan, D., Srinivasan, K., Evaluation of disc brake materials for squeal reduction, Tribology Transactions 54 (4) (2011) 644–656.

[21] Omolayo, P. M., Ogheneortega, O. J., Numerical simulation of thermoelastic contact problem of disc brake with frictional heat generation, New York Science Journal 5 (10) (2012) 39–43.

[22] Öztürk, B., Arslan, F., Öztürk, S., Effects of different kinds of fibers on mechanical and tribological properties of brake friction materials, Tribology Transactions 56 (4) (2013) 536–545.

[23] Palmer, E., Fieldhouse, J., Mishra, R., Optimisation of pin shape and its configuration for a pin type vente brake disc using CFD, FISITA, Yokohama, 2006.

[24] Rajagopal, T. K. R., Ramachandran, R., James, M., Gatlewar, S. Ch., Numerical investigation of fluid flow and heat transfer characteristics on the aerodynamics of ventilated disc brake rotor using CFD, Thermal Science 18 (2) 2014 667–675.

[25] Reimpel, J., Technologie de freinage, Vogel Verlag, Würzburg, 1998.

[26] Voldřich, J., Frictionally excited thermoelastic instability in disc brakes-transient problem in the full contact regime, International Journal of Mechanical Sciences 49 (2007) 129–137.

[27] Voldřich, J., Morávka, S., Student, J., Transient temperature field in intermittent sliding contact at temperature dependent coefficient of friction, Proceedings of Computational Mechanics 2006, Nečtiny, 2006, pp. 697–704.

[28] Wallis, L., Leonardi, E., Milton, B., Air flow and heat transfer in ventilated disc brake rotors with diamond and tear-drop pillars, Proceedings of International Symposium on Advances in Computational Heat Transfer, Australia, 2002, pp. 643–655.

[29] Yi, B. Y., Barber, J. R., Zagrodzki, P., Eigen value solution of thermoelastic instability problems using Fourier reduction, Proceedings of the Royal Society of London, 2000, pp. 279–282.

[30] Zagrodzki, P., Thermoelastic instability in friction clutches and brakes-transient modal analysis revealing mechanisms of excitation of unstable modes, International Journal of Solids and Structures 46 (2009) 2 463–2 476.

[31] Zhang, J., Xia, C., Research of the transient temperature field and friction properties on disc brakes, Proceedings of the 2nd International Conference on Computer and Information Application, 2012, pp. 201–204.

[32] Zhu, Z. C., Peng, U. X., Shi, Z. Y., Chen, G. A., Three-dimensional transient temperature field of brake shoe during hoist's emergency braking, Applied Thermal Engineering 29 (2009) 932–937.

11

Optimization methodology for beam gauges of the bus body for weight reduction

R. Jain[a,*], P. Tandon[a], M. Vasantha Kumar[b]

[a] *Mechanical Engineering Discipline, PDPM Indian Institute of Information Technology, Design and Manufacturing Jabalpur Jabalpur-482005, Madhya Pradesh, India*

[b] *Altair Engineering, Bengaluru, India*

Abstract

During service, a bus is subjected to various loads that cause stresses, vibrations and noise in the different components of its structure. It requires appropriate strength, stiffness and fatigue properties of the components to be able to stand these loads. Besides, quality and optimum weight of the vehicle, for efficient energy consumption, safety and provision of the comfort to the user are highly desired. The present work proposes a methodology to minimize the bus weight by modifying its beam gauges to optimum thickness. The bus performance is evaluated by multiple iterations on the basis of parameters like frequency, distortion, stress and stiffness. The algorithm performs gauge optimization of the bus by analyzing and satisfying its structural strength through linear static analysis on a laden bus. It also performs structural stiffness analysis and vibration analysis for safety of the bus structure. This work unfolds an integrated methodology to the bus manufacturers to optimize the structural weight for improving the fuel efficiency, static and dynamic safety, and robust design. The work is implemented by creating a finite element model of the bus and optimizing in HyperWorks environment. The results are verified for a full length 11 m, 65 seats bus. The methodology helps in weight reduction along with improvement in performance parameters.

Keywords: bus body, gauge optimization, weight reduction, linear static analysis, stiffness, vibration analysis

1. Introduction

Due to recent advancements in technologies and need for a sustainable environment, the latest trend in automobile industry is towards development of light weight vehicles for reduced gas emissions and better fuel economy. It is either done by optimizing the existing auto body structure using advanced computational techniques or foraying into design improvements. Innovations in existing designs are profitable in comparison to new product designs due to existence of established manufacturing and production units, and acceptability by the market. Thus, the aim of this work is to reduce the weight of the bus structure by optimizing the beam thicknesses, achieving in process, environmental and manufacturing benefits, without compromising the structural performance.

Computer Aided Engineering (CAE) analysis, if done prior to fabrication of the product, is a time and cost saving tool in product design cycle. For engineering products, it is executed at earlier phases of design, after conceiving the concept designs in CAD environment. The CAE analysis done in this research is based on Finite Element Analysis (FEA), assisted by well-advanced HyperWorks software tools such as HyperMesh for pre-processing, Optistruct

*Corresponding author. e-mail: richa20.jbp@gmail.com.

for structural analysis and optimization, and HyperView for post-processing and result visualization.

Optimization techniques applied for structural design is a major research area to reduce the weight of vehicles and form refinement. The performance of the bus is estimated by linear static analysis [5,9], evaluating the strength and deformation of the bus structure when subjected to service loads while the bus is on road. Many times, the structure of the bus body is designed that is sufficiently strong yet unsatisfactory due to insufficient stiffness. Designing for acceptable stiffness [3,11,19] is, therefore, often more critical than designing for sufficient strength. Dynamic analysis [2,9,16] of the structure is a concern for the safety of the structure and it is performed to evaluate the natural frequency and mode shapes of the structure. These analysis setups facilitate an iterative optimization [12,14,16], proving the performance criteria for the converged design solution of the bus to be reasonable under bending, torsion, buckling and dynamic loads. The design solution satisfies all the manufacturing and weight constraints and enhances performance in terms of mass, stiffness, frequency and deformation.

The primary objective of this work is to reduce the weight of passenger bus without compromising the performance of the vehicle; and if possible enhancing the structural performance of the bus body through employing FEA, structural analysis and optimization techniques to optimize beams of the bus.

2. Methodology

To optimize the bus body, an analytical study of the requirement is done and based on that, the following methodology is proposed, Fig. 1.

From the meshed model shown in Fig. 2, the geometrical model has to be extracted. Thus, the first step is to import the geometrical model of the bus from the free domain to the desired CAD environment or model it from scratch, which is a tedious task. The geometry of the bus body is divided into six frames, i.e., top, bottom, two sides, rear and front, for the ease of handling.

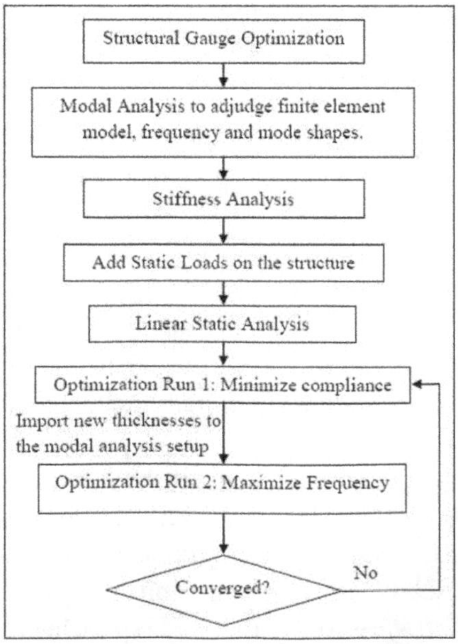

Fig. 1. Proposed algorithm

Fig. 2. Meshed bus model (from free domain)

Secondly, the median surfaces of the beams are extracted, which is followed by generation of finite element mesh on it. The meshed frames are joined replicating the weld joints between the beams. The whole meshed model is scrutinised under quality check to ensure no discontinuity or failure of elements. A finite element chassis model is prepared to support the bus structure and set up boundary conditions on the axle. Chassis model is resistant to any variation throughout the experiment.

Once the finite element model of the bus is prepared, it is subjected to (a) vibration analysis, (b) stiffness analysis and (c) linear static analysis. At first, vibration analysis is done to verify the meshed model and calculate first bending and torsion frequency. The model is verified with the established data which should be analogous to the natural frequency of a simply supported hollow beam, then the bending and torsion stiffness of the structure is calculated. This is followed by linear static analysis, for which the bus is loaded with external weights and equipment weights. The service loads are now imposed, categorized under bending, torsion, combined bending and torsion, lateral loading and longitudinal loading. These are done to analyze and visualize the cases for deformation and stress concentration, spotting the weak zones in the beams of the structure.

After procuring and fixing all the performance indices of the baseline model, the established model is to be optimized. The response parameters are frequency (to be maximized), displacement (to be minimized) and compliance (to be minimized), with an objective to reduce the mass. The design variables are the thickness of the various beams. Later, a comparison of the baseline and iterative optimized model is done. The model which complies with all the desired values of variables, responses and objective is set as optimum.

The gauge optimization of the bus body includes the following steps:

1. Modeling: The modeling of the bus body is a two-step process comprising of, first a geometric model of the bus and second, a finite element model. It would be good if both the models are developed in the same CAD environment. This helps in extraction of geometry of interconnected beams.

2. Pre-processing: This helps in defining the material, thickness and other essential properties of the bus. Pre-processor also allows the solver in the next step to predict the action of these elements and analyze the reaction to external forces and interactions.

3. Finite Element Analysis: FEA of the base model is carried out for strength, stiffness and safety. It includes linear static analysis to identify strength criteria that include maximum stresses and deflections; vibration analysis for dynamic behaviour; and bending and torsion analysis for stiffness benchmarking.

4. Vibration Analysis: Vibration analysis is done to 'Design for Safety'. Every structure or body has its own natural frequency and when this frequency coincides, resonance occurs.

To prevent resonance, the natural frequency of the bus has to be high so that its frequency does not coincide with natural frequency of human, humps, engine vibration and other parts.

5. Stiffness analysis: The bending stiffness and torsion stiffness accounts for the overall strength of the bus structure. To evaluate the baseline stiffness of the bus, the structure is assumed to be a beam and its deformation on application of forces depends on its stiffness.

6. Optimization: In this work, structural optimization in the form of topology, shape and size are carried out. All three have a purpose in the design phase and the choice or combination depends on desirables. The purpose of parameter optimization is to find the minimum thickness of the inter-related beams to minimize the weight.

The basic idea of gauge and size optimization is to modify gauge properties so that the residual structure evolves towards an optimum solution. Gauge optimization [12] is a particular form of size optimization where design variables t_i are in the form of 2D shell elements only.

The element property p is derived as a function of design variable and constants, i.e., $p = f(DV_j, C_j)$, where DV_j is design variable and C_j is constant. The responses to be evaluated for optimum sizing of beam gauges [2,6] would be done by iterative optimization to increase frequency on modal and decrease displacement on linear static analysis setups.

3. Implementation

In this work, the implementation of the proposed methodology is carried out with HyperWorks suite taking help from [10,11,15]. The HyperMesh interface facilitates both geometric modelling as well as the pre-processing of the model. FEA and gauge optimization of the bus structure is done with Radioss as solver. After solving and retrieving the results, visualization of the results of the analysis is supported by HyperView.

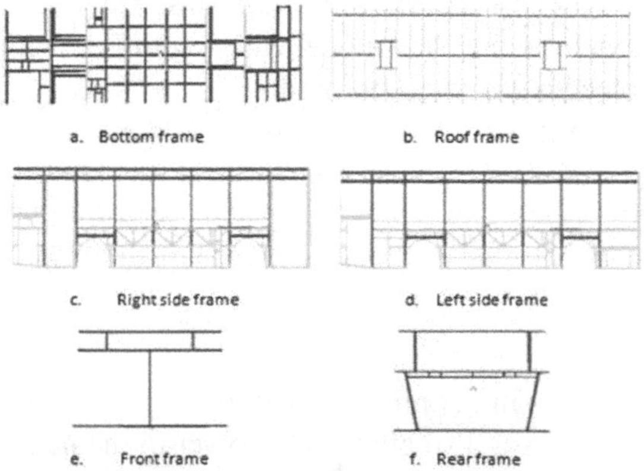

Fig. 3. Geometric models of the frames

The geometric models of various frames of the bus body are shown with the help of Fig. 3. Finite element model is represented by interconnection of a mesh of 2-D quad elements with average size 10 on the median surface of the beams. Both aluminium (Al) and steel were considered as possible materials for beams [3,20], but in the present work, steel is preferred over Al, which in spite of being light, strong and stiff is expensive. Moreover, the results of the

Table 1. Material properties

Element type	2D shell	Beam
Material	Steel (IS:4923)	High strength steel
Young's modulus	$2.1 \cdot 10^5$ MPa	$2.1 \cdot 10^{10}$ MPa
Poisson's ratio ν	0.300	0.300
Modulus of rigidity G	$8.1 \cdot 10^4$ MPa	$8.1 \cdot 10^9$ MPa
Density	$7.9 \cdot 10^{-9}\ \mathrm{N \cdot mm^{-3}}$	$7.9 \cdot 10^{-9}\ \mathrm{N \cdot mm^{-3}}$

bus body made of steel can be validated through the published data. The material properties are defined in Table 1.

The mass of the bus structure after assigning material and thickness properties comes out to be 1.521 kg. To carry out the strength analysis of the bus structure, the bus body has to be statically loaded for the worst possible loads while on road. This is done using rigid elements (Rbe3) as shown in Fig. 4.

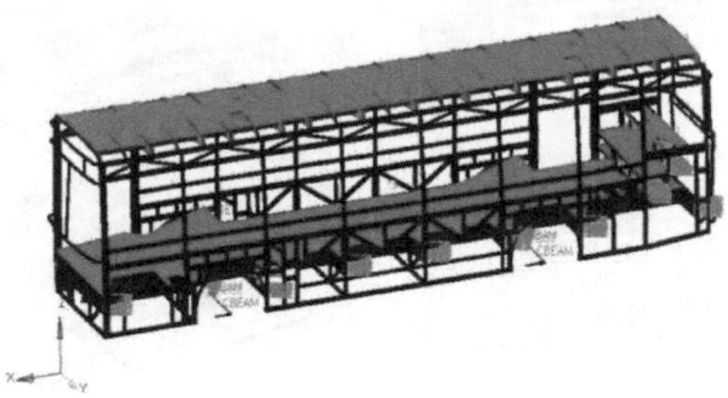

Fig. 4. Various loads on the bus body lumped using rigid (Rbe3) elements

The static loads as referred in [5, 13, 16] include the loads imposed, e.g., by passengers, engine, body coatings, windows and doors and are given in Table 2.

Table 2. Static loads

Mass category	Mass value [tonnes]
Bus structure	1.521
Chassis structure	1.153
Engine (rear mount)	0.5
Body coating	2.599
Windows and doors	0.575
Total passenger mass	6.00
Passenger mass · No. of person	$0.068 \cdot 65$
Seat and luggage	1.58
Other accessories	1.00
Total	13.416

3.1. Vibration analysis

The road induced frequencies are normally less than 5 Hz [1]; therefore, it is necessary to keep the natural frequency of the body higher than this. The natural frequency and modes are evaluated on solving the eigenvalues equations associated with the design model, and presented with the help of Table 3 and Fig. 5, respectively.

Table 3. First ten modes of free modal analysis

Modes	Frequency [Hz]	Mode shape
1	$4.26 \cdot 10^{-3}$	Rigid body mode
2	$7.79 \cdot 10^{-4}$	Rigid body mode
3	$3.26 \cdot 10^{-4}$	Rigid body mode
4	$1.22 \cdot 10^{-3}$	Rigid body mode
5	$1.25 \cdot 10^{-3}$	Rigid body mode
6	$1.40 \cdot 10^{-3}$	Rigid body mode
7	5.95	1st torsion mode
8	8.79	1st bending mode
9	10.36	2nd bending mode
10	11.74	Combined torsion and bending

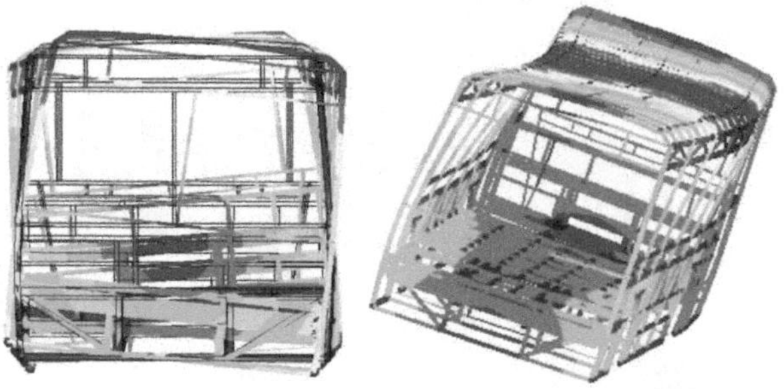

Fig. 5. First torsion and first bending mode

For free body modal analysis, the first six modes are rigid body modes. On visualisation of mode shapes, the seventh mode shows tendency of torsion, whereas the eighth mode comes out to be the first bending mode. Since the first torsion frequency is already greater than five, the objective is to enhance it further.

3.2. *Stiffness analysis*

The bus setup as shown in Fig. 6, where 1, 2, 3 represent the x, y, z directional constraints, respectively, is assumed to be a simply supported beam. It deforms upon application of certain force depending on the bending stiffness of the structure. Deformation is visualized along the length and for the bottom beam is presented in the form of a graph in Fig. 7.

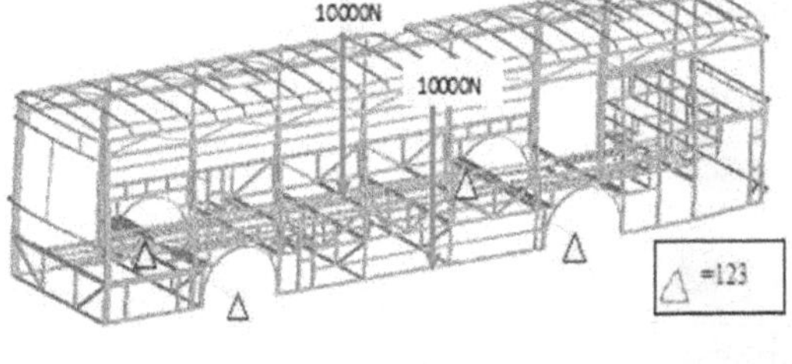

Fig. 6. FEA setup to calculate bending stiffness

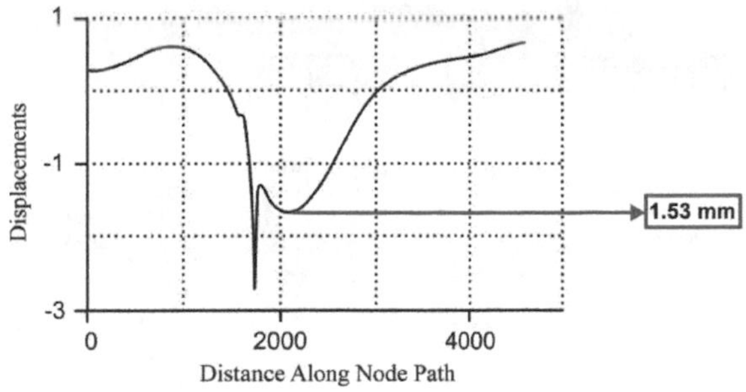

Fig. 7. Graph between the distance along the frame length and vertical displacement for the bottom beam

Torsion stiffness is a parameter to check the twist/deflection of the bus structure upon the application of torque. It should be high enough to resist the bus from twisting under various torsion loadings subjected to the vehicle. The torsional stiffness (TS) is defined with the help of the following equation:

$$\mathrm{TS} = \frac{F\,t}{\arctan\left(\frac{z_1+z_2}{t}\right)}, \tag{1}$$

where F is the force, t is the track width and z_1, z_2 are the displacements of the two spindles of the front axle. Fig. 8 shows the FEA setup to calculate the torsion stiffness, while Fig. 9 shows the displacements on the two spindles.

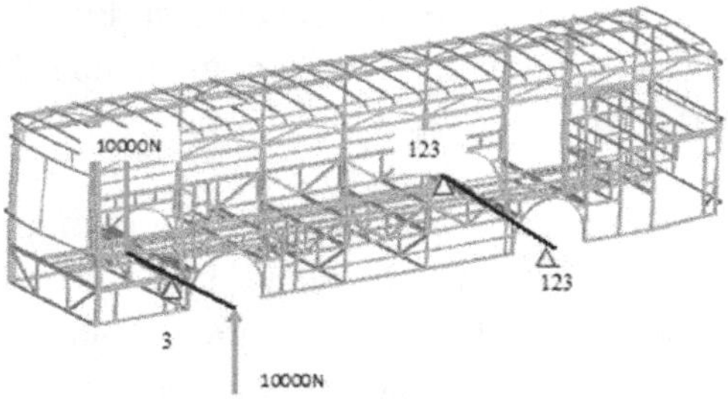

Fig. 8. FEA setup to calculate torsion stiffness

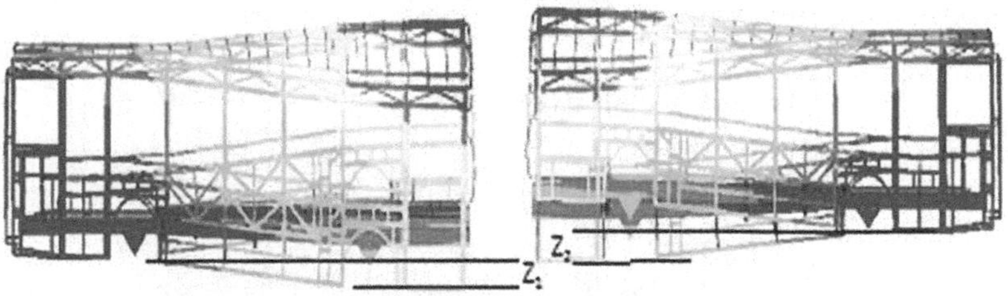

Fig. 9. Deflections z_1 and z_2

3.3. *Linear static analysis*

The loads imposed on the chassis and the body structure of a bus as the vehicle traverses uneven ground or when the driver performs various manoeuvres, lead to stresses and deformations. They have to be substantiated. The boundary conditions of loads are defined at the spindle locations of front and rear axles as mentioned in [7,8,11].

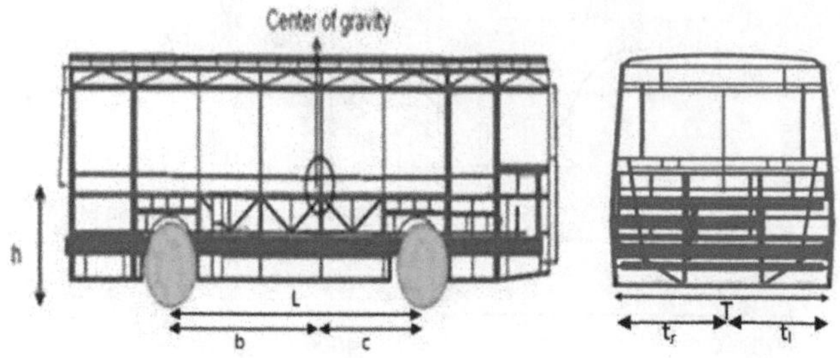

Fig. 10. Bus body dimensions

Table 4. Dimensions of the bus. Here, the abbreviation C.G. stands for the center of gravity

Variable	Definition	Measure
L	wheel base length	6 000 mm
b	distance from the C.G. to the front axle	3 566.5 mm
c	distance from the C.G. to the rear axle	2 433.5 mm
h	height from the ground to the C.G.	1 227.8 mm
T	track width of the bus	2 490 mm
t_r	lateral distance from the C.G. to the right frame	1 233 mm
t_l	lateral distance from the C.G. to the left frame	1 257 mm

Fig. 10 shows the bus model with proper dimensions, supported by its corresponding configuration in Table 4. To design for structural strength, all standard loads are grouped into five basic load categories, defined below:

(a) Bending case: It depends on the weights of the major components of the vehicle and the payload. Weight of the driver, engine, passenger, structure and chassis under an effect of gravity produces sag in the frame. Frame is assumed to act as simply supported beam and the four wheels as supports tend to produce reactions vertically upward at the axles as shown in Fig. 11. The relations for static load on the two axles and two wheels are given as

$$\begin{aligned} &\text{static loads on the front axle:} && W_{fs} = W\,\frac{c}{L}, \\ &\text{static loads on the rear axle:} && W_{rs} = W\,\frac{b}{L}, \\ &\text{static loads on the right wheel:} && W_{rw} = W_a\,\frac{l_t}{T}, \\ &\text{static loads on the left wheel:} && W_{lw} = W_a\,\frac{l_r}{T}, \end{aligned} \tag{2}$$

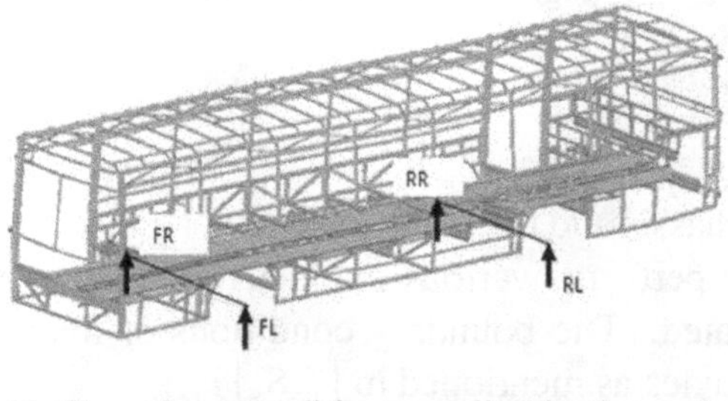

Fig. 11. Boundary conditions applied in the static case

where W is the weight of the loaded bus and W_a is the weight on axles. The percentage of weights on rear and front axles are 59.44 % and 40.56 %, respectively.

(b) Torsion: The maximum torsion is critical at the lighter loaded axle, and its value given as

$$\frac{R_f}{2}T_f = \frac{R_{r'}}{2}T_r, \tag{3}$$

where T_f and T_r are the front and rear end track widths, respectively, and R_f and $R_{r'}$ are the coupling forces on the front and rear axles, respectively, is the load on that axle multiplied by the wheel track width, Fig. 12.

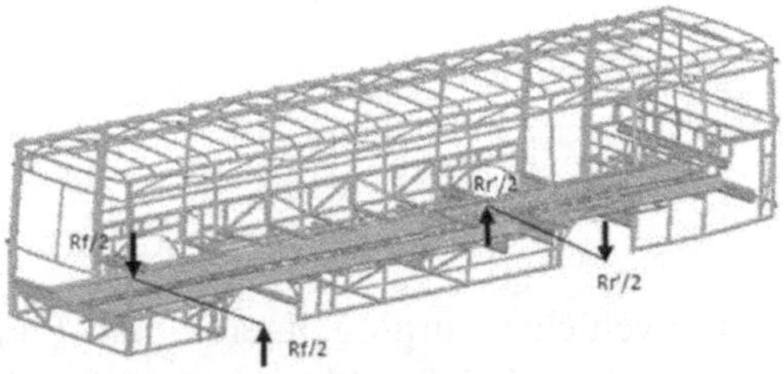

Fig. 12. Applications of boundary loads for torsion case

(c) Combine bending and torsion: The condition of pure torsion cannot exist on its own because vertical loads always exist due to gravity. Therefore, to get the realistic scenario, combined bending and torsion case is analysed as shown in Fig. 13. This figure shows a situation when one wheel of the lighter loaded axle is raised on a hump of sufficient height to cause the other wheel on the same axle to leave the ground (vertical bump case). In Fig. 13, R_r is the normal reaction force.

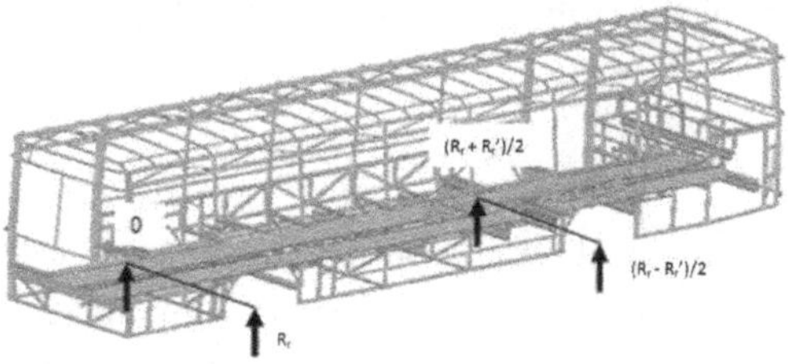

Fig. 13. Boundary loads for combine bending and torsion (vertical bump case)

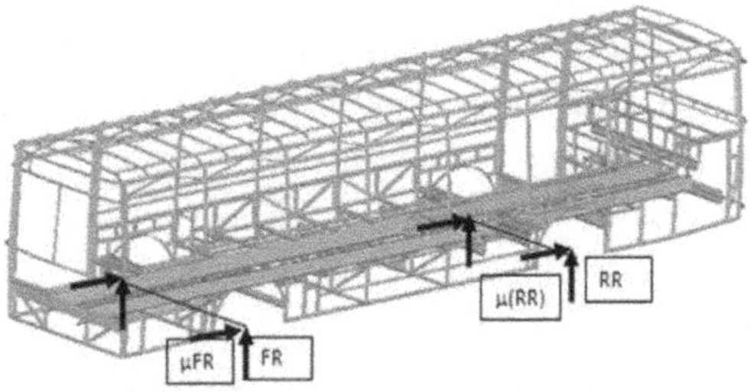

Fig. 14. Application of braking loads

(d) Longitudinal loading: When a vehicle accelerates (or decelerates), inertial forces are generated, Fig. 14. During acceleration, the weight is transferred from the front axle to the rear axle and vice versa in case of braking or decelerating conditions. The maximum load during acceleration is $A_x = 1g$ and the braking load is $D_x = 0.75g$ along the longitudinal axis. The distribution of axle loads is given as follows:

$$\begin{aligned}
&\text{load on the front axle during braking:}\\
&W_f = W\frac{c}{L} + D_x\frac{hW}{Lg} = W_{\text{fs}} + W_d,\\
&\text{load on the rear axle during braking:}\\
&W_r = W\frac{b}{L} - D_x\frac{hW}{Lg} = W_{\text{rs}} - W_d,\\
&\text{dynamic load transfer:}\\
&W_d = D_x\frac{hW}{Lg}.
\end{aligned} \tag{4}$$

(e) Lateral loading: When the vehicle is turning, lateral loads (A_y) are generated at the tyre-ground contact patches, which are balanced by the centrifugal force, thus,

$$\begin{aligned}
&\text{lateral weight transfer:}\\
&W_{lt} = W\frac{A_y h}{gT},
\end{aligned}$$

when turning left,

$$\begin{aligned}
&\text{load on the left axis spindle:} \qquad W_l = W\frac{t_r}{T} + W_{lt},\\
&\text{load on the right axis spindle:} \qquad W_r = W\frac{t_l}{T} - W_{lt}.
\end{aligned} \tag{5}$$

3.4. *Optimization of beam gauges*

In the present work, the design variables for optimization are the thickness of the beams constrained within a defined lower and upper bound, i.e., 1.5 mm and 3 mm, respectively. Response parameters are the properties based on the design variables, like stress, displacement, frequency and mass, which are used to evaluate the performance of the structure. The objective functions, here, are three in number: to minimize mass, to minimize compliance and to maximize frequency. A few boundary conditions or constraints are used on the response and design variables to make sure that the properties of the structure are within an allowed interval. The present work includes two optimization runs, on the basis of displacement criteria and modal criteria, respectively:

(a) Displacement criteria:

When displacement is considered, a structure is considered stiffer if the force to achieve that displacement is high, thus, minimizing compliance corresponding to maximum stiffness. This means that a lower compliance means a stiffer structure and lower displacement. The constraint is to have the mass of designed beams less than 1.521 kg (as given in Table 2) and the objective is to minimize the compliance.

(b) Modal criteria:

This is a measure of the dynamic behaviour of the model, in which higher frequencies mean better ride for the passengers (to avoid resonance). Since the first two bending and torsion mode frequencies, see Table 3, contribute most to the structure stiffness, they are to be maximized with an objective to minimize the weight. The design variable properties are imported from the results of the previous (displacement) run.

In this case, the constraint is the mass taken from the previous run, i.e., mass less than 1.521 kg. The objectives are to maximize the 7th and 8th frequencies of the free modal analysis. Two iterative runs are consecutively executed and studied to achieve an optimum result.

4. Experimental results

4.1. Result of linear static analysis

During the linear static analysis, the highest stresses observed are those of 279 MPa and 283 MPa for combined bending and combined longitudinal and lateral loadings, respectively. The locations of these highest stresses are shown in Fig. 15.

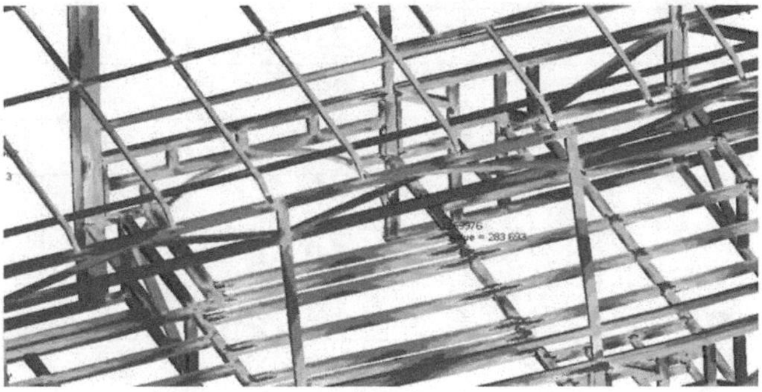

Fig. 15. Beam subjected to high stresses

When evaluating the performance of the base model, emphasis is placed on the magnitude and location of stresses and deformations occurring on the body. As circled in Fig. 16, the largest stresses are localized near the wheels and the floor adjacent to the doors.

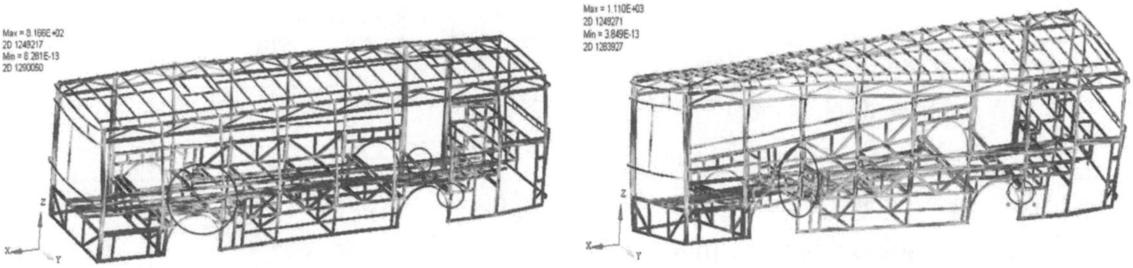

a) b)

Fig. 16. a) Stresses for left cornering with braking; b) stresses in the torsion case

4.2. Result of optimization

The iterative optimization process results into multiple outcomes because of two different runs. The results in the form of varied thicknesses can be imported from one setup to another so that the optimized results are in good agreement. The beams are categorized on the basis of thickness and symmetry into small sets, denoted by $T_1, T_2, \ldots, T_{12}$. In the optimization mode, runs 1 and 2 are carried out to minimize displacement, while runs 3, 4, and 5 are carried out to maximize frequency, subjected to variation in beam gauges. The beam gauges are presented in Table 4. The analysis of the gauge variations [16, 18] during runs 1–5 led to selection of specifications of beams' thickness as per Indian market standards. On the basis of data selected for the cases A, B and C, the performance characteristics are mapped and summarized in Table 5.

The results of runs are normalized and plotted in Fig. 17. The case C is the optimum result with modified specifications.

Table 5. Parametric values after each optimization run

Beam thickness [mm]	Specifications of the base model	Product specifications as per market availability		
		A	B	C
T_1	2.0	2.0	2.0	1.50
T_2	3.0	2.5	3.0	3.00
T_3	3.0	2.5	2.5	2.50
T_4	2.0	2.0	2.0	2.00
T_5	1.0	1.5	1.5	1.50
T_6	1.5	2.0	1.5	1.75
T_7	2.0	2.0	2.0	2.00
T_8	3.0	3.0	3.0	3.00
T_9	2.0	2.5	1.5	1.50
T_{10}	2.0	1.5	1.5	2.00
T_{11}	1.5	1.5	1.5	1.50
T_{12}	2.0	2.0	2.0	1.50
		Performance characteristics		
Frequency in torsion mode [Hz]	5.950	6.180	5.980	6.050
Worst case displacement during left turning and braking [mm]	22.400	23.400	22.500	21.900
Worst case displacement during combined torsion and bending [mm]	67.300	63.830	67.260	64.800
Mass [tonnes]	1.521	1.488	1.505	1.501

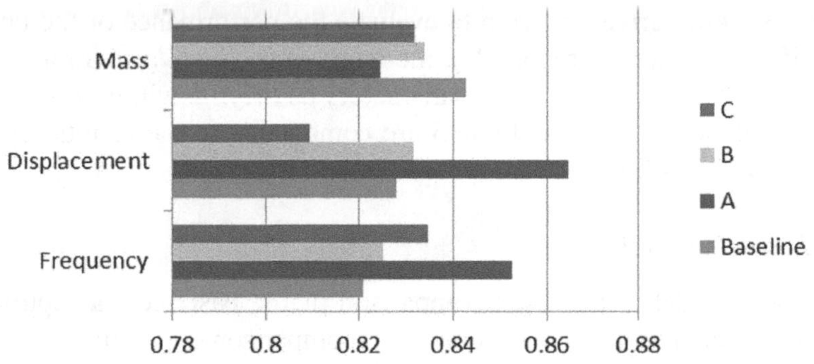

Fig. 17. Analytical chart showing performance comparison of the iterative runs

5. Discussion

5.1. Baseline model analysis

Similar to the stress raising members (e.g., holes, joints etc.), the elements at the two beams junctures of the bus body would have very high stresses, as per Neuber's criteria for elasto-plastic analysis. In the present work, these junctures are not considered due to the inherent complexity. The yield stress of the steel used in the bus structure is 400 MPa and a safety factor of 2.6 is considered so that the permissible stress [1,5,17] is less than 150 MPa. Further, the high stresses at specific location can be accounted to the design limitation, i.e., to the unavailability of a suspension system which acts as damper to stresses.

The bus subjected to stresses would also suffer deflection. The deflections are minimized to account for high stiffness of the structure. The maximum deflection varies according to the severity of the loading. It is highest for the torsion case, i.e., for the vertical bump when the bus rises on a hump. When deflections are not continuous and high as circled in Fig. 18, unstable and weak zones are implied. There is no limiting value to be generalized, but the objective is to minimize deflection as much as possible or to maximize stiffness/minimize compliance.

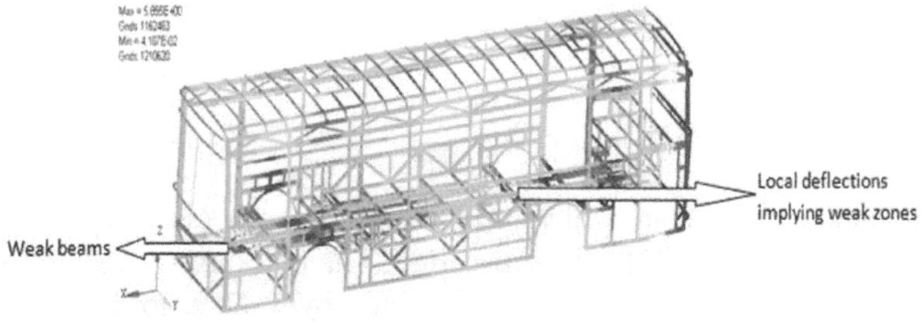

Fig. 18. Deflections under bending loads

The frequencies of the baseline model are above a limiting value of 5 Hz, which is further improved during the course of this work. The mass of the bus structure is also lower than the one reported in literature [3,5]. Since the objective of this work is to improve the performance parameters along with weight reduction, the optimization is carried out with the minimal possible mass as a constraint.

The stiffness is an important criterion to evaluate the performance of the bus structure. In literature, the stiffness values are reported in the range of 5–15 kN/mm for a bus structure [9]. The stiffness values of baseline model are satisfactory as they lie within this range. The worst case displacement values as given in Table 5 are comparable to the results reported in literature [4,16] and, hence, justified.

5.2. *Optimized model analysis*

To validate the new model, a tabulated comparison of the baseline and optimized model is made in Table 6. This table does not include the comparison of maximum stresses induced on beams due to indefinite load cases. The range of maximum stresses for the worst cases, which are turning with braking and combined torsion with bending, has reduced from a range of 250–280 MPa to the range of 200–220 MPa. Besides, for other load cases, the stresses on the beams of the optimized model are within the permissible range of 150 MPa [5]. Stresses on some of the weak beams still exceed the permissible limit and, hence, are identified for remodeling or artificial stiffeners have to be attached at identified locations.

Table 6. Comparison between baseline and optimized model

Parameters	Baseline model	Optimized model	Difference
Structure mass [tonnes]	1.521	1.500	−1.33 %
Torsion frequency [Hz]	5.950	6.050	+1.66 %
Bending frequency [Hz]	8.780	8.900	+1.43 %
Torsion stiffness [kNm/deg]	17.920	18.230	+1.7 %
Bending stiffness [kN/mm]	13.110	12.940	−1.20 %
Displacement [mm] (left turn and braking case)	22.400	21.900	−2.20 %
Displacement [mm] (vertical bump case)	67.000	64.880	−3.56 %

6. Conclusion

The gauge optimization on the bus structure is an iterative process and the mass of the final optimized model is by 1.33 % less than the baseline. The first torsion frequency has increased by 1.66 % to 6.05 Hz and the first bending frequency by 1.43 % to 8.906 Hz. The torsion stiffness is now by 1.7 % higher than that of the baseline model, whereas the bending frequency is almost the same for the two models. The deformations have moderately come down by 2.2 % for the combined turning with braking case, while it shows a decline of 3.56 % for the torsion case. Thus, one can conclude that by following the proposed algorithm, the bus body may be optimized along with minor decrease in weight by optimizing beam gauges. This has led to significant improvement of performance characteristics. The work is verified by conducting detailed analysis on a full length 11 m, 65 seats bus and helps in reducing structural mass by 20 kg along with enhanced stiffness, vibration and stress performances.

Acknowledgements

The proposed work is done during the Project Based Internship (PBI) of the undergraduate students at PDPM IIITDM Jabalpur. We acknowledge the support provided by the Institute and Altair Engineering Pvt. Ltd. to facilitate the conduct of this work.

References

[1] Belure, S. B., Kadam, S. S., Wadkar, S. B., Criticality analysis of passenger bus body structural design using finite element method, Proceedings of COMPUTIME, National Conference on Computational Methods in Mechanical Engineering, Hyderabad, India, 2005.

[2] Biradar, J. M., Vijay, B. V., Jat, K., Automotive chassis sizing optimisation for modal and distortion criteria, SASTech 7 (2) (2008) 1–7.

[3] Boada, B. L., Gauchia, A., Boada, M. J. L., Diaz, V., A genetic-based optimization of a bus structure as a design methodology, Proceedings of the 12th IFToMM World Congress, Besancon, 2007.

[4] Chinnaraj, K., Prasad, M. S., Rao, C. L., Experimental analysis and quasi-static numerical idealization of dynamic stresses on a heavy truck chassis assembly, Applied Mechanics and Materials 13–14 (2008) 271–280.

[5] Croccolo, D., Agostinis, M. D., Vincenzi, N., Structural analysis of an articulated urban bus chassis via FEM: A methodology applied to a case study, Strojniški vestnik – Journal of Mechanical Engineering 57 (11) (2011) 799–809.

[6] Fornace, L. V., Weight reduction techniques applied to formula SAE vehicle design: An investigation in topology optimization, Master thesis, University of California, San Diego, 2006.

[7] Happian-Smith, J., An introduction to modern vehicle design, 2nd edition, Elsevier, 2001, pp. 125–155.

[8] Heibing, B., Ersoy, M., Chassis handbook: Fundamentals, driving dynamics, components, mechatronics, perspectives, Vieweg + Teubner Verlag, 2010.

[9] Helsen, J., Cremers, L., Mas, P., Sas, P., Global static and dynamic car body stiffness based on a single experimental modal analysis test, Proceedings of the International Conference on Noise and Vibrations Engieneering — ISMA, Leuven, Belgium, 2010, pp. 2 505–2 522.

[10] Kumar, S., Optimization of BIW, chassis and casting at Mahindra and Mahindra, Proceedings of the Altair CAE Users Conference, Bangalore, India, 2005.

[11] Lafreniere, M., Design of a formula SAE race car chassis: Composite analysis utilizing Altair Engineering OptiStruct software, Bachelor thesis, University of Toronto, 2007.

[12] Lin, Y. C., Nian, H. C., Structural design optimization of the body section using the finite element method, SAE Technical paper 2006-01-0954, 2006.

[13] Manokruang, S., Butdee, S., Methodology of bus-body structural redesign for lightweight productivity improvement, AIJSTPME 2 (2) (2009) 79–87.

[14] Proos, K. A., Steven, G. P., Querin, O. M., Xie, Y. M., Multicriterion evolutionary structural optimization using the weighting and the global criterion methods, AIAA Journal 39 (10) (2001) 2 006–2 012.

[15] Reed, C., Application of Optistruct optimization to body in white design, Altair Engineering Ltd., 2002.

[16] Ruíz, O., Ramírez, E. I., Jacobo, V. H., Schouwenaars, R., Ortiz, A., Efficient optimisation of the structure of a passenger bus by iterative finite element models with increasing degrees of complexity, Proceedings of the 3rd International Conference on Engineering Optimization, Rio de Janeiro, Brazil, 2012.

[17] Shinabuth, D., Benyajati, C., Phuchamnong, A., Pimsam, M., Okuma, M., A numerical and experimental measurement in a dynamic strain response of an electric bus body structure, Proceedings of the 2nd TSME International Conference on Mechanical Engineering, Krabi, Thailand, 2011.

[18] Singh, R. V., Structural performance analysis of a formula SAE car, Jurnal Mekanikal 31 (2010), 46–61.

[19] Xingwang, Z., Zhen, T., A study on shape optimization of bus structure based on stiffness sensitivity analysis, Proceedings of the 10th International Conference on Computer-aided Industrial Design & Conceptual Design, Wenzhou, China, 2009, pp. 1 225–1 229.

[20] Zehnder, J., Pritzlaff, R., Lundberg, S., Gilmont, B., Aluminium in commercial vehicle, European aluminium association AISBL, Brussels, 2011.

12

Development of aerodynamic bearing support for application in air cycle machines

J. Šimek[a,*], P. Lindovský[b]

[a]TECHLAB Ltd., Sokolovská 207, 190 00 Praha, Czech Republic
[b]První brněnská strojírna V. Bíteš, a. s., Vlkovská 279, 595 12 Velká Bíteš, Czech Republic

Abstract

Air cycle machines (ACM) are used in environmental control system of aircrafts to manage pressurization of the cabin. The aim of this work is to gain theoretical and experimental data enabling replacement of rolling bearings, which require lubrication and have limited operating speed, with aerodynamic bearing support. Aerodynamic bearings do not pollute process air and at the same time allow achieving higher operating speed, thus enabling to reduce machine mass and dimensions. A test stand enabling the verification of aerodynamic bearing support properties for prospective ACM was designed, manufactured and tested with operating speeds up to 65 000 rpm. Some interesting features of the test stand design and the test results are presented. A smaller test stand with operating speed up to 100 000 rpm is in design stage.

Keywords: air cycle machine, aerodynamic bearings, foil bearings, titling pad journal bearings, elastically supported pads, spiral groove thrust bearings, rotor stability

1. Introduction

Former types of air cycle machines (ACM) had rotors supported in rolling bearings, which could not operate without lubrication. Traces of lubricant polluted the air, thus, endangering the crew and passengers. Moreover, working life of rolling bearings as well as their operating speed is limited, which prevents achieving higher speed and at the same time smaller ACM dimensions. In order to improve competitive advantages of machines produced by PBS V. Bíteš, a project of development ACM with aerodynamic bearings was started in 2012.

Most air cycle machines use foil bearings [1–3], which have some unique properties, namely possibility of adaptation to operating conditions and excellent dynamic properties due to additional damping caused by friction between the bearing and supporting foils (see Fig. 1) and between the supporting foil and the bearing casing. Both bearing and supporting foils are deformed by the generated aerodynamic pressure. As the bearing gap and at the same time bearing properties change with speed, it is relatively difficult to calculate bearing characteristics.

However, even better dynamic properties have tilting pad journal bearings with elastically supported pads (Fig. 2) described in detail in [4]. They combine excellent stability of tilting pad bearings, as a result of very small cross coupling stiffness terms, with positive properties of foil bearings, namely possibility to adapt itself to changed operating conditions and additional damping due to friction of elastic elements on bearing casing.

*Corresponding author. e-mail: j.simek@techlab.cz.

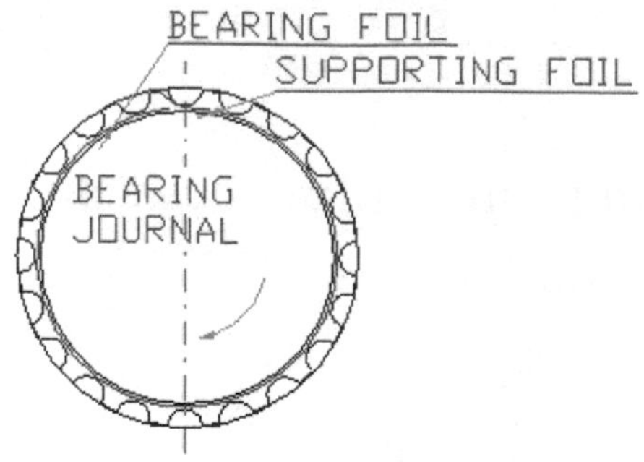

Fig. 1. Typical foil bearing

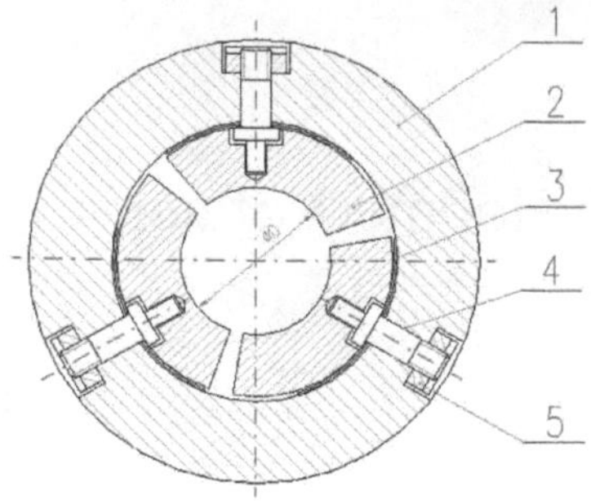

Fig. 2. Tilting pad bearing (ESTP)

As can be seen in the original design of ACM in Fig. 3, the rotor with the turbine (left) and blower (right) impeller was supported in two angular contact rolling bearings preloaded by the spring. Rolling bearings were replaced by aerodynamic journal bearings with elastically supported tilting pads (ESTP) and aerodynamic spiral groove thrust bearings.

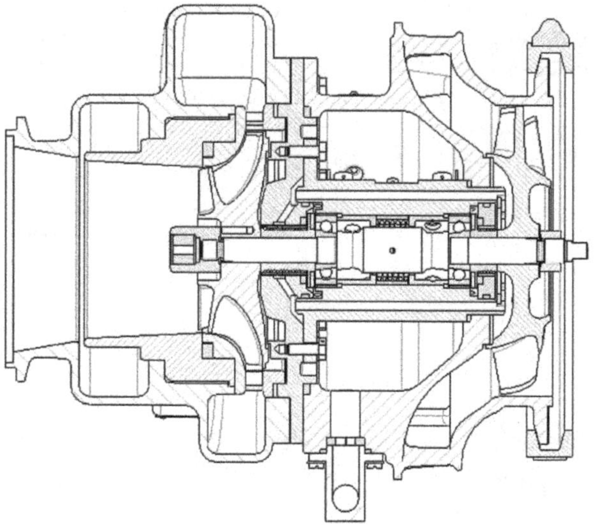

Fig. 3. Original ACM with rotor supported in rolling bearings

ACM designed as a test stand for verification of bearing properties [5] is presented in Fig. 4. The rotor is supported in two aerodynamic tilting pad bearings, and axial forces are taken up by double-sided aerodynamic spiral groove thrust bearing 4, 5. The bearing pads 3 are supported on the elastic elements 6, which are deformed to required shape by means of pins 7 and nuts 8. The difference between the inner radius of the bearing casing 2 and the outer radius of the pad enables the rolling of pads on elastic element inner surface, so that they can tilt in circumferential direction. The elastic elements are preloaded to such an extent that maximum pad load capacity exceeds the force necessary for elastic element deformation. In case that the bearing clearance is reduced to a dangerously low value, elastic elements make it possible to restore the bearing clearance back to its safe value. Friction between the elastic elements and bearing body contributes to the damping of a gas film, similarly as in the foil bearings. Moreover, the overall damping is further increased by the squeeze effect of gas pushed out of the gap between the elastic elements and the bearing body.

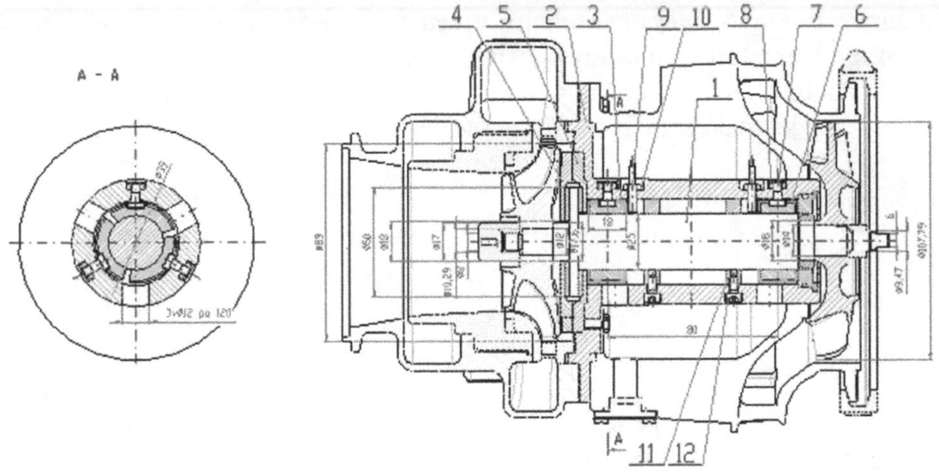

Fig. 4. Test stand with aerodynamic bearing support

The aerodynamic thrust bearings 4 a 5 substitute to some extent original labyrinth seal, because they prevent the air to flow from the turbine (left impeller) to the blower (right impeller) side. However, this was the cause of unexpected problems with the axial force magnitude. The maximum axial force directed to the turbine is higher than the axial load acting in the opposite direction and therefore, it is beneficial that the thrust bearing 4 could have bigger sliding surface (lower inner diameter) than the other one.

2. Calculation of bearing characteristics and rotor dynamics

The theoretical basis for the calculation of aerodynamic tilting pad bearing function properties is briefly described in [4]. Static and dynamic bearing characteristics are calculated with the computer programs "SATPJB" and "DATPJB".

The journal bearings have the diameter of 25 mm and the width to diameter ratio of 0.72. Manufacturing clearance in the range 0.035 to 0.040 mm was selected due to relatively low speed. The Manufacturing clearance defined as $c_S = R_S - R_C$ is an important design parameter together with the bearing preload $\delta = 1 - c_L/c_S$, where $c_L = R_L - R_C$ (variable designation in Fig. 5).

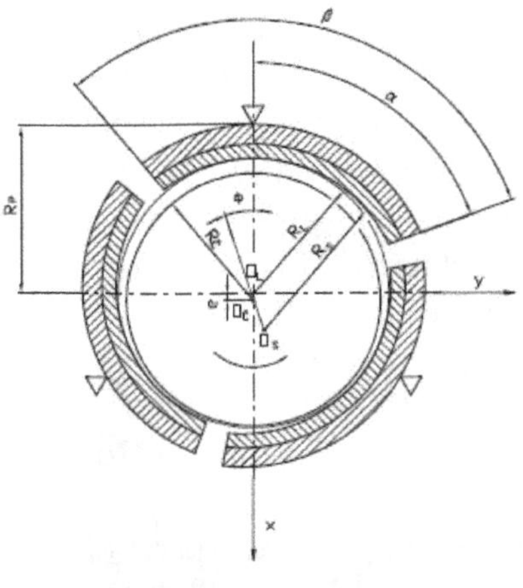

Fig. 5. TPJ bearing geometry

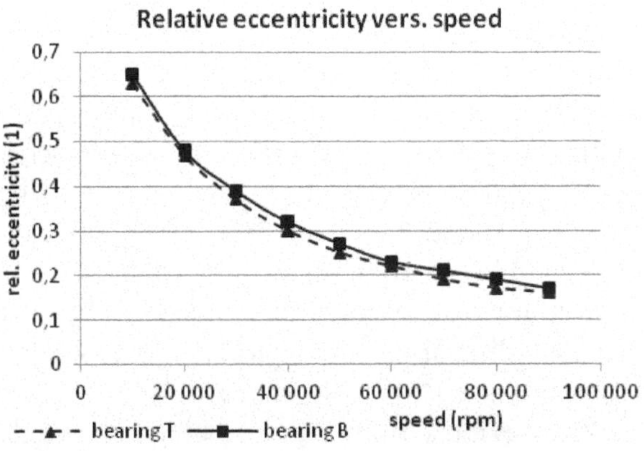

Fig. 6. Relative eccentricity in journal bearings

The diagram in Fig. 6 shows the relative eccentricity (eccentricity relative to the bearing clearance) of the journal in both bearings, which is almost the same because of not much differing load. It can be seen that with increasing speed, the relative eccentricity decreases to very small values — the journal centre is almost in the centre of the bearing.

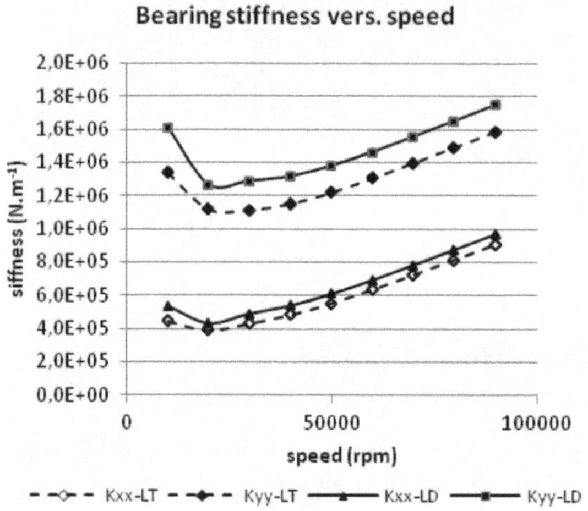

Fig. 7. Bearing stiffness coefficients

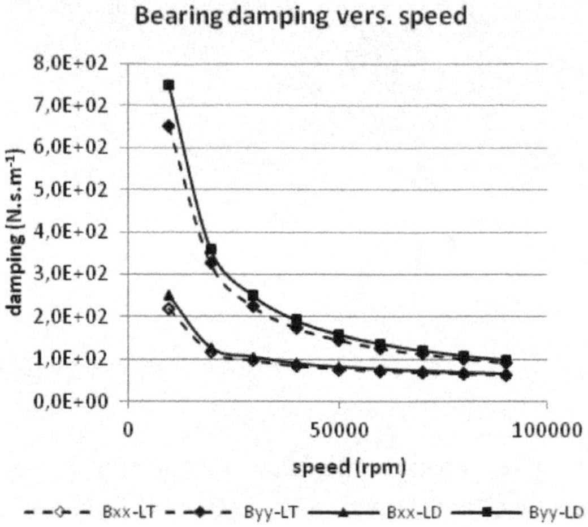

Fig. 8. Bearing damping coefficients

The bearing dynamic properties represented by the stiffness and damping coefficients calculated by the program "DATPJB" are shown in the diagrams in Figs. 7 and 8. Principal bearing stiffness and damping coefficients are shown in dependence on speed. It is evident that in contrast to the stiffness coefficients of hydrodynamic tilting pad bearings, aerodynamic bearings have the element K_{xx} lower than the K_{yy} one. It is due to the bearing geometry with only 3 pads, whereas the hydrodynamic bearings have 4 or 5 pads. With the exception of the lowest speeds, the bearing stiffness increases with increasing speed. The bearing damping is decreasing monotonously with increasing speed, which is standard for all sliding bearings. The calculated bearing stiffness and damping coefficients are used in the calculation of rotor dynamics.

Due to low damping, rotors in aerodynamic bearings cannot pass through the bending critical speed. Moreover, the 1st bending critical speed should be at least 60 % above the maximum operating speed; otherwise one could expect problems with excessive vibrations due to the start of rotor bending. This requirement and the achievement of rotor stability are, therefore, the two most important aspects of dynamic analysis [6]. The analysis showed four "bearing" critical speeds, i.e., critical speeds of rigid rotor on gas film, which are situated in the range from 6 900 to 11 900 rpm and are well damped. The rotor exhibits somewhat unorthodoxly positioned bending critical speeds, because apart from the 1st bending critical speed of counter-rotating precession at 113 700 rpm, the next critical speed is counter-rotating precession with the 2nd bending mode at 197 100 rpm. The 1st bending critical speed of co-rotating precession is at 261 800 rpm, i.e., more than 400 % above the operating speed. According to the Campbell diagram in Fig. 9, the critical speeds of 113 700 and 261 800 rpm are two branches of one eigenvalue split by the gyroscopic moments. The vibration mode at 197 100 rpm (not shown in the Campbell diagram) is the typical 2nd one (sine wave), while the vibration mode at 261 800 rpm is typically the 1st one.

As is evident from Fig. 10, which shows the eigenvalue damping, real parts of the four lowest eigenvalues are negative, which means that the rotor operation is stable. The stability

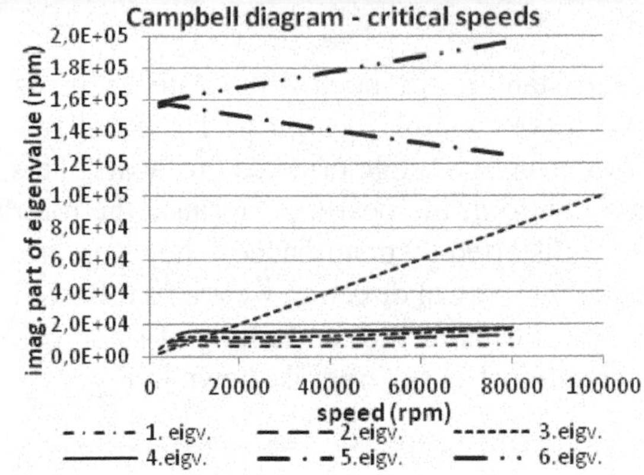

Fig. 9. Campbell diagram of critical speeds

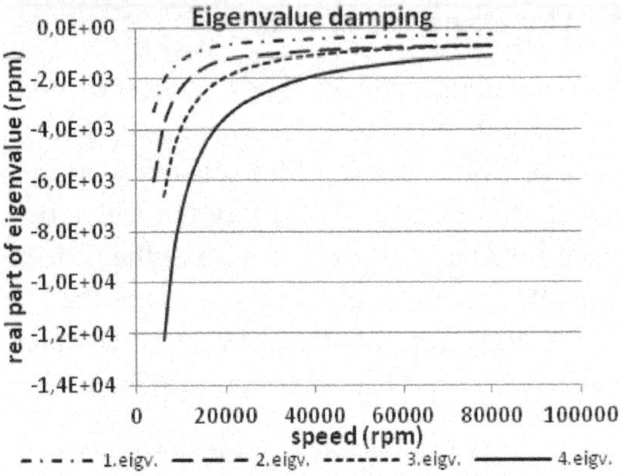

Fig. 10. Damping of critical speeds

reserve of the rotor is defined as

$$\chi = -2\mathrm{Re}(\lambda)/\mathrm{Im}(\lambda) \cdot 100\ [\%],$$

where $\mathrm{Re}(\lambda)$ and $\mathrm{Im}(\lambda)$ are the real and imaginary parts of the eigenvalue, respectively.

At the speed of 60 000 rpm, the two lowest eigenvalues have the stability reserve of 10 % and 13.8 %, respectively, which should be quite sufficient for safe operation. The stability reserve of 10 % is generally considered to be safe for stable rotor operation. However, in some previous applications, rotors operated securely with stability reserve around 5 %. Rotor operability was also confirmed with a calculated response to unbalance. The residual unbalance for a rotor with the maximum operating speed of 60 000 rpm according to the class G2.5 of ISO 1940 standard is 0.38 g.mm. This unbalance was divided between the turbine and the blower impellers in ratio of their masses. The rotor response to static unbalance (unbalances in-phase) and dynamic unbalance (unbalances out-of-phase) in dependence on speed is shown in Figs. 11 and 12. It can be seen that up to 80 000 rpm, there is no resonance peak with the exception of the speed interval around 15 000 rpm with well damped critical speeds of rigid rotor on a gas film.

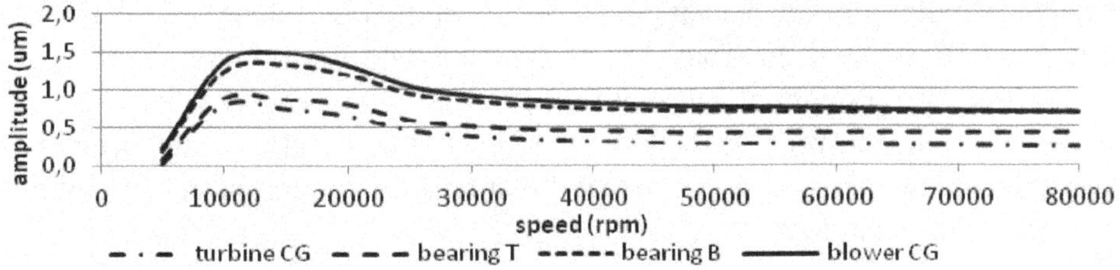

Fig. 11. Rotor response to static unbalance

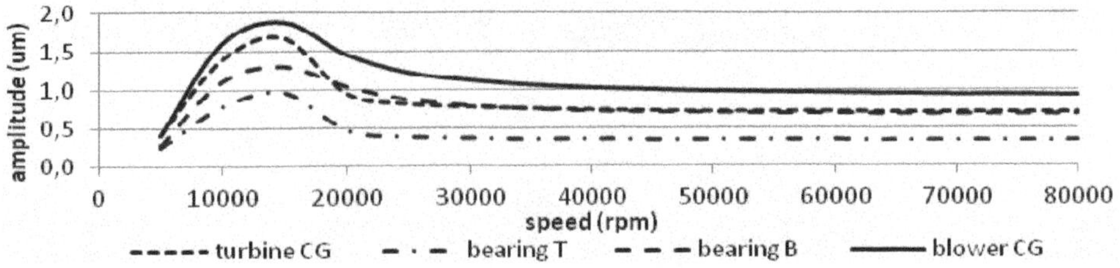

Fig. 12. Rotor response to dynamic unbalance

3. The 1st period of tests

As is evident from the cross section in Fig. 4, the test stand is equipped with relative sensors for the detection of rotor vibrations 9 located next to both journal bearings. There are two pairs of Micro-epsilon S04 sensors working on eddy current principle oriented 90° apart. This orientation enables to display the trajectory of rotor centre in the bearing. Because the rotor vibrations had very low level and the trajectories were distorted by disturbances, they were not evaluated. Beside radial vibrations of the rotor, excursions in axial direction were also followed by the eddy current sensor. Certain correlation between the axial shift and tilting of rotor axis can be observed [7]. Rotor run-up from about 9 000 to almost 17 000 rpm is shown in Fig. 13.

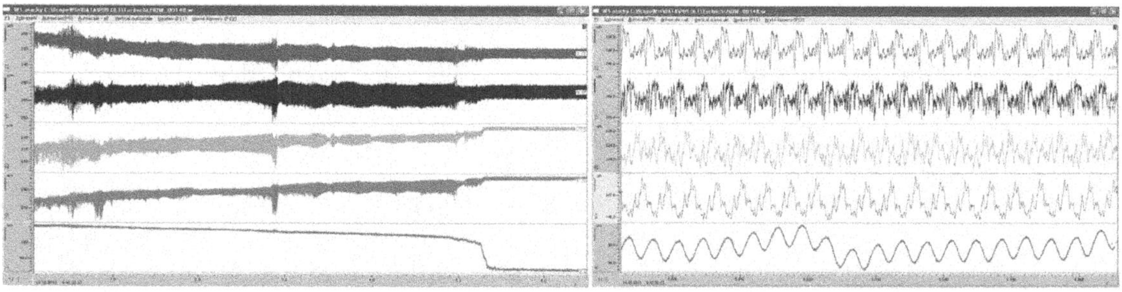

Fig. 13. Run-up of the rotor, generation of the full aerodynamic film at about 16 000 rpm

Top down in Fig. 13 and all following records of vibrations are signals:

- turbine bearing – horizontal direction,
- turbine bearing – vertical direction,
- blower bearing – horizontal direction,
- blower bearing – vertical direction,
- axial direction.

Expanded signals in the right diagram of Fig. 13 show a situation with a fully developed air film. Small disturbances can be seen in all vibration signals in the radial direction, which can be attributed to material properties changing around the shaft periphery. Fig. 14 presents rotor acceleration from 35 500 to 42 200 rpm. The change of axial force direction took place at about 39 000 rpm, which also brought about relatively great shift in the radial direction, such as the one in the bearing at the blower side. It shows that due to the pad elastic support with relatively low stiffness, the rotational axis can tilt so that thrust bearing and thrust runner sliding surfaces align to be parallel to each other.

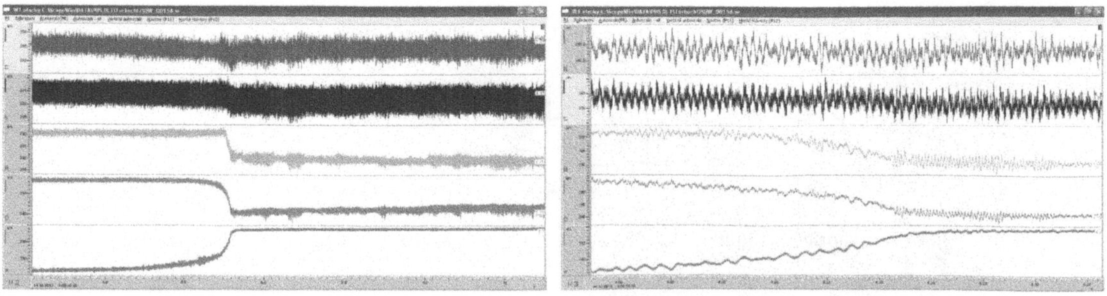

Fig. 14. Rotor acceleration from 35 500 to 42 200 rpm with change of axial force direction

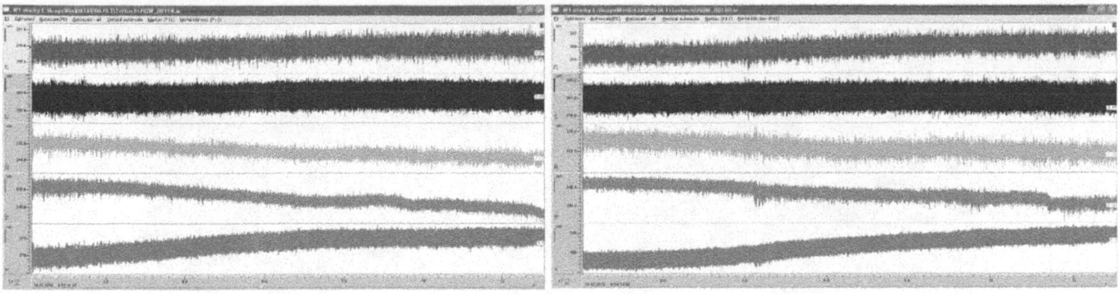

Fig. 15. Acceleration from 42 500 to 47 000 rpm and from 47 500 to 53 000 rpm

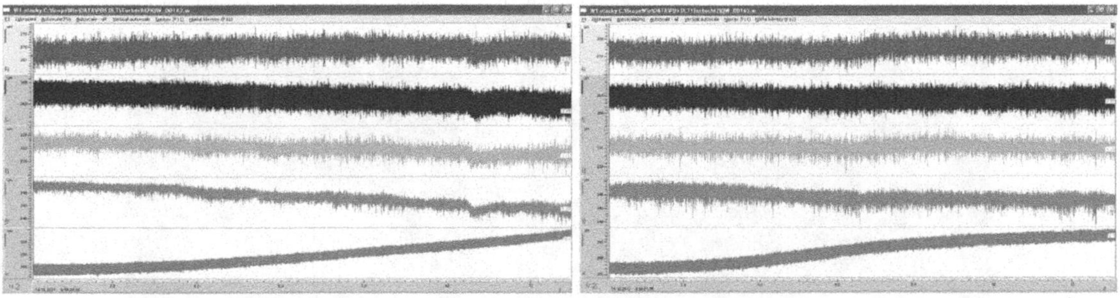

Fig. 16. Acceleration from 55 000 to 59 000 rpm and from 59 400 to 63 700 rpm

This is most important from the standpoint of thrust bearing load carrying capacity, which is very dependent on the paralelism of sliding sufaces. Tilting of rotatinal axis acompanying the change of axial force/axial shift can be also seen in Fig. 15 illustrating rotor acceleration from 42 500 to 47 000 rpm and from 47 500 to 53 000 rpm. The maximum radial shift occurs in horizontal direction. Similar rotor behaviour follows then the speed is increased from 55 000 to 59 000 rpm and from 59 400 to 63 700 rpm as shown in Fig. 16. Small increase in vibration amplitude can be seen in signals from the bearing on blower side, which could indicate that the thrust bearing load capacity is nearing its limit.

Although the operation speed was already achieved, it was decided to increase the speed once more. Results for acceleration from 62 800 to 66 600 rpm are shown in Fig. 17. Apart from further shift in the axial direction and a slight increase in vibration amplitude in blower bearing, there were no indications of potential problems. However, after a relatively short time, the vibration amplitude dramatically increased, as shown in the right part of Fig. 17. At this point, the machine was stopped and disassembled. Essential parts of the aerodynamic bearing support after failure are shown in Figs. 18 and 19.

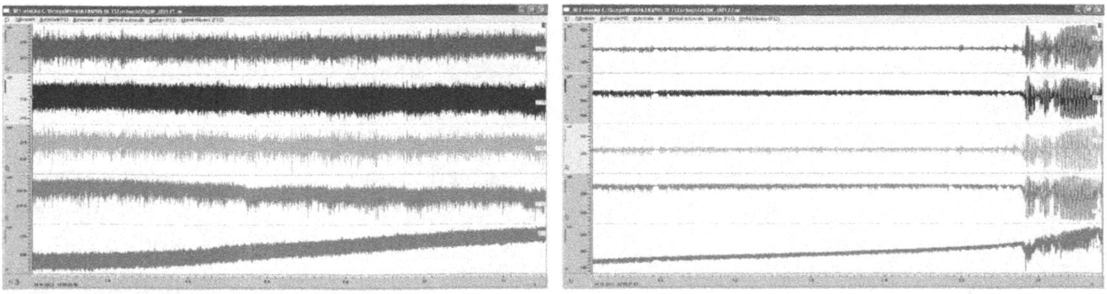

Fig. 17. Acceleration from 62 800 to 66 600 rpm; Beginning of failure at 65 000 rpm

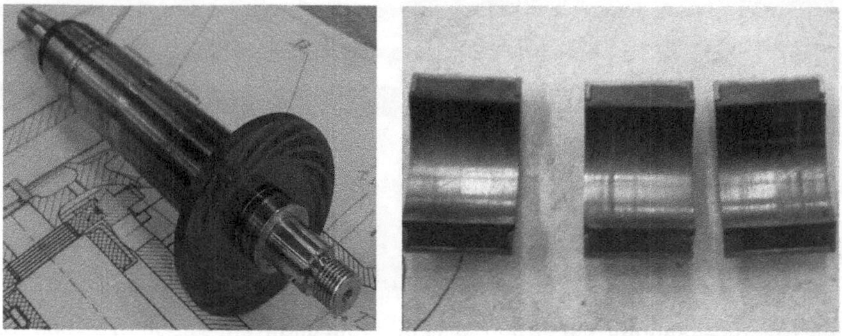

Fig. 18. Rotor with thrust disk and pads from journal bearing at blower side after failure

Fig. 19. Auxiliary (left) and main (right) thrust bearing after failure

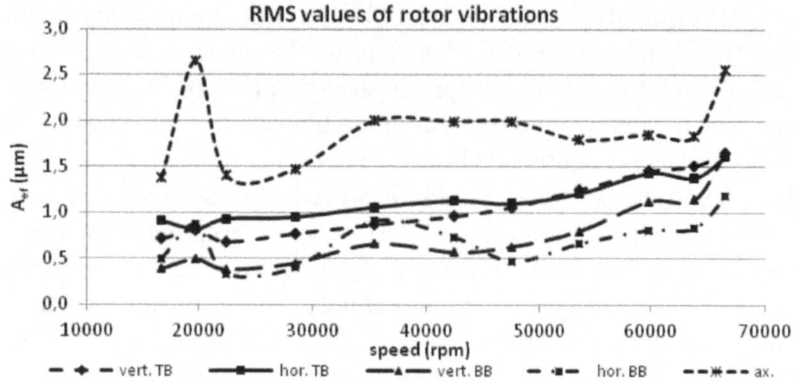

Fig. 20. RMS value of rotor vibration amplitude

It can be seen that the main thrust bearing with graphite lining as well as steel thrust disk on the rotor are heavily damaged by the mutual contact at sliding speed exceeding $170\ \mathrm{m \cdot s^{-1}}$. On the other hand, the bearing pads and opposed sliding surfaces on the rotor were quite undamaged. The failure was apparently caused by the overloading of the main thrust bearing and the subsequent contact of sliding surfaces, which resulted in substantially increased vibrations in the journal bearings within the whole clearance (including clearance of elastically supported pads in bearing casing).

As is evident from diagram in Fig. 20, the rotor vibrations up to 67 000 rpm were quite low. Curves in Fig. 20 represent the RMS values of relative vibrations in horizontal, vertical and axial directions in bearings on blower (BB) and turbine (TB) side. No RMS value of rotor vibration amplitude in radial direction exceeded 2 μm. Vibrations in axial direction were lower than 3 μm in the whole speed range. These data reflect the very good level of rotor balancing as well as the proper function of both the journal and thrust bearings. The only indication of thrust bearing overloading was the very slight vibration increase in the journal bearing at blower side.

According to calculation, the main thrust bearing was operating with gas film thickness lower than 3 μm, which is an extremely low value for a bearing 50 mm in diameter. The axial force acting on the thrust bearing is a resultant of pressures acting on the impellers and the unloaded part of the thrust disk. Change of axial force direction at about 40 000 rpm was caused by pressure increase at the turbine side, which was then amplified by the pressure acting on the other (now unloaded) side of the thrust disk. The development of axial force with speed is shown in diagram in Fig. 21. As can be seen from the diagram, at the speed of 65 000 rpm axial force exceeded value of 150 N, scheduled for bearing design, almost three times. The change of axial force direction around 40 000 rpm was confirmed by the measured data.

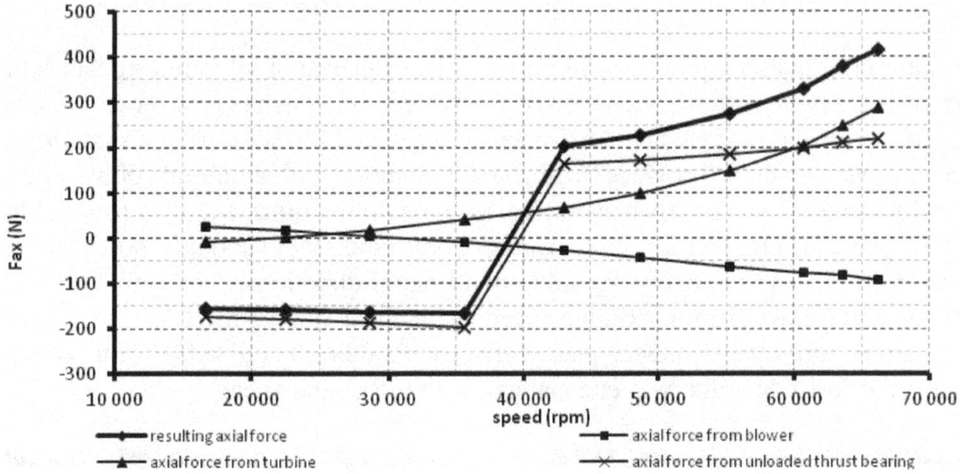

Fig. 21. Calculated axial force

4. Test stand modifications and new series of tests

An axial force analysis showed that it is necessary to reduce the thrust bearing load [8]. Labyrinth seal was, therefore, installed behind the turbine impeller (see Fig. 22). The tests continued by sequentially increasing the speed with similar rotor vibrations as in the previous tests. Sam-

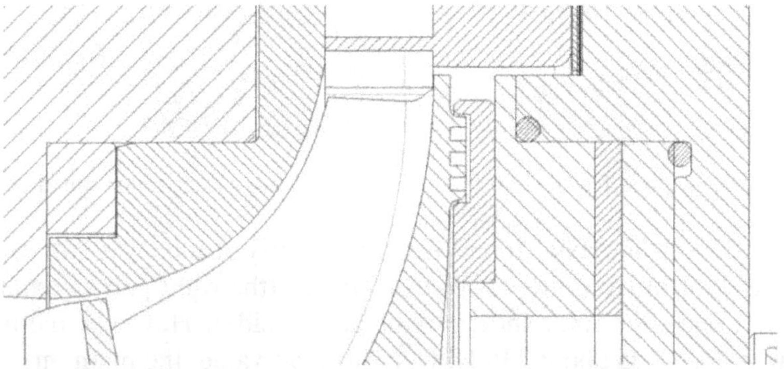

Fig. 22. Labyrinth seal behind the turbine impeller

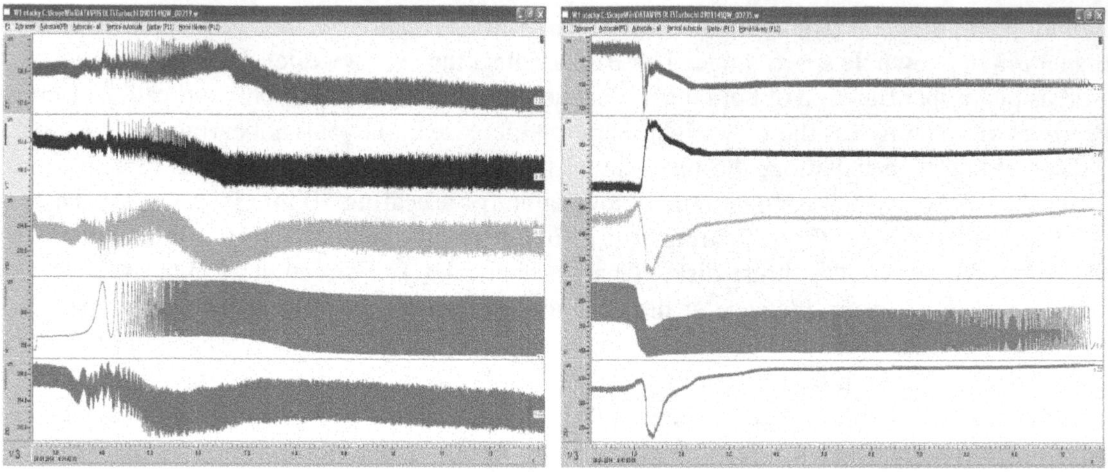

Fig. 23. Run up to 23 000 rpm and run-down from 65 000 rpm

ples of vibration signals are presented in Fig. 23. The sequence of vibration signals is the same as in the previous tests, with the exception of axial direction, which is now the 2nd one from below. The axial signal is distorted by the damage of measured surface during disassembly. There is an eccentric hole, which "provides" virtual axial vibration with amplitude of about 70 μm. It is practically impossible to assess real amplitude of axial vibrations, but axial shifts of the rotor can be still observed. Fig. 23 shows rotor run-up from stand-still to about 23 000 rpm with axial shift around 40 μm and relatively big shifts in all radial directions. The right part of Fig. 23 presents rotor run-down from maximum speed to practical stand-still. It can be seen that the change of axial force direction takes place between 59 000 rpm and 48 000 rpm. However, no change of axial force direction was encountered with increasing speed.

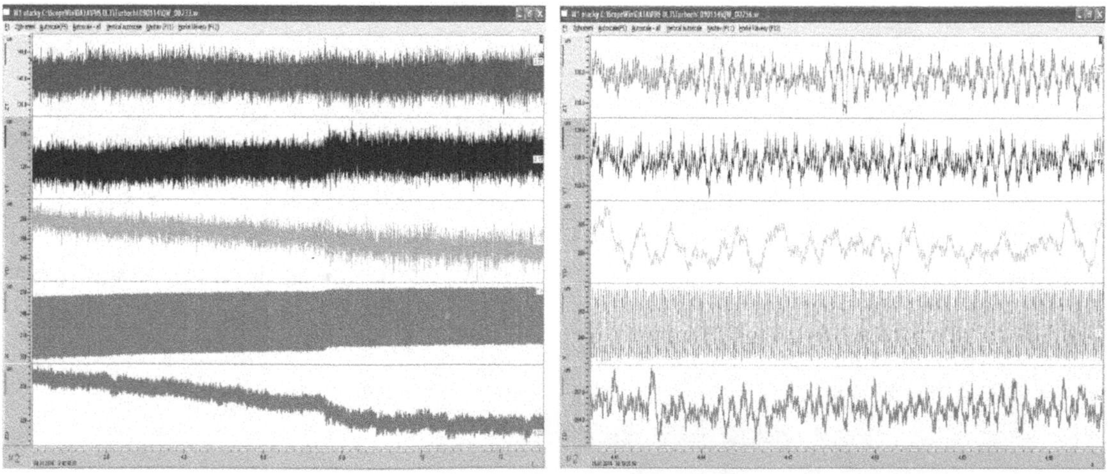

Fig. 24. Increasing speed from 57 500 to 60 200 rpm; Stable maximum speed of 65 000 rpm

Due to labyrinth seal, the axial force was significantly reduced; it still increases with increasing speed, as evident from the left part of Fig. 24 (the right part of axial force signal is higher than the left one – the rotor shifted to the blower side). However, the maximum calculated value of axial load was about 235 N; with this load value, the main thrust bearing works with a film thickness of 5 μm. The right part of Fig. 24 shows vibration signals at the stable maximum speed of about 65 000 rpm. The peak to peak vibration amplitudes of all signals in radial direction did not exceed 5 μm. The RMS values of vibration amplitudes in radial direction in dependence on speed are presented in Fig. 25. Sensor at the blower side in horizontal direction (dotted line in Fig. 25) was damaged during the rotor failure in the 1st test and its characteristics cannot be considered as quite correct. The other 3 sensors show RMS values of vibration amplitude lower than 1.5 μm in the whole speed range, i.e., extremely small values for such a high speed.

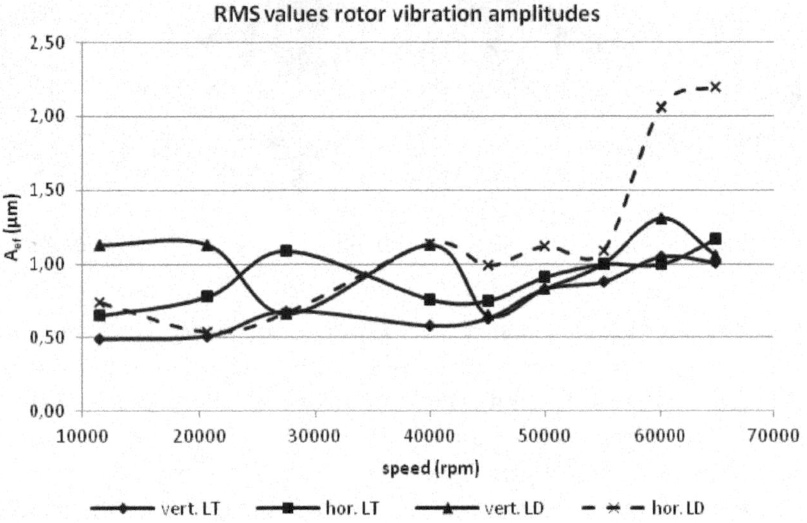

Fig. 25. RMS values of rotor vibration amplitude in radial direction vs. speed

5. Final tests and machine disassembly

Final tests consisted of run-up and run-down cycles and of cycling from minimum speed to maximum speed. In the 1st part of the test, the machine completed 1 000 cycles from standstill to the maximum speed of 60 000 rpm with a 1 minute-long run followed by a run-down to stand still. The 2nd part of the test consisted of 500 cycles of 5 minute run at 30 000 rpm, increasing speed to 60 000 rpm, 5 minute run at this speed, and again decreasing speed to 30 000 rpm etc. After finishing the 2nd test, the machine was disassembled. Two important facts were found out when examining the bearing parts:

1. A small amount of oil from the system of pressurized air driving the machine penetrated into main thrust bearing, as documented in Fig. 26. The oil mixed with the graphite particles loosened from the bearing lining significantly increased the starting torque during the rotor run-up. The presence of oil at the graphite lining is by no means desirable and does not reduce the wear of sliding surfaces during the run-up and run-down; it has rather the opposite effect.

2. Some of the nuts securing the pad position and by that also basic bearing clearance were completely or partly loosened. Pertinent pads were then pressed by their elastic supports to the shaft so that the bearing support lost its biggest advantage in comparison with the foil bearings, in the form of a low starting torque. On the other hand, it showed that the tilting pad bearings with elastically supported pads can operate in the same way as the foil bearing, i.e., without basic bearing clearance. Similarly as in the foil bearings, the pads are separated from the shaft surface and a full gas film is generated when the rotor achieves certain speed.

Fig. 26. Main thrust bearing and thrust disk polluted by the mixture of oil and graphite particles

It is quite natural that during the start up with traces of oil in the thrust bearing gap, the rotor exhibited high level of vibration before a full air film was established. It is documented in the left part of Fig. 27, showing the run-up from standstill to about 20 000 rpm. The vibration amplitudes around 200 μm p-p in radial direction (practically in the range of the whole bearing clearance including the clearance of pads in bearing casing) are apparently caused by the contact of sliding surfaces in the auxiliary thrust bearing, which is preloaded by the main thrust bearing operating in the "hydrodynamic" regime (i.e., with oil as a process medium). As soon as the oil is thrown out of the thrust disk and a complete gas film is established in the auxiliary thrust bearing, the vibration level drops to the usual level of around 4 μm p-p. With a further speed increase, there is no increase in vibration level, only shifts in the radial direction as the rotor aligned with the main thrust bearing sliding surface. Almost steady shift in the axial direction to the turbine impeller (last but one signal) indicates growing axial force with increasing speed.

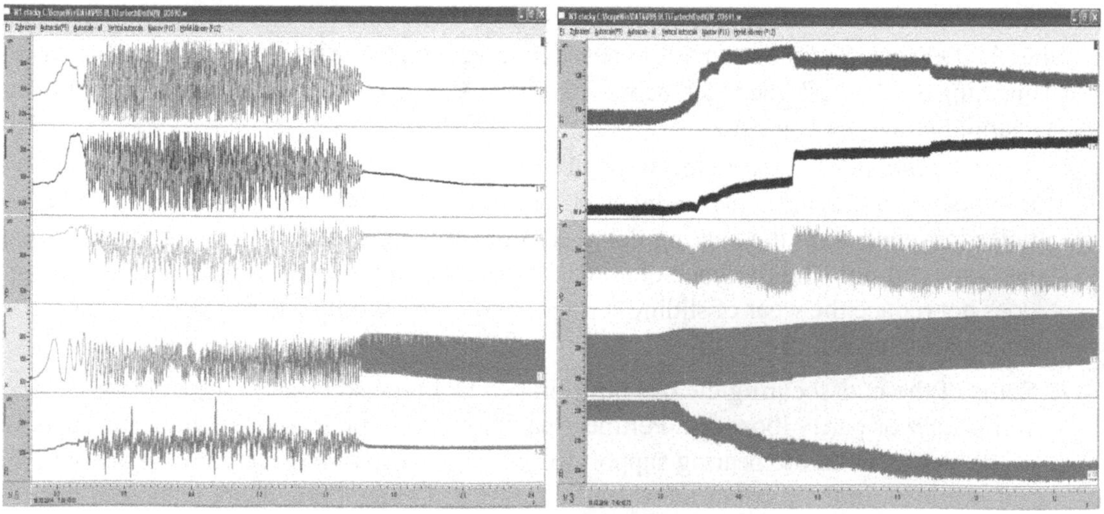

Fig. 27. Run-up from standstill to about 20 000 rpm (left); Speed increase from about 20 000 to 50 300 rpm (right)

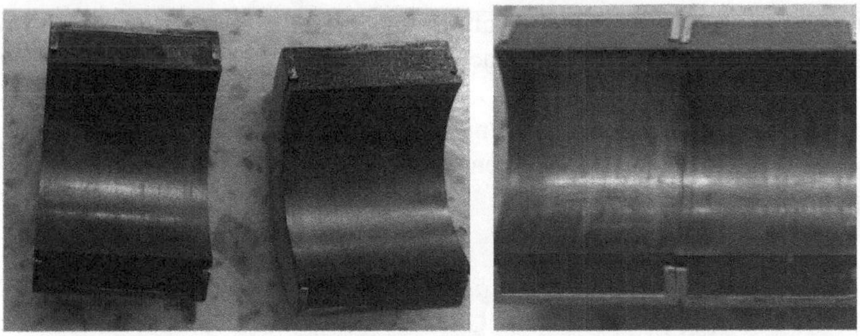

Fig. 28. Bearing pads at the turbine side (left) and blower side (right) after disassembly

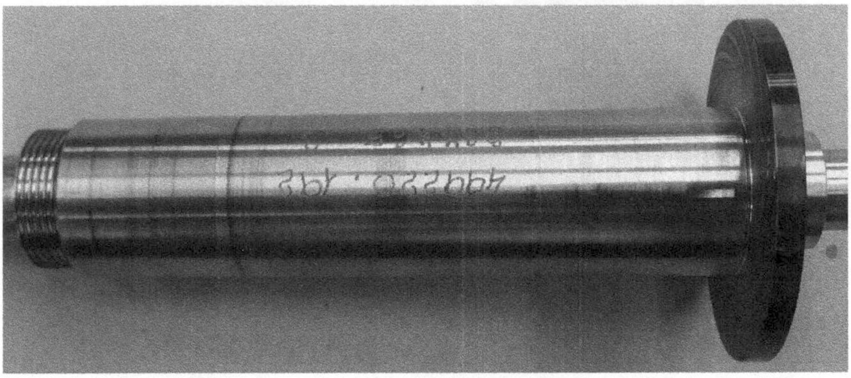

Fig. 29. Cleaned rotor

As can be seen from Fig. 28, the bearing pads at the turbine side (next to the thrust bearing) were also somewhat polluted by oil in contrast to the pads from the bearing at the blower side, which stayed quite clean. After shaft cleaning, it is evident that the sliding surfaces are not damaged and neither on the thrust disk and neither on the shaft at the journal bearing locations (see Fig. 29). Traces at the outer diameter of the thrust disk indicate that there was a slight and very short contact of sliding surfaces, but the damage is very small and the rotor and bearings are able to operate further quite satisfactorily.

6. Conclusions

An air cycle machine (ACM) with rolling bearing rotor support was reconstructed and used as a test stand for the verification of air bearing properties. The 1st part of the test proved operability of the air bearing support, consisting of elastically supported tilting pad journal bearings and spiral groove thrust bearings, up to maximum speed of 65 000 rpm. The failure occurred due to excessive axial force, which exceeded the nominal value for thrust bearing design almost three times.

Installation of a labyrinth seal behind the turbine impeller proved effective in reducing the axial load to its maximum value about 235 N, with which main thrust bearing works with film thickness about 5 μm. Up to the maximum speed of 65 000 rpm, the rotor runs with RMS values of vibration amplitudes up to 1.5 μm, which corresponds to peak-to-peak values lower than 4.5 μm. Such a small values of rotor vibration confirm correct function of bearing support and a very good level of rotor balancing.

The aerodynamic bearing support operability and reliability was further confirmed by a serie of 1 000 starts and stops and 500 cycles of acceleration from 30 000 rpm to 60 000 rpm and back to 30 000 rpm.

A smaller ACM with maximum operating speed of 95 000 rpm was designed and is prepared for manufacture (Fig. 30), as machine dimensions and weight are the most important parameters for flight applications.

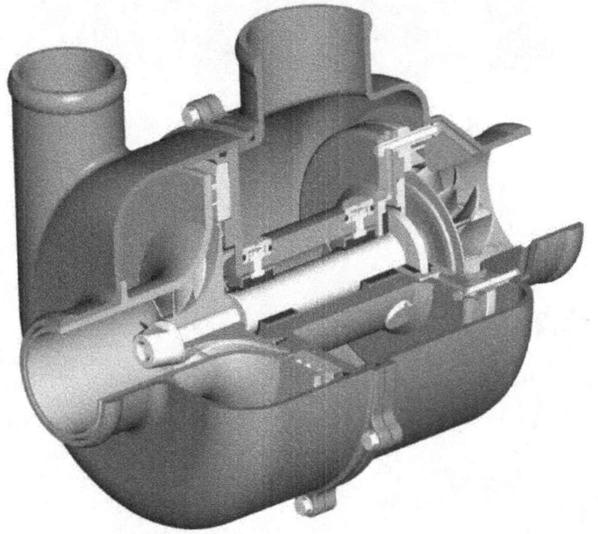

Fig. 30. Design of a smaller ACM variant

Acknowledgements

This work was supported by the Technology Agency of the Czech Republic under project PID TA02011295 "Verification of air bearing technology".

References

[1] Agrawal, L., Foil air/gas bearing technology — an overview, ASME publication 97-GT-347.

[2] Howard, S. A., Bruckner, R. J., DellaCorte, CH., Radil, K. C., Gas foil bearing technology advancement for Brayton cycle turbines, NASA/TM-2007-214470.

[3] Howard, S. A., Bruckner, R. J., Radil, K. C., Advancement towards oil-free rotorcraft propulsion, NASA/TM-2010-216094.

[4] Šimek, J., Application of a new type of aerodynamic bearing in power gyroscope, Engineering Mechanics 19 (5) (2012) 359–368.

[5] Šimek, J., Design of aerodynamic bearing support of the test stand, Technical report TECHLAB No. 12–417, 2012. (in Czech)

[6] Šimek, J., Design of the test stand for verification of air bearing technology, Technical report TECHLAB No. 12–420, 2012. (in Czech)

[7] Šimek, J., Experimental verification of air bearing properties, Technical report TECHLAB No. 13–413, 2013. (in Czech)

[8] Šimek, J., Experimental verification of air bearing properties. Phase 2 — after reduction of axial force, Technical report TECHLAB No. 14–401, 2014. (in Czech)

Application of the Monte Carlo method for investigation of dynamical parameters of rotors supported by magnetorheological squeeze film damping devices

J. Zapoměl[a,*], P. Ferfecki[a], J. Kozánek[a]

[a] *Department of Dynamics and Vibrations, Institute of Thermomechanics, Department of Mechanics, Dolejškova 1402/5, 182 00 Praha 8, Czech Republic*

Abstract

A flexible suspension with damping devices is an efficient technological tool for reducing forces transmitted between the rotor and its frame. To achieve optimum performance of the damping elements, their damping effect must be adaptable to the current operating conditions. In practical rotordynamic applications this is offered by magnetorheological squeeze film dampers. Some of parameters, which determine behaviour of rotors, may have uncertain character. Then a probabilistic approach is needed for analysis of such systems. In this paper there is investigated the vibration amplitude of a rigid rotor damped by two magnetorheological squeeze film dampers and magnitude of the force transmitted to the stationary part during the steady state operating regime. The uncertain parameters of the studied system are the rotor unbalance and speed of its rotation. The Monte Carlo method was employed for this analysis.

Keywords: uncertain parameters of rigid rotors, magnetorheological dampers, force transmission, Monte Carlo method

1. Introduction

Unbalance of rotating machines is the main source of increase of time variable forces transmitted to the rotor frame. A frequently used technological solution which makes it possible to reduce their magnitude is represented by application of a flexible suspension with added damping elements. Both the theory and practical experience show that to achieve optimum performance of the damping devices their damping effect of must be controllable. This is offered by semiactive magnetorheological squeeze film dampers.

Values of some geometric, operational or technological parameters of rotating machines may be uncertain or their magnitudes may slightly vary during the operating regime. Then the approaches based on stochastic principles should be utilized for their investigation. The worst scenario method [3], the theory of fuzzy sets and interval mathematics [1,4,6,11], the probability methods [2] and variational procedures belong to them. The worst scenario approach is based on searching for the worst combination of the input uncertain parameters. Its drawback is that it does not take into account probability of occurrence of such a case. The theory, confirmed by practical experience, shows that the fuzzy set approach overestimates influence of uncertain parameters on behaviour of mechanical systems. Therefore, some correction procedure is needed to minimize this undesirable effect. The probabilistic method of the Monte Carlo type requires

*Corresponding author. e-mail: zapomel@it.cas.cz.

to perform a large number of repeated computational simulations for randomly generated values of uncertain quantities. Application of this method seems to be suitable for investigation of practical problems as it is not complicated and may bring reliable and accurate predictions.

If there is specified an allowed value of some parameter (e.g. the vibrations amplitude, magnitude of the transmitted force, state of stress, etc.), then the Monte Carlo method enables to determine the system reliability ψ

$$\psi = 1 - \frac{N_{DIS}}{N_{SAT} + N_{DIS}}, \tag{1}$$

where N_{SAT}, N_{DIS} are the numbers of simulations when the required condition is satisfied and is not satisfied respectively.

The force transmitted to the rotor casing during the steady state operating regime depends on unbalance and on speed of the rotor rotation. In this paper, values of both these parameters are considered to be uncertain and their influence on the force transmission between the rotating and stationary parts and on amplitude of the rotor vibration is the subject of investigations.

2. The investigated rotor system

The rotor of the studied rotating machine (Fig. 1) is rigid and consists of a shaft and of one disc. Two magnetorheological squeeze film dampers are used to mount the rotor with the bearing housings flexibly coupled with the rotor frame. The whole system can be considered as symmetric relative to the disc middle plane perpendicular to the shaft axis.

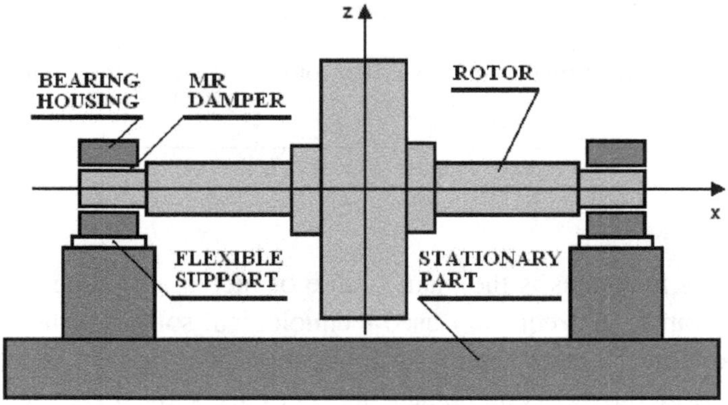

Fig. 1. The investigated rotor system

The rotor turns at constant angular speed, is loaded by its weight and excited by the disc unbalance. The squirrel springs are prestressed to eliminate their deflection (relative to the bearing housings) caused by the rotor weight.

The principal parts of each magnetorheological damper (Fig. 2) are two rings between which there is a layer of magnetorheological fluid. The outer ring is fixed to the damper's body. The inner ring is coupled with the shaft by a rolling element bearing and with the stationary part by a squirrel spring. The damping effect is produced by squeezing the lubricating layer produced by the rotor lateral vibrations. The damper is equipped with an electric coil generating magnetic flux passing through the layer of the magnetorheological liquid. As resistance against its flow depends on magnetic induction, the change of the current can be used to control the damping force.

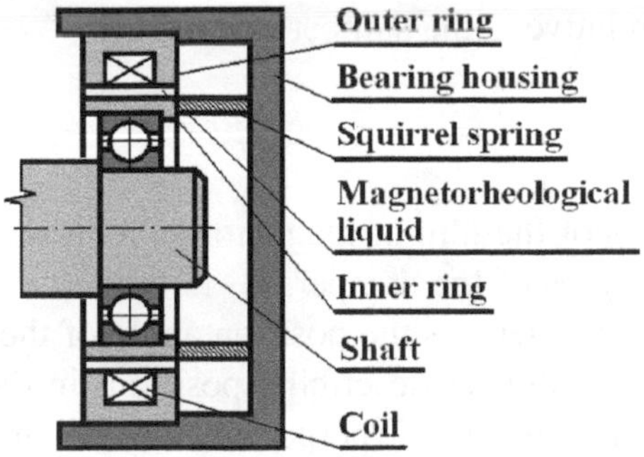

Fig. 2. Magnetorheological damper

The mathematical model of the squeeze film magnetorheological damper is based on assumptions of the classical theory of lubrication except those for the lubricant. As the magnetorheological oils are liquids with the yielding shear stress, the lubricant is represented by bilinear theoretical material whose properties depend on magnetic induction. In addition, it is assumed that the geometric and design parameters of the damper enable to consider it as short.

In the computational model, the rotor, the bearing housings and the frame are considered as absolutely rigid and the magnetorheological squeeze film dampers are represented by springs and force couplings.

Taking into account the system symmetry, the vibration of the rotating machine is governed by a set of four nonlinear equations of motion

$$m_R\ddot{y}_R + b_P\dot{y}_R + 2k_D y_R - 2k_D y_B = m_R e_T \omega^2 \cos(\omega t + \psi_R) + 2F_{mry} + 2F_{psy}, \quad (2)$$

$$m_R\ddot{z}_R + b_P\dot{z}_R + 2k_D z_R - 2k_D z_B = m_R e_T \omega^2 \sin(\omega t + \psi_R) + 2F_{mrz} + 2F_{psz} - m_R g, \quad (3)$$

$$m_B\ddot{y}_B + b_B\dot{y}_B - k_D y_R + (k_D + k_B)\, y_B = -F_{mry} - F_{psy}, \quad (4)$$

$$m_B\ddot{z}_B + b_B\dot{z}_B - k_D z_R + (k_D + k_B)\, z_B = -F_{mrz} - F_{psz} - m_B g, \quad (5)$$

where m_R, m_B are masses of the rotor and the bearing housing, b_P is the coefficient of the rotor external damping, k_D is the squirrel spring stiffnesses, k_B, b_B are the stiffness and damping coefficients of the bearing housing support, e_T is eccentricity of the rotor centre of gravity, g is the gravity acceleration, t is the time, y_R, z_R, y_B, z_B are displacements of the rotor and bearing housings centres in the horizontal and vertical directions, ω is angular speed of the rotor rotation, ψ_R is the phase shift, F_{mry}, F_{mrz}, F_{psy}, F_{psz} are the y and z components of the magnetorheological damping and prestress forces and (.), (. .) denote the first and second derivatives with respect to time.

To obtain the steady state solution of the equations of motion, a trigonometric collocation method was applied.

3. Determination of the magnetorheological damping forces

Magnetorheological fluids belong to the class of liquids with yielding shear stress. This implies that in the area where the shear stress between two neighbouring layers is less than critical one a core is established. In this region the magnetorheological oils behave almost like solid bodies.

Further attention is focused only on dampers whose geometric and design parameters make it possible to treat them as short. The thickness of the oil film depends on positions of the centres of the damper rings relative to the damper body [5,7]

$$h = c - e_H \cos(\varphi - \gamma), \tag{6}$$

where h denotes the thickness of the film of magnetorheological oil, c is the width of the gap between the inner and outer rings of the damper, e_H is the rotor journal eccentricity, φ is the circumferential coordinate and γ denotes the position angle of the line of centres (Fig. 3), X, Y, and Z read for the local coordinates describing positions in the magnetorheological oil in the circumferential X, radial Y and axial Z (perpendicular to X and Y) directions respectively. Axes y and z define directions of the fixed frame of reference in which the rotor vibration is investigated.

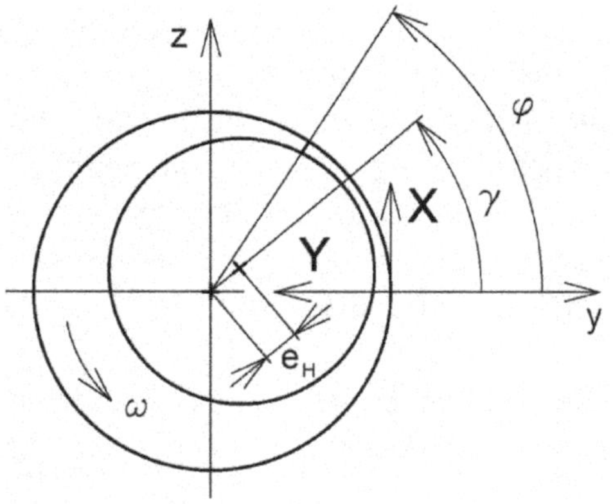

Fig. 3. Coordinate systems of the damper

Character of the flow in the lubricating film is depicted in Fig. 4. The pressure distribution in the layer of the full lubricating film is governed by the Reynolds equation adapted for bilinear material. Its derivation starts from the equation of continuity, the equation of equilibrium of the infinitesimal element specified in the lubricating layer and from the constitutive relationship of the magnetorheological oil [10].

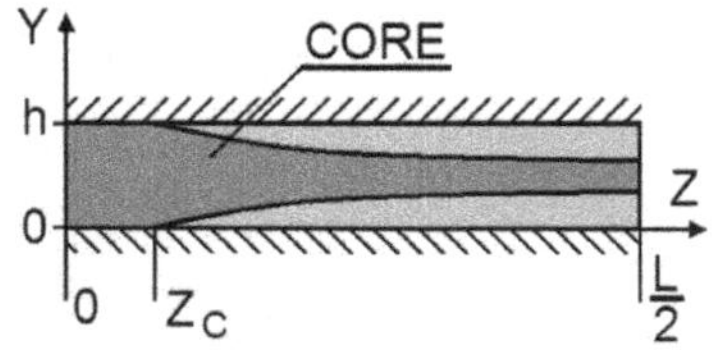

Fig. 4. The core formation in the lubricating film

After performing some manipulations the resulting relations for the pressure distribution p in the oil film read

$$p = \frac{6\eta_C \dot{h}}{h^3}\left(Z^2 - Z_C^2\right) + p_C \qquad \text{for} \qquad 0 \le Z < Z_C,\ Z_C < \frac{L}{2}, \tag{7}$$

$$a_3 p'^3 + a_2 p'^2 + a_1 p' + a_0 = 0 \qquad \text{for} \qquad Z_C \le Z \le \frac{L}{2}, \tag{8}$$

where

$$a_3 = 0.5\, h^3 \eta_C, \tag{9}$$

$$a_2 = -3h^2 \left(\tau_C - \tau_y\right) \eta_C - \left(12\dot{h}Z + C\right) \eta\, \eta_C, \tag{10}$$

$$a_1 = -6h\tau_C^2 \eta_C, \tag{11}$$

$$a_0 = 4\tau_C^3 \eta_C - 12\tau_y \tau_C^2 \eta_C - 8\tau_C^3 \eta, \tag{12}$$

or if only the core is formed in the gap

$$p = -\frac{6\eta_C \dot{h}}{h^3}\left(\frac{L^2}{4} - Z^2\right) + p_A \qquad \text{for} \qquad 0 \leq Z < \frac{L}{2}, \ \frac{L}{2} \leq Z_C. \tag{13}$$

The derived relations for the pressure distribution (7), (8) and (13) are valid only for the dampers symmetric to their middle plane perpendicular to the shaft axis. τ_y is the yielding shear stress, τ_C is the shear stress at the core border, η is the dynamic viscosity of the oil if no magnetic field is applied, η_C is viscosity of the oil in the core, Z is the axial coordinate, Z_C is the axial coordinate of the location where the core border touches the lower and upper walls of the gap and C in (10) is the integration constant. Its value is determined from the boundary condition expressing that for $Z = Z_C$ the pressure gradient is $p' = p'_C$. L is the axial length of the damper and p_A denotes pressure in the ambient space (atmospheric pressure).

The relation between the shear stresses τ_C and τ_y and the axial coordinate Z_C are given by the following expressions [9]

$$\tau_C = \frac{\eta_C}{\eta_C - \eta}\tau_y, \tag{14}$$

$$Z_C = -\frac{\tau_C h^2}{6\eta_C \dot{h}}. \tag{15}$$

Relation (8) represents a cubic algebraic equation for unknown value of the pressure gradient in the axial direction p'. The root which has the physical meaning must satisfy two conditions:

- is real (not complex),
- $p' < -\frac{2\tau_C}{h}$.

When the pressure gradient p' is known, the pressure profile in the oil film is consequently calculated by the integration

$$p = \int_{\frac{L}{2}}^{Z} p'\, \mathrm{d}Z + p_A. \tag{16}$$

Relations (7), (8) and (13) are valid only for the case when $0 \leq Z$ and $\dot{h} < 0$.

In cavitated areas (in areas where the thickness of the oil film rises with time) pressure of the medium is assumed to be constant and equal to the pressure in the ambient space. The nonlinear damping forces are calculated by integration of the pressure distribution in the oil film (taking into account the different pressure distributions in non-cavitated and cavitated areas) around the circumference and along the length of the damper

$$F_{mry} = -R \int_{-\frac{L}{2}}^{\frac{L}{2}} \int_0^{2\pi} p_d \cos\varphi \,\mathrm{d}\varphi\, \mathrm{d}Z, \tag{17}$$

$$F_{mrz} = -R \int_{-\frac{L}{2}}^{\frac{L}{2}} \int_0^{2\pi} p_d \sin\varphi \,\mathrm{d}\varphi\, \mathrm{d}Z, \tag{18}$$

where p_d is the pressure distribution in the damper clearance, R is the radius of the damper inner ring and φ is the circumferential coordinate defining positions in the oil film.

Results of theoretical analyses and measurements done by researchers and producers of magnetorheological liquids show that the dependence of the yielding shear stress on magnetic induction can be approximated by a power function

$$\tau_y = k_y B^{n_y}, \tag{19}$$

where k_y and n_y are material constants of the magnetorheological oil (n_y is often equal to 2) and B is magnetic induction.

In the simplest design case when the inner and outer rings of the damper can be considered as a divided core of an electromagnet the relation for the yielding shear stress can be expressed [8]

$$\tau_y = k_C \left(\frac{I}{h}\right)^{n_y}, \tag{20}$$

where I is the electric current and k_C denotes the design coefficient whose value depends on the number of the coil turns, on material parameters of the damper body and on arrangement of its individual parts.

4. Results of the computational simulations

The technological parameters of the investigated rotor system are: mass of the rotor 450 kg, mass of the bearing housing 100 kg, the rotor external damping coefficient 50 N · s/m, the squirrel spring stiffness 2.0 MN/m, the stiffness and damping coefficient of the bearing housing support 50.0 MN/m, 1 000 N · s/m.

The task was to find out if the vibration amplitude and the maximum force transmitted to the rotor frame in the vertical direction were less than the allowed values 100 μm, 3 850 N respectively at the nominal rotor running speed of 170 rad/s.

The unbalance and the rotor angular velocity are uncertain parameters in the investigated case. Therefore, they are given by their mean values and estimated deviations 0.036 ± 0.009 kg·m and 170 ± 6.0 rad/s, respectively. The Monte Carlo method was adapted to perform the analysis. The unbalance probability function is assumed to be uniform because it can take any value from some interval given by accuracy of the balancing procedure. As the speed of the rotor rotation is set by a controller which forces the rotor to achieve the nominal angular velocity, the speed probability function is assumed to have a normal distribution.

In Figs. 5 and 6 there are drawn the frequency response characteristics and dependence of amplitude of the variable component of the force transmitted to the rotor frame in the vertical direction calculated for mean values of the unbalance, nominal angular velocity and magnitudes of the applied current of 0.0, 0.5, 1.0, 1.5 and 2.0 A. The results show that to suppress the vibration amplitude below the allowed value the current must be at least 1.5 A. The orbits of the rotor centre and of the centres of the bearing housings are drawn in Fig. 7 and the corresponding time histories of the force transmitted to the rotor frame in the horizontal and vertical directions can be seen in Fig. 8. Both these results are related to the nominal speed of the rotor rotation.

The results show that the requirements put on performance of the rotating machine without taking into account the system uncertainties are met.

To study influence of the uncertain parameters on behaviour of the investigated rotor, the Monte Carlo method was employed and 200 000 computational simulations were carried out. The probability functions of the rotor unbalance and speed of its rotation are drawn in Figs. 9 and 10.

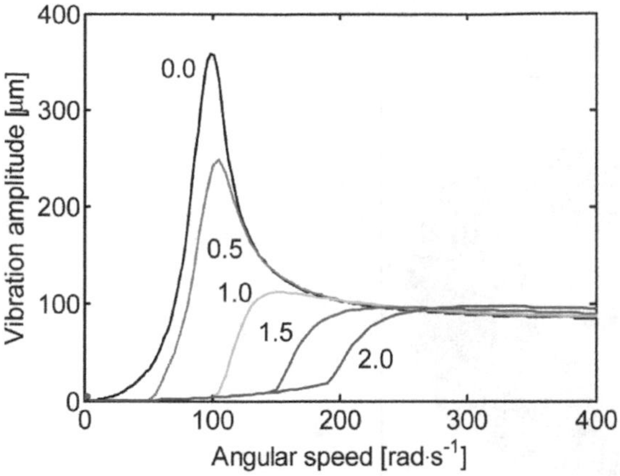

Fig. 5. Frequency response characteristic

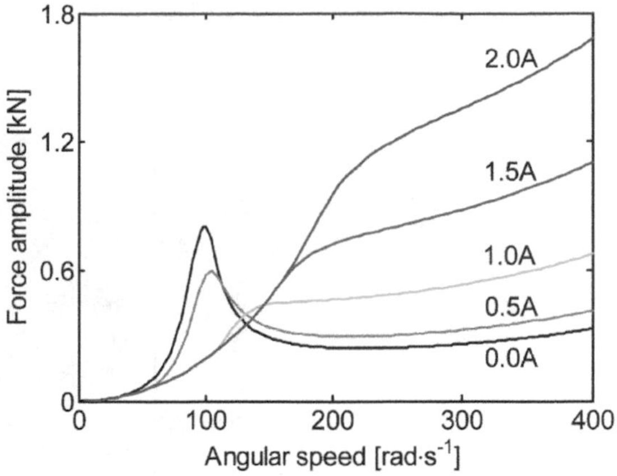

Fig. 6. Force-frequency relationship

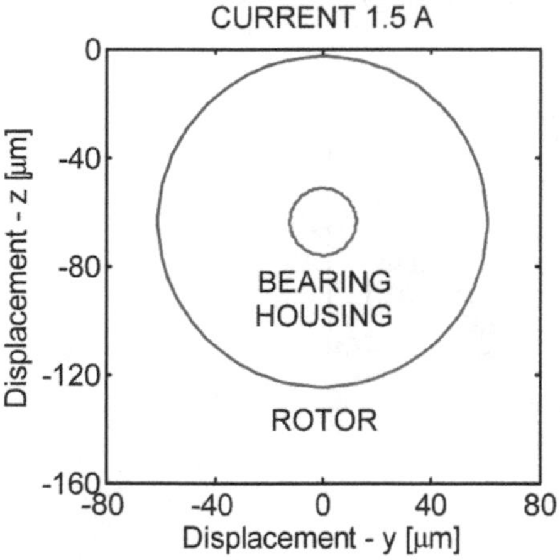

Fig. 7. Rotor and bearing housing centre orbits

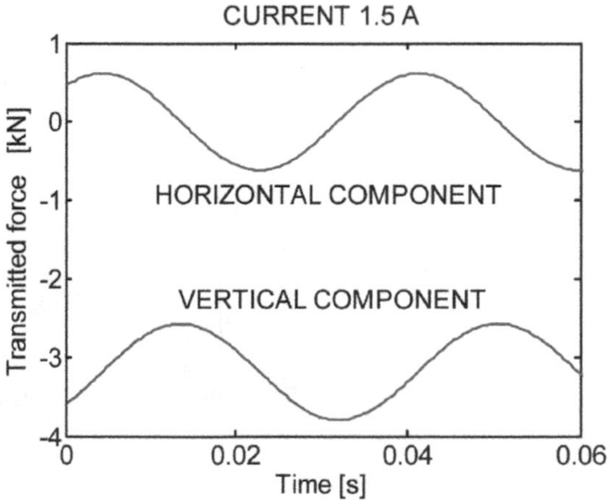

Fig. 8. Transmitted force components (nominal speed)

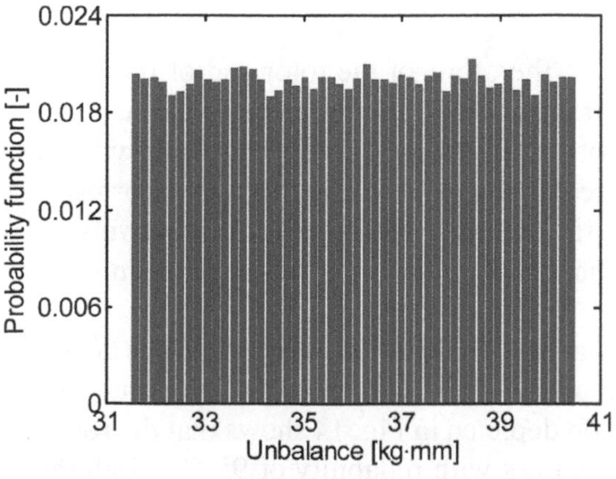

Fig. 9. Probability function – unbalance

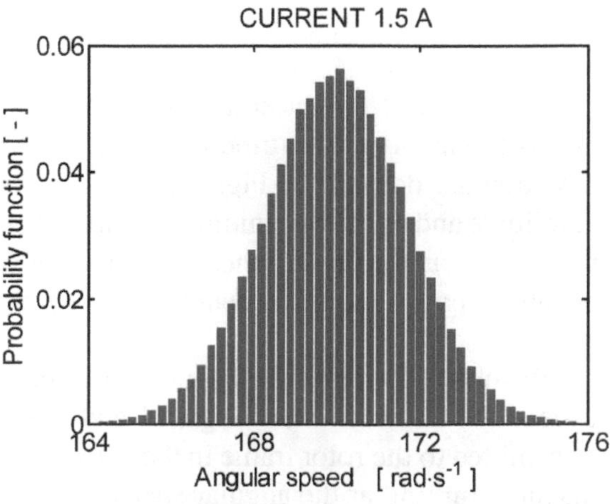

Fig. 10. Probability function – angular speed

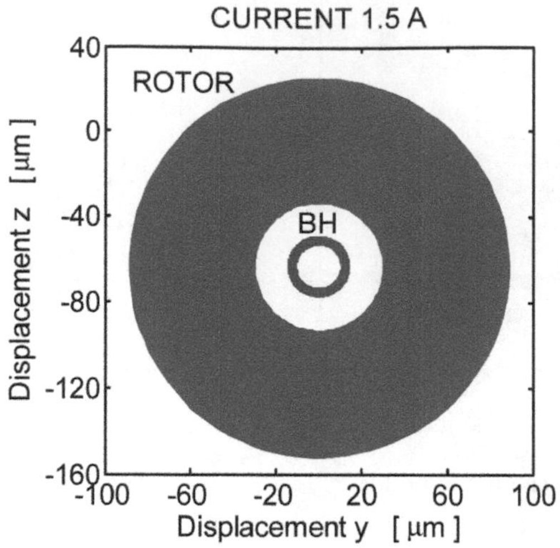

Fig. 11. Rotor orbit (BH-bearing housing)

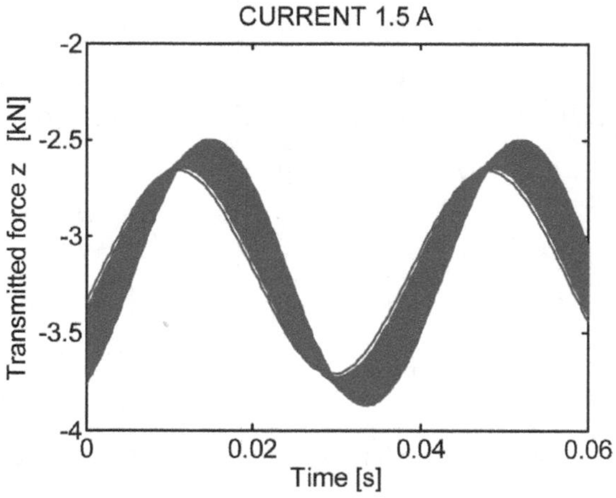

Fig. 12. Maximal transmitted force in the vertical direction

The effect of the uncertain parameters on size of the orbits of the rotor and of the bearing housing centres and the time history of the force transmitted to the rotor frame in the vertical direction are depicted in Figs. 11 and 12. The probability functions of the resulting vibration amplitude and of the maximum transmitted force are drawn in Figs. 13 and 14 which correspond to the estimated values of these quantities of 61 ± 15 μm and $-3\,802 \pm 38$ N respectively. The negative sign of the force denotes that it acts in the direction from the bearing housing to the frame.

Results of the performed simulations confirm that amplitude of the rotor vibrations does not exceed the allowed value (Figs. 11 and 13). The probability function of the maximum force transmitted to the rotor frame in the vertical direction depicted in Fig. 14 shows that the rotating machine turning at the angular speed of 170 rad/s works with reliability of 95 %. Both these results are related to the applied current of 1.5 A.

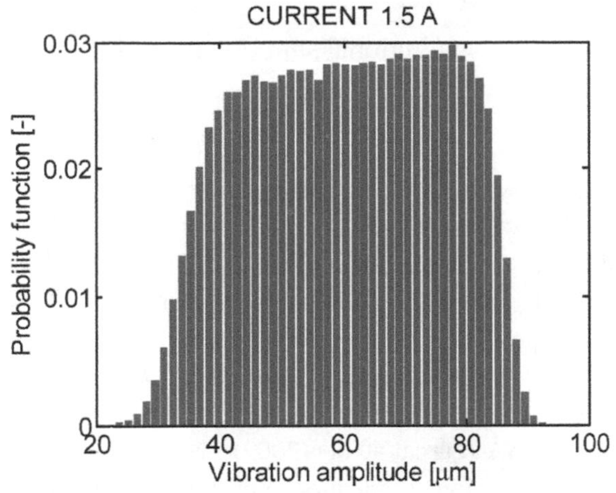

Fig. 13. Probability function – vibration amplitude

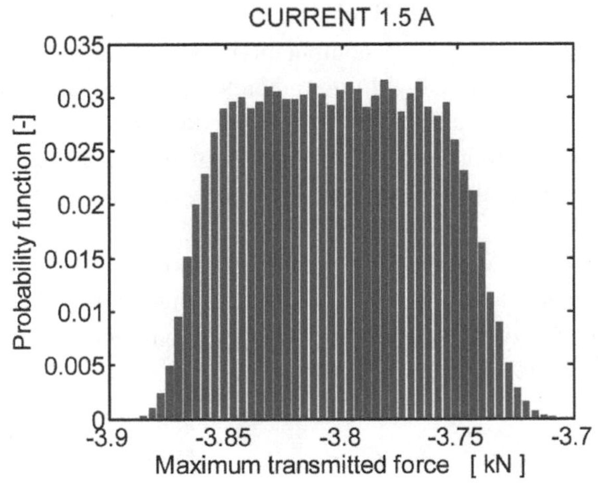

Fig. 14. Probability function – transmitted force

5. Conclusions

The probabilistic Monte Carlo method is a strong tool for dynamical investigation of mechanical systems whose behaviour is effected by geometric, design or operational parameters of uncertain values. The results are expressed in the form of interval numbers and characterized by probabilistic quantities. It follows from the theory, confirmed by experience, that to obtain the reliable results a large number of computational simulations must be performed which makes this method considerably time consuming. Advantage of the Monte Carlo method is that the number of simulations needed for achieving the required accuracy of the results does not depend on the number of uncertain system parametres. Some its modifications (e.g. stratified sampling, importance sampling, antithetic variates) based on reducing the variance of uncertain quantities make it possible to decrease the number of simulations which speeds up the calculations and rises efficiency of the procedure.

Acknowledgements

This work has been done with the institutional support RVO:61388998.

References

[1] Chen, L., Rao, S. S., Fuzzy finite-element approach for the vibration analysis of imprecisely-defined systems, Finite Elements in Analysis and Design 27 (1) (1997) 69–83.

[2] Kalos, M. H., Whitlock, P. A., Monte Carlo methods, Wiley-VCH, 2008.

[3] Liberdová, J., Zapoměl, J., Effect of uncertainty of the support stiffness, damping and unbalance excitation on dynamical properties of the turbine rotor of an aircraft engine, Transactions of the VSB-Technical University of Ostrava 56 (2) (2010) 289–294.

[4] Moens, D., Vandepitte, D., Fuzzy finite element method for frequency response function analysis of uncertain structures, AIAA Journal 40 (1) (2002) 126–136.

[5] Szeri, A. Z., Tribology: Friction, lubrication, and wear, Hemisphere Publishing Corporation, Washington, New York, London, 1980.

[6] Zadeh, L. A., Fuzzy sets, Information and Control 8 (3) (1965) 338–353.

[7] Zapoměl, J., Computer modelling of lateral vibration of rotors supported by hydrodynamical bearings and squeeze film dampers, VSB-Technical University of Ostrava, Ostrava, 2007 (in Czech).

[8] Zapoměl, J., Ferfecki, P., Forte, P., A computational investigation of the steady state vibrations of unbalanced flexibly supported rigid rotors damped by short magnetorheological squeeze film dampers, Journal of Vibration and Acoustics 135 (6) (2013).

[9] Zapoměl, J., Ferfecki, P., Mathematical modelling of magnetorheological oil in short squeeze film dampers for rotordynamic applications, Modelling and Optimization of Physical Systems 12 (2013) 59–64.

[10] Zapoměl, J., Ferfecki, P., Mathematical modelling of magnetorheological oil in squeeze film dampers for rotordynamic applications, Modelling in Engineering 42 (2013).

[11] Zimmermann, H.-J., Fuzzy set theory and its applications, Kluwer Academic Publishers, Boston/Dordrecht/London, 1996.

14

Dry friction damping couple at high frequencies

L. Půst[a,*], L. Pešek[a], J. Košina[a], A. Radolfová[a]

[a] *Institute of Thermomechanics, AS CR, v.v.i., Dolejškova 5, 182 00 Prague, Czech Republic*

Abstract

The contribution deals with the application of dry friction couples for noise and vibration damping at high frequency of several kHz what brings new problems connected with the small amplitudes of relative slipping motion of contact surfaces. The most important information from the experimental results is knowledge that the value of evaluated friction coefficient can have different physical sense according to the magnitude of excitation force and to the frequency of applied vibrations. If amplitudes of motion are very small, then the external harmonic force produces only elastic micro-deformations of contacting bodies, where no slip occurs and then the traction contact force is proportional only to elastic deformation of the sample.

Keywords: dry friction, damping, high frequencies, experimental data

1. Introduction

Undesirable vibrations and noise of machines and means of transport have very often high frequency, which must be quenched by means of introducing some kind of structural damping, very often based on dry friction connection between vibrating elements. Characteristics of dry friction have been both experimentally and analytically investigated for more than two centuries, but mainly for comparatively low frequencies and also at sufficiently large amplitudes. Let us mention general friction problems in [1,3,8,9], application to turbine blade dampers [2,7] etc.

The application of dry friction couples for noise and vibration damping at high frequency of several kHz brings new problems connected with the small amplitudes of relative slipping motion of contact surfaces [4–6]. The tangential forces in these surfaces produce tangential micro-deformations that are of the same level as slip motions, and in some cases these elastic micro-deformations are the only response on the external harmonic excitation. Such kinds of contact couples do not work as energy damage, but as an elastic spring without any loss of energy, i.e., without any damping properties.

Topic of this contribution is an experimental analysis of these friction contacts.

2. Experimental sets for dry friction characteristic measurements

The tribologic literature oriented on the description of experimental investigation of friction properties at vibrations contains information usually only up to 50 Hz, exceptionally up to 200 Hz. We have found no information about properties of experimental research of contact friction behavior in the range of several kHz. Therefore, we developed a simple experimental set, which would be suitable for ascertaining with sufficient exactness in measurements of

*Corresponding author. e-mail: pust@it.cas.cz.

Fig. 1. Experimental set for dry friction measurements

friction forces F_t, normal trust forces F_N and relative tangential motion and velocity in friction surface, Fig. 1.

The request of measurements of friction properties up to 3 kHz we tried to fulfil by reinforcement of some parts of experimental set and also by theirs lightness, but in spite of these reconstructions, several eigenfrequencies of experimental rig are under the upper frequency boundary 3 kHz. If they are under the lower boundary frequency 1 kHz, these frequencies did not disturb fortunately the measured signals. However, there appears another problem.

Dynamical behavior of dry friction at vibrations is much complicated than of friction with constant relative velocity between contacting surfaces. Friction velocity changes its sign twice in a period from the positive value to the negative one and vice versa and the friction force changes its sign jump-wise as well. Harmonic decomposition of this force time history contains infinite set of higher harmonic components, some of which can resonantly coincide with some of experimental set eigenfrequencies and can essentially disturb records of force and motion. The low pass filter for cut off these higher disturbing frequencies cannot be applied because it cut of not only disturbing frequencies but also the higher harmonic components of the basic jump-wise force signal. One way is to increase essentially the frequency spectrum of the whole experimental rig by decreasing its dimensions at holding all physical similarity laws. The second way is the developing of a new method for evaluation of records, which enable to gain the needed values with the good technical accuracy. This problem has been solved in paper [7] and we use it also for evaluation of the following presented results.

Measurement and excitation apparatuses completed the experimental set in Fig. 1. The accelerometer B&K 4374 measures relative motion in the friction surface between contacting bodies. Signal from this accelerometer was after double integration as a voltage proportional to the displacement $x(t)$ led on one input of digital oscilloscope YOKOGAWA. Friction force F_t was picked up by the force transducer B&K 8200 and its signal was led into Conditioning Amplifier B&K 2626 and then on the second input of the digital oscilloscope.

The moving coil exciter LDS V400 fed by digital HP 332A Synthesized generator through PA100E power amplifier was applied for vibration excitations. Ampere meter measured the feeding current. Digital generator enables measurements in linear sweep regime prescribed in given frequency range (e.g., 1–3 kHz) and in prescribed time interval (e.g., 5 s).

Time histories of friction force $F_t(t)$ and of oscillating relative motion $x(t)$ were during measurements recorded in the memory and on the screen of digital oscilloscope. At the end of measurements, these records were saved on Card Reader and off-line evaluated by means of system Xviewer.

A record of one measurement length 5 s at interval 500 Hz contains more than 10 000 periods of friction process, prepared in the memory for evaluation. However, for our purposes, it suffices to elaborate only 10 selected events approx. at 50 Hz.

In the presented examples, the sweep measurements were repeated always at two normal loads $F_N = 11.56, 23.12$ N, which act on the length of 5 cm of the line contact surface. Contact pressures are $p = 231.22$ and 462.44 N/m. The entire frequency interval 1–3 Hz was divided for simple evaluation into four subintervals 1–1.5, 1.5–2, 2–2.5, 2.5–3 kHz. Example of a record at the lowest frequency boundary 2 050 Hz and amplitude 6 μm is in Fig. 2.

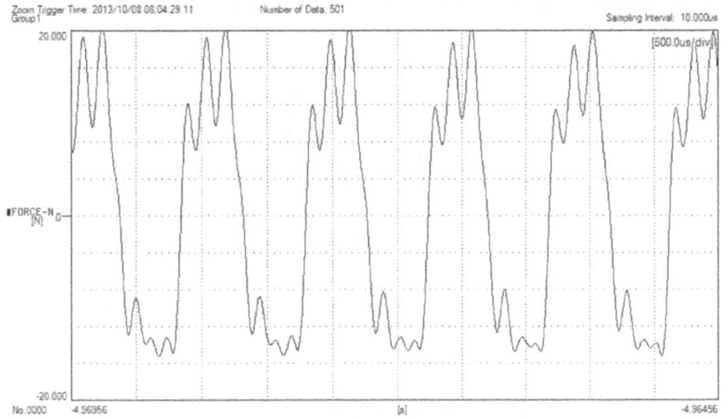

Fig. 2. Distortion of friction force record

Jumps of friction force in the extreme positions are recorded as nearly vertical lines; the friction forces between these jumps are distorted by the higher resonance oscillations of the subsystem connected the friction surface with force meter B&K 8200. These friction forces can be ascertained only as an average value of the waved upper and lower parts of records. The smaller are relative vibrations amplitudes, the more pronounced is the influence of tangential elastic micro-deformations and the slope of jump lines are smaller.

3. Examples of friction characteristics

3.1. Records of low frequency events

Let us see first of all on the records of friction forces at very small frequency, e.g., at 12 Hz, when the inertia forces and dynamic compliances of contacting bodies are negligible. The time history shown in Fig. 3 presents the harmonic course of relative slip motion in contact surface and

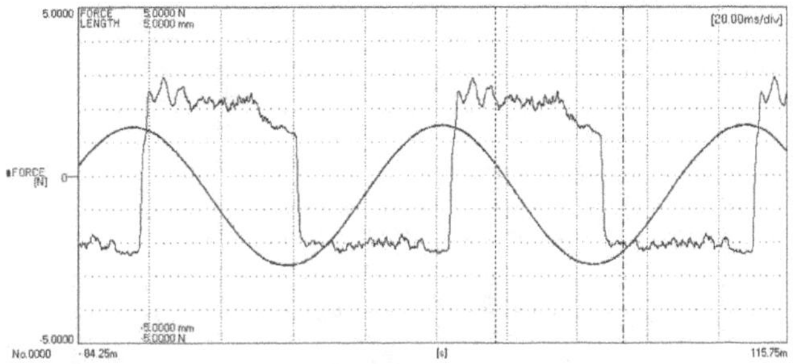

Fig. 3. Harmonic motion and friction force at 12 Hz

roughly rectangular form of friction force. The jumps between positive and negative values of friction force in the return points are nearly vertical, but the sliding parts of the force history are recorded by very noisy horizontal curves. This is caused by the micro-unevenness of contacting surfaces and theirs abrasive wear.

The mapping of the same friction process in the force-displacement coordinates is shown in Fig. 4. The closed rectangular form proves stationary properties of friction process. The very small elastic tangential micro-deformations in reverse points are very small in comparison to the displacement amplitude (approx. 2 mm) and cause only very gentles inclinations of vertical jump lines. The area of this hysteretic loop is proportional to the friction energy loss during one period of vibration.

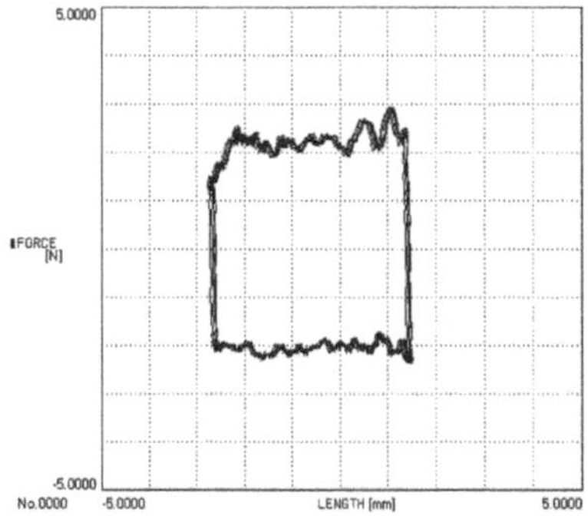

Fig. 4. Hysteretic loop at 12 Hz

3.2. Records of high frequency events

The random disturbances caused by the micro-unevenness of contacting surfaces are contained also in the friction processes realized at much higher frequency vibrations, in the range of one or more kHz, but the more important record's distortion is usually the influence of eigenfrequencies of the experimental rig, excited by the higher harmonic components of the jumps in the rectangular course of friction force, as has been mentioned in the previous chapter. However by means of special evaluations procedure and careful readings it is possible to ascertain the friction coefficient also in such cases, but the accuracy of these values is worse, sometimes with error up to 10 % or more.

Let us see on the results measured in the frequency range 1 kHz up to 3 kHz. Measured friction couple consists of a steel cylinder ⊘ 14 mm (stainless steel, CSN 17240), length 25 mm contacting on the surface straight line of steel cylindrical groove (ER7 – cat. 2), Fig. 5. Surface finish of both elements was Ra 3.2. Specific pressure on 1 m length of contact line was $p =$ 231.24 N/m.

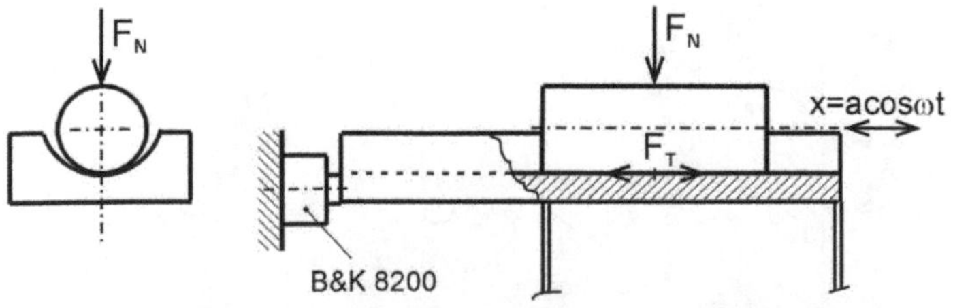

Fig. 5. Friction couple for high frequency measurements

Dry friction coefficient f at vibrations in the lowest range 1–1.5 kHz is shown in Fig. 6, where values at three groups of amplitudes are plotted. Vibrating motion of cylinder was realized by an electro-dynamic vibrator feed by alternating currents $I = 2, 4, 6$ A (in Fig. 6 labeled by □, △, ◇, respectively). Corresponding amplitudes in the given frequency range are plotted in Fig. 7.

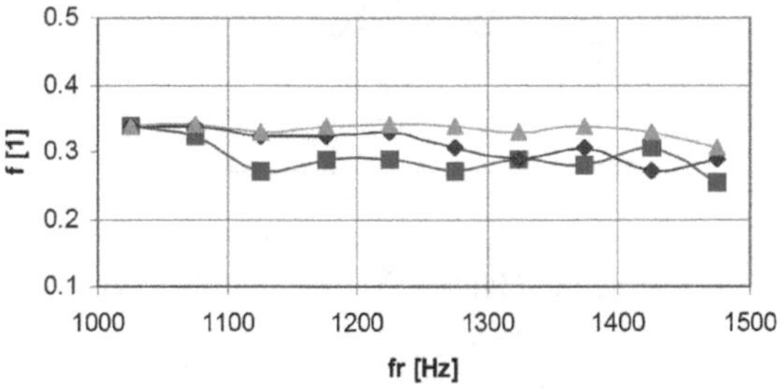

Fig. 6. Dry friction coefficient f in frequency range 1–1.5 kHz, contact pressure $p = 231.24$ N/m

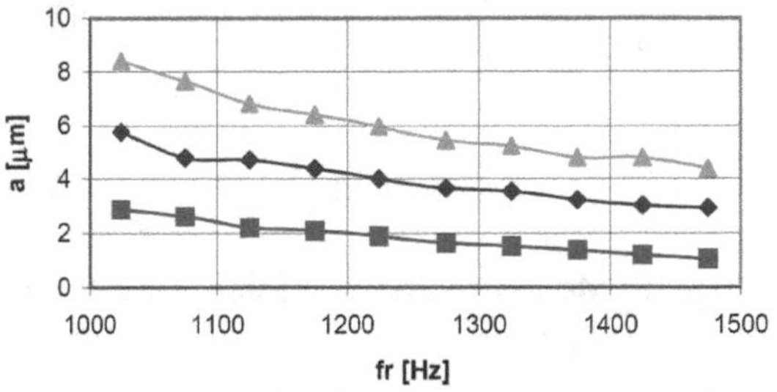

Fig. 7. Amplitudes of relative vibrations, $I = 2, 4, 6$ A, (□, △, ◇)

From the last two figures it is evident that the dry friction coefficient in the whole range of amplitudes up to 9 μm is constant and equals $f = 0.3$. The measured coefficients of dry friction at the twice higher pressure $p = 462.48$ N/m and at the same feeding currents $I = 2, 4, 6$ A are plotted in Fig. 8. It is seen that also at these higher normal forces, friction coefficients stay roughly constant $f \approx 0.3$ in the entire range of frequencies 1–1.5 kHz, with the exception of the smallest forcing motion $I = 2$ A (□) near 1 500 Hz, where $f \approx 0.2$.

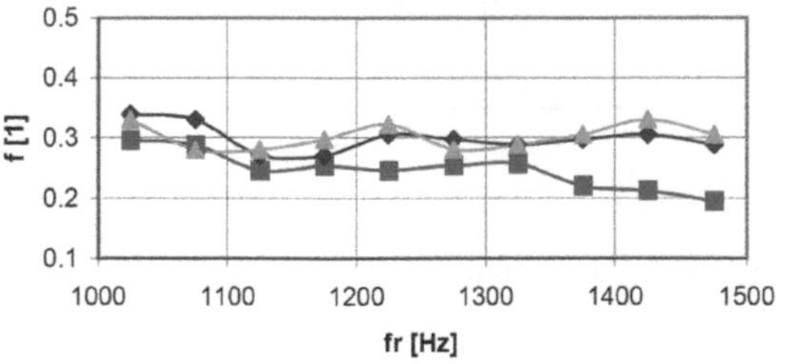

Fig. 8. Dry friction coefficient f in frequency range 1–1.5 kHz, contact pressure 462.48 N/m

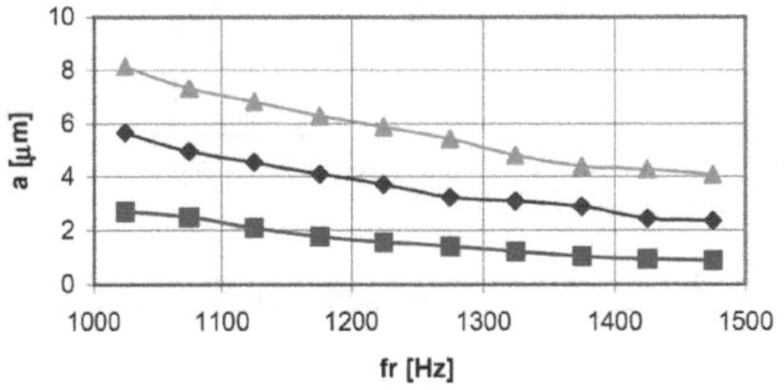

Fig. 9. Amplitudes of relative vibrations, contact pressure $p = 462.48$ N/m

Courses of amplitudes vibrations do not change essentially with the increase of normal thrust on double, as seen from the comparison of curves in Figs. 7 and 9.

The distortion of force records in this range of frequency is shown in Fig. 10 picked up at 1 249 Hz. Repeated jumps excite at this high frequency unexpected oscillations of the signal trace between origin of the friction force and the force transducer BaK 8200 so that the upper and lower parts of records have the forms of wave curve, instead of horizontal lines corresponding to the constant friction forces.

Hysteretic curve of force-displacement record at the same frequency 1 249 Hz is plotted in Fig. 11. The high eigenfrequency oscillations of the trace between source of the friction force and the force transducer cause also the wavelike distortion of the upper and lower parts of hysteretic curve, where the straight lines could be expected, if the friction force should be measured direct in the contact surfaces.

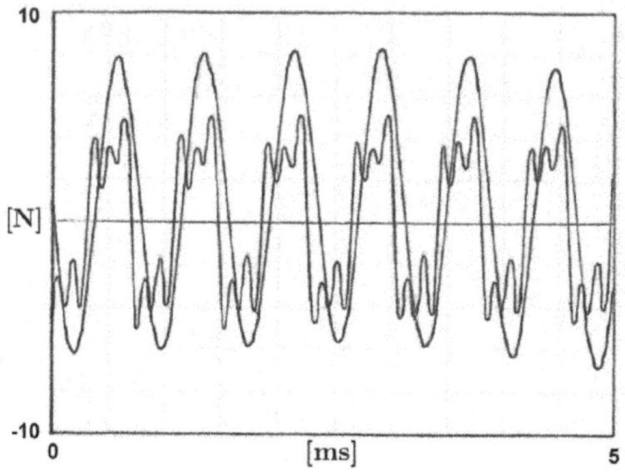

Fig. 10. Motion and friction force at 1 249 Hz

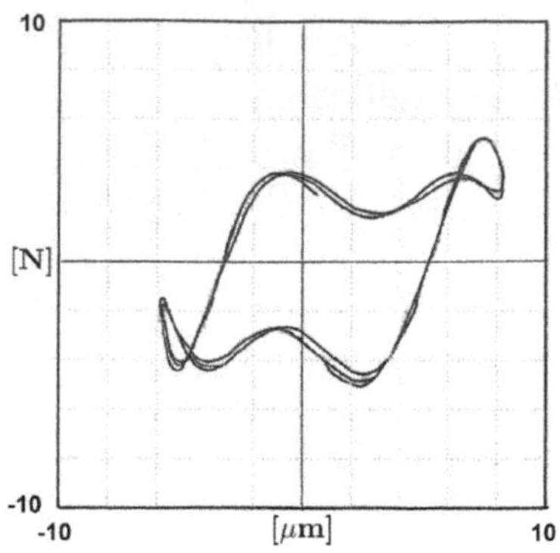

Fig. 11. Hysteretic loop at 1 249 Hz

The area of this hysteretic curve is again proportional to the energy lost during one cycle, but it must be taken into account, that two small marginal loops have opposite circulation and therefore determine not lost but gained energy.

Increase of frequency range on 1.5–2 kHz at the same contact pressure $p = 462.48$ N/m acting on contact line of length 5 cm and at the same feeding currents $I = 2, 4, 6$ A causes the small decrease of dry friction coefficient as seen in Fig. 12. The lowest coefficient values $f \approx 0.2$ are again at the smallest forcing motion created by feeding current $I = 2$ A (□).

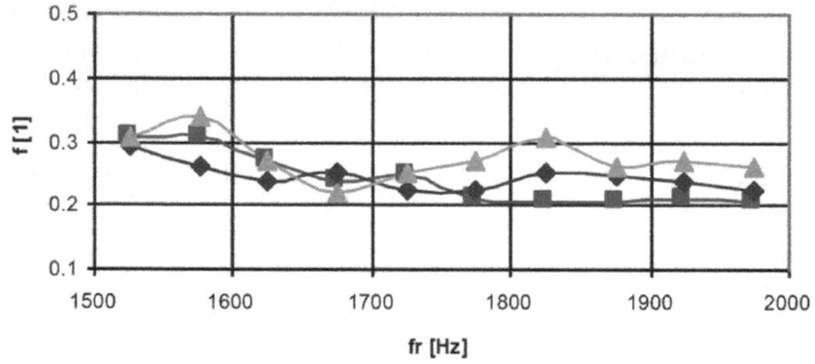

Fig. 12. Friction coefficient f in frequency range 1.5–2 kHz, contact pressure 462.48 N/m

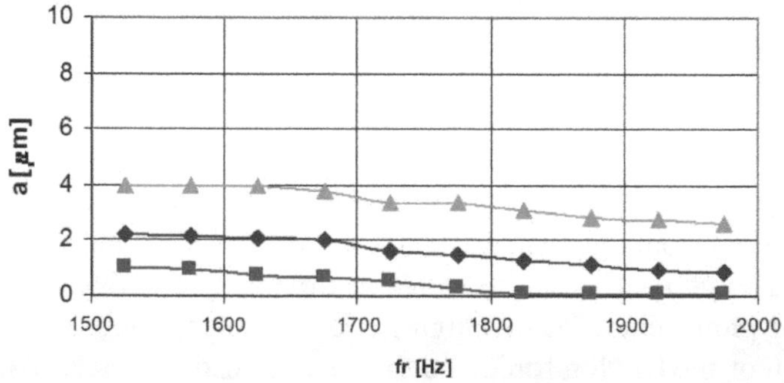

Fig. 13. Amplitudes of relative vibrations in frequency range 1.5–2 kHz

Courses of amplitudes vibrations decrease in this higher frequency range very essentially as seen from comparison of Figs. 9 and 13. This decrease is strongest evident at the smallest feeding current $I = 2$ A (□), where amplitudes in the frequency range 1.8–2 kHz are near to zero. This behavior can be explained by the insufficient excitation force, which does not reach the boundary dry friction force and can produce only elastic tangential (stick) micro-deformations in the contact line surface. These elastic micro-deformations influence also motions at higher excitation-moving-coil forces (feeding currents $I = 4$ A $\triangle$, 6 A $\diamondsuit$), where the slip motions are only parts of measured amplitudes, which results in the lower corresponding amplitude-frequency records in Fig. 13.

The details of force and motion records at frequency 1 790 Hz is shown in Fig. 14. Similar as in the previous case, the signal of friction force picked up by force transducer BaK 8200 contains distortion due to higher eigenfrequency (approx. 5 800 Hz) of the signal trace between origin of the friction force and the force transducer. These distorted vibrations have three periods in one period of relative motion as distinct from the record in Fig. 10, where are five periods of distorted vibrations per one period of relative motion. Hysteretic curve at frequency 1 790 Hz is plotted in Fig. 15. Its horizontal width is smaller than in Fig. 11 and corresponds to the amplitude $a = 3.6$ μm (see Fig. 13 curve 6 A).

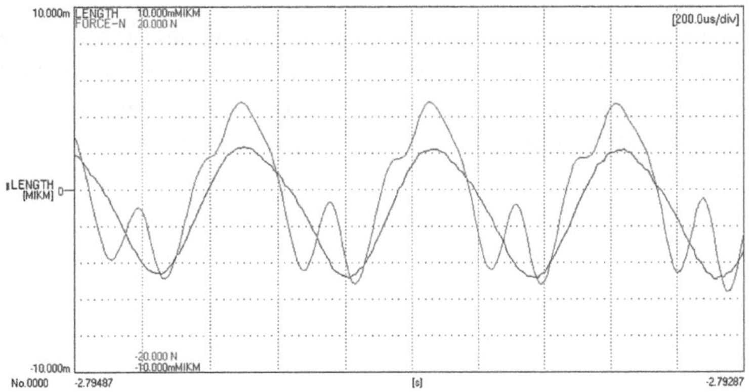

Fig. 14. Motion and friction force at 1 790 Hz

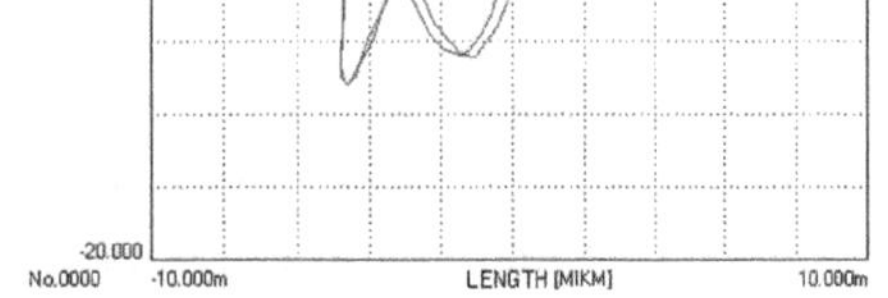

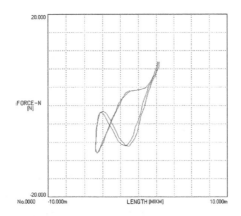

Fig. 15. Hysteretic loop at 1 790 Hz

The higher are the excitation frequencies at the limited vibration excitation force, the smallest are amplitudes of relative tangential motions in contact surface and the strongest is influence of elastic tangential stick micro-deformations. This case occurred, e.g., in the highest frequency range 2.5–3 kHz. Measured dependences of friction coefficient — frequency and amplitude — frequency for the same linear contact pressure $p = 462.48$ N/m and the same currents $I = 2, 4, 6$ A feeding moving-coil vibrator, as in the previous examples, are plotted in Figs. 16 and 17.

Curves in Fig. 16 demonstrate nearly constant, with frequency moderately decreasing values of coefficient f, but they are different ones for various excitation forces ($I = 2$ A □, 4 A △, 6 A ◇) and roughly proportional to these forces in the ratios $2 : 4 : 6$.

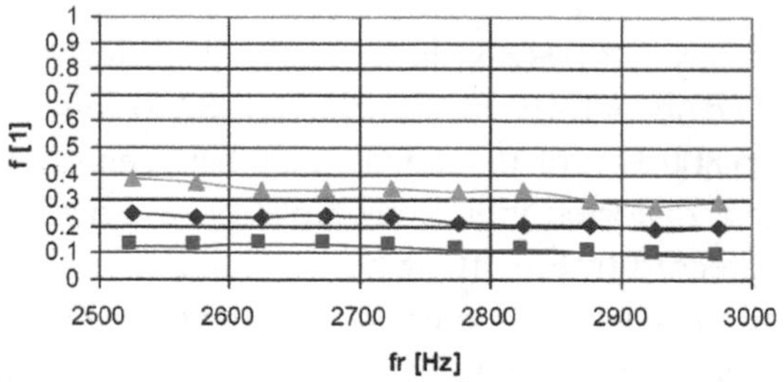

Fig. 16. Coefficient f in frequency range 2.5–3 kHz, contact pressure 462.48 N/m

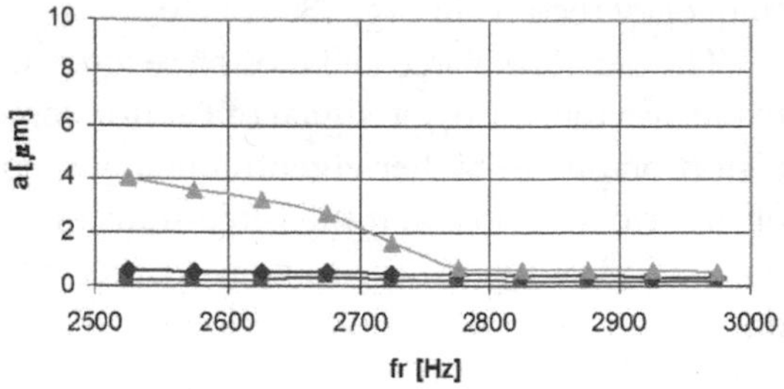

Fig. 17. Amplitudes of relative vibrations in frequency range 2.5–3 kHz

Corresponding amplitudes, plotted in Fig. 17, have approximately zero values (under 1 μm) with the exception of the highest forcing force ($I = 6$ A, ◇) in range 2 500–2 750 Hz, where the force of vibrator reaches over the friction force and evokes greater vibrations. No slip in the contact surface occurs in the other cases, but only elastic tangential micro-deformations with amplitudes in the ratios of excitation forces.

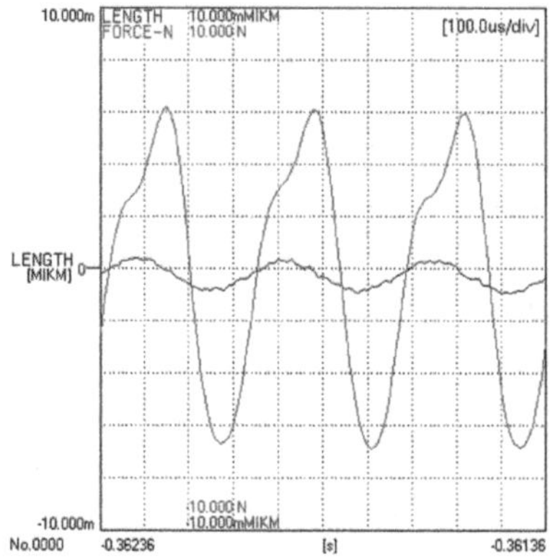

Fig. 18. Motion and friction force at 2 850 Hz

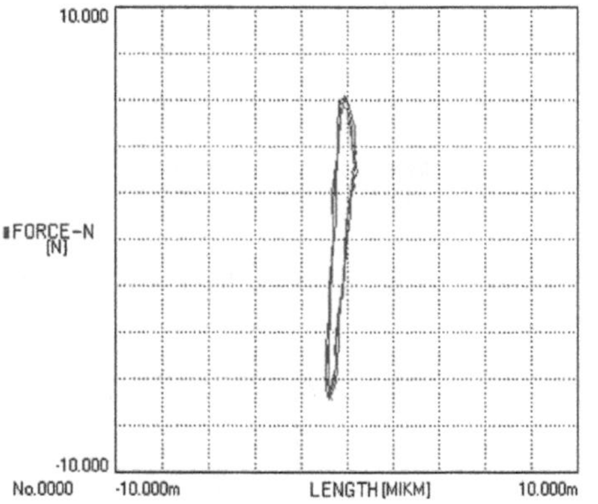

Fig. 19. Hysteretic loop at 2 850 Hz

The example of force and motion records in this range of frequency are picked up at excitation frequency 2 850 Hz and at feeding current $I = 6$ A. Three periods of this record are shown in Fig. 18. The smallest sinusoidal curve indicates recorded motion $x(t)$ of upper cylindrical body (see Fig. 5) with amplitudes $a \approx 0.5$ μm, a little distributed by the noise of applied apparatus. The quasi-sinusoidal curve with higher amplitude (approx. 6 N) describes the friction force. It is evident that this measured tangential force causes only elastic deformation and therefore it has quasi-harmonic form without any slips and it equals to the acting force of electro-dynamic vibrator. A small distortion of form is due to the resonance effect caused by the nearby eigenfrequency of experimental rig.

Fig. 19 presents hysteretic loop (force – displacement). Its small area indicates that in some parts of contact surfaces, marginally and low pressed, friction micro-slips can occur. Moderate inclination of this loop is caused by the elastic compliance properties of the force transfer trace between friction surface and position of force measurement.

The further example of force and motion records in this range of frequency, picked up at 2 950 Hz, is shown in Fig. 20. The feeding current was $I = 6$ A and pressure $p = 452.48$ N/m.

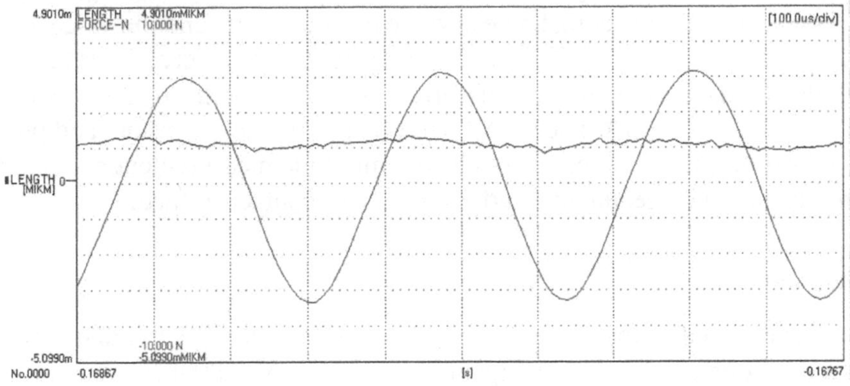

Fig. 20. Motion and friction force at 2 950 Hz

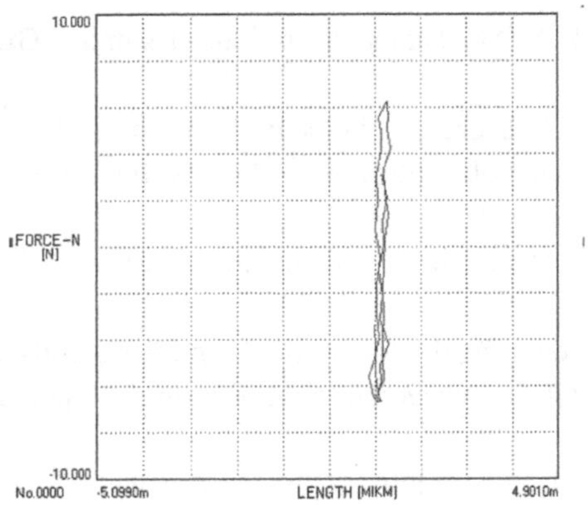

Fig. 21. Hysteretic loop at 2 950 Hz

The measured tangential force causes again only elastic deformation and therefore it has a pure harmonic form corresponding to the force of electro-dynamic vibrator.

The record of relative motion $x(t)$ is much smaller than in previous case. Corresponding hysteretic loop force-displacement degrades to a straight, moderately leaned line (Fig. 21) with no area and also with no damping properties.

4. Conclusion

The dry friction properties at high frequency vibration of contacting bodies are strongly influenced both by the elastic tangential compliances of these bodies near the contacting surface and also by the dynamic frequency spectrum of the entire measuring set, particularly of its part lying between friction contact (force-signal source) and the force meter position. In spite of these distorting influences, the dry friction coefficient $f = 0.3$ at 1 kHz with a moderate decrease to higher frequencies has been ascertained.

One of the important information gained from the presented results of measurements is knowledge that the value f ascertained as ratio of measured tangential force F_T to the perpendicular force F_N can have different physical sense according to the magnitude of excitation force and to the frequency of applied vibrations. At sufficiently large amplitudes a of motions, F_T gives the friction force and ratio $F_T/F_N = f$ ascertains dry friction coefficient.

If amplitude a of motion is very small, then the external harmonic force produces only elastic micro-deformations of contacting bodies, where no slip occurs and the value $f = F_T/F_N$ is usually smaller than dry friction coefficient. The tangential friction force F_T is then proportional to the amplitude of excitation motion (or of current) from the electro-dynamic vibrator and expresses static friction ('stiction'). The ratio F_T/a is approximately constant and proportional to the stiffness [N/m] of trace between the points of amplitude measurement and tangential force measurement, including tangential micro-deformation in contact surface.

Acknowledgements

This work was elaborated in Institute of Thermomechanics AS CR v.v.i. and it was supported by conceptual development of research organizations RVO: 61388998.

References

[1] Brepta, R., Půst, L., Turek, F., Mechanical vibrations, Technical Guide 71, Sobotáles, Praha, 1994. (in Czech)

[2] Charleux, D., Gibert, C., Thouverez, F., Dupeux, J., Numerical and experimental study of friction damping in blade attachments of rotating bladed disks, International Journal of Rotating Machinery 1 (2006) 1–13.

[3] Juliš, K., Brepta, R. et al., Mechanics II – Dynamics, Technical Guide 88, SNTL, Praha, 1987. (in Czech)

[4] Pešek, L., Půst, L. et al., Identification of friction conditions and microslips of friction ring within slot for optimization of railway wheel damping, Research Report Z – 1487/12, IT AS CR, Praha, 2012. (in Czech)

[5] Pešek, L., Půst, L., Košina, J., Radolfová, A., Properties of damping couple at very high frequencies, Proceedings of the conference Dynamics of Machines 2014, Prague, IT AS CR, 2014, pp. 133–141.

[6] Půst, L., Pešek, L., Radolfová, A., Records-distortion at discontinuous forces measurements, Proceedings of Interaction and Feedbacks 2013, Prague, pp. 43–50.

[7] Rao, J. S., Turbomachine blade vibration, New Age, New Delhi, 1991.

[8] Schwingshackl, C. W., Petrov, E. P., Ewins, D. J., Validation of test rig measurements and prediction tools for friction interface modelling, Proceedings of ASME Turbo Expo 2010, Glasgow, pp. 1–10.

[9] Sextro, W., Dynamical contact problems with friction, 2nd edition, Springer, Berlin, 2007.

15

Wall effects on Reiner-Rivlin liquid spheroid

B. R. Jaiswal[a,*], B. R. Gupta[a]

[a]*Department of Mathematics, Jaypee University of Engineering and Technology, 473226 Guna, M. P., India*

Abstract

An analysis is carried out to study the flow characteristics of creeping motion of an inner non-Newtonian Reiner-Rivlin liquid spheroid $r = 1 + \sum_{k=2}^{\infty} \alpha_k G_k(\cos\theta)$, here α_k is very small shape factor and G_k is Gegenbauer function of first kind of order k, at the instant it passes the centre of a rigid spherical container filled with a Newtonian fluid. The shape of the liquid spheroid is assumed to depart a bit at its surface from the shape a sphere. The analytical expression for stream function solution for the flow in spherical container is obtained by using Stokes equation. While for the flow inside the Reiner-Rivlin liquid spheroid, the expression for stream function is obtained by expressing it in a power series of S, characterizing the cross-viscosity of Reiner-Rivlin fluid. Both the flow fields are then determined explicitly by matching the boundary conditions at the interface of Newtonian fluid and non-Newtonian fluid and also the condition of impenetrability and no-slip on the outer surface to the first order in the small parameter ϵ, characterizing the deformation of the liquid sphere. As an application, we consider an oblate liquid spheroid $r = 1 + 2\epsilon G_2(\cos\theta)$ and the drag and wall effects on the body are evaluated. Their variations with regard to separation parameter, viscosity ratio λ, cross-viscosity, i.e., S and deformation parameter ϵ are studied and demonstrated graphically. Several well-noted cases of interest are derived from the present analysis. Attempts are made to compare between Newtonian and Reiner-Rivlin fluids which yield that the cross-viscosity μ_c is to decrease the wall effects K and to increase the drag D_N when deformation is comparatively small. It is observed that drag not only varies with λ, but as η increases, the rate of change in behavior of drag force increases also.

Keywords: Reiner-Rivlin fluid, Gegenbauer function, stream functions, liquid spheroid, drag force, wall correction factor, spherical container

1. Introduction

Since the pioneering work [31], a considerable body of knowledge has accrued on the drag force experienced by a range of shapes of particles settling in immovable fluid media or held stationary in moving fluids. A spherical particle is unique in that it presents the same projected area to the oncoming fluid irrespective of its orientation. For nonspherical particles, on the other hand, the orientation must be known before their terminal settling velocity or the drag force acting on them can be calculated. On the contrary, under appropriate circumstances, non-spherical particles are vulnerable to accomplish a preferred or most stable orientation irrespective of their preliminary orientation. These entire phenomenons are strongly influenced not only by the shape of the particle, its size and density, fluid properties, but also by the shape and size of confining boundaries and the fluid motion as well. The literature, not as much extensive as that of spherical particles, available on regularly or irregularly shaped nonspherical particles in incompressible Newtonian and non-Newtonian fluid media has been reviewed by few researchers in the recent past. Ramkissoon, independently, examined the problems of symmetrical flow of Newtonian fluid past a Newtonian fluid spheroid [22] and non-Newtonian Reiner-Rivlin fluid

*Corresponding author. e-mail: jaiswal.bharat@gmail.com.

spheroid [19] and evaluated drag on these fluid bodies. The problems of Stokes flow of a micropolar fluid past a Newtonian fluid spheroid by [15], a Reiner-Rivlin fluid sphere with no-spin condition for micro rotations by [20] and later this problem was reinvestigated with spin condition for micro rotations by [6]. The paper [5] solved the problem for the case of an approximate sphere in micropolar fluid. The problem of a viscous incompressible fluid past a spheroid which departs but little in shape from a sphere with mixed slip-stick boundary conditions was investigated by [14] and drag exerted on oblate spheroid is calculated which is found in the present case to be less than that of the Stokes resistance for a slightly oblate spheroid.

The problems of motion of a particle at the instant it passes the center of the spherical container serves as a model of interaction in multi-particle systems. The simplest geometry permits one to study the impressive and effective shape of fluid particles on their settling velocity and the drag resistance on approximate sphere. This class of problems is important because it provides some information on wall effects. Wall effects for a sphere at the instant it passes the center of the spherical container have been studied by several investigators and these investigations were abbreviated and summarized by [4, 10, 13] and [7]. A survey of literature regarding the fluid flows past a solid or a liquid body of different shapes indicates that while abundant information is available for flows in an infinite expanse of fluid, but very few studies of wall effects for approximate spheres are available for flows in enclosures. The authors of [2] and [32], independently, considered the motion of a solid sphere in a spherical container and presented the solutions for the case of inner solid sphere. [3] have made an equivalent or analogous study for the motion of an inner classical fluid sphere. The paper [16] investigated the motion of inner Reiner-Rivlin (non-Newtonian) fluid sphere in a spherical container and they deduced that cross viscosity μ_c is to reduce the wall effects. Two years later, [17] also investigated the problem of a solid spherical particle in a spheroidal container. They obtained the expression for the drag on the inner sphere and examined the wall effects and concluded that as the deformation of the spheroidal container increases the wall effects also increase. In all the above mentioned problems, the authors utilized no-slip condition on the surface of the inner sphere. [18] studied the motion of solid spheroidal particle in a spherical container using a slip condition at the surface of the inner particle and evaluated the expression for drag on it and examined the wall effects. They concluded that the wall effects increase as the spheroidal particle becomes more spheroidal. Maul and Kim [11, 12] investigated a point force in a fluid domain bounded by a rigid spherical container and the image of a point force in a spherical container and its connection to the Lorentz reflection formula by using Stokes equation. By applying the boundary conditions of continuity of the velocity, pressure and tangential stresses at the porous-liquid interface, the quasisteady translation and steady rotation of a spherically symmetric composite particle composed of a solid core and a surrounding porous shell located at the center of a spherical cavity filled with an incompressible Newtonian fluid is studied by [8]. The quasisteady translation and steady rotation of a spherically symmetric porous shell located at the center of a spherical cavity filled with an incompressible Newtonian fluid is investigated analytically by [9]. They evaluated the hydrodynamic drag force and torque exerted by the fluid on the porous particle and found that the boundary effects of the cavity wall on the creeping motions of a composite sphere can be significant in appropriate situations. The work [30], independently, has studied the motion of a porous sphere in a spherical container using Brinkmans model in the porous region and concluded that the wall correction factor increases and drag coefficient decreases with permeability parameter. The motion of a spheroidal particle at the instant it passes the centre of a spherical envelope filled with a micropolar fluid is investigated by [23] using the slip condition at the surface of the particle. The flow problem of an incompressible axisymmetrical quasisteady translation and steady rotation of a porous spheroid in a concentric spheroidal container is studied analytically by [24]. Srinivasacharya and Prasad have study the problems on the motion of a porous spherical shell in a bounded medium [27], slow steady rotation of a porous sphere in a spherical container [29], steady rotation of a composite sphere in a concentric spherical cavity [28], creeping motion of a porous approximate sphere with an impermeable core in spherical container [26] and axisymmetric motion of a porous approximate sphere in an approximate spherical container [25]. They found that, in all the cases, the wall correction factor increases and drag decreases as the size of the container increases. The outcomes and motivations of these research findings lead us to discuss the present problem which includes some previous results.

In this paper, we consider the creeping motion of a Reiner-Rivlin liquid spheroid in spherical container and the flow considered is axisymmetric in nature. This, we should mention, is an extension of previous works by [16], where the Reiner-Rivlin droplet was assumed to be perfectly spherical, and [19], where Reiner-Rivlin liquid spheroid was considered in an infinite expanse of Newtonian fluid. Here, we assume that the inner sphere is moving with uniform velocity U in the positive z-direction. Both the internal and external flow fields, flows inside the liquid spheroid and within a spherical container, are determined to the first order in the small parameter characterizing the deformation by using the Stokes approximation and expanding the internal stream function in a power series of a dimensionless parameter S. The drag experienced by the inner liquid spheroid is evaluated, and this enables us to examine the wall effects numerically and graphically as well.

2. Mathematical formulation and solution of the problem

We consider a non-Newtonian Reiner-Rivlin liquid spheroid of radius, $r = 1 + \alpha_k G_k$, and to examine the drag force at an instant it passes the center of a large spherical vessel of radius b containing a steady and incompressible Newtonian fluid (liquid region I). The problem of course, is equivalent to the inner liquid spheroid remains at rest while the outer spherical container moves, say, with constant velocity U in the positive z-direction. Further, we assume that the inner flow (liquid region II) is also steady, incompressible and the liquid spheroid to be stationary. A physical model illustrating the problem under consideration is shown in Fig. 1. The

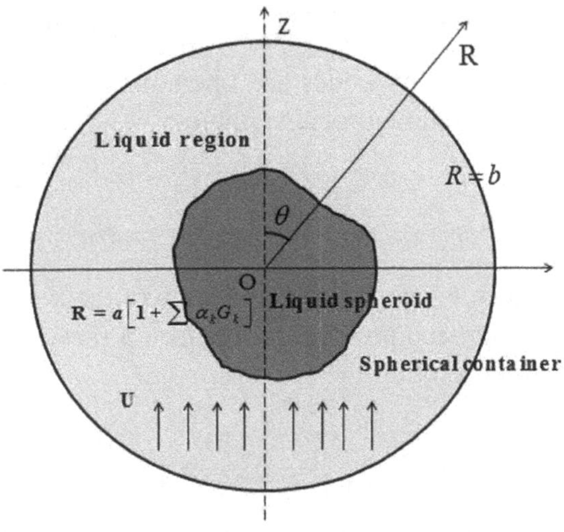

Fig. 1. Physical model and co-ordinate system illustrating the axisymmetric flow past a Reiner-Rivlin liquid spheroid in spherical container

parameters pertaining to the exterior and the interior of the liquid spheroid to be distinguished respectively by the index in the superscripts under the bracket of an entity $\chi^{(i)}, i = 1, 2$.

2.1. Basic equations

Within the spherical container (liquid region I), the corresponding flow-field equations describing the motion of a viscous fluid (Newtonian) in region I are governed by the Stokes equation and equation of continuity respectively as

$$\operatorname{grad} \tilde{p}^{(1)} + \mu_1 \operatorname{curlcurl} \tilde{q}^{(1)} = 0 \tag{1}$$

and

$$\operatorname{div} \tilde{q}^{(1)} = 0, \tag{2}$$

where $\tilde{q}^{(1)}$ is the velocity, μ_1 is the coefficient of viscosity and $\tilde{p}^{(1)}$ is the pressure.

The constitutive equation for isotropic non-Newtonian Reiner-Rivlin liquid in the region II takes the form

$$\tilde{\tau}_{ij} = -\tilde{p}^{(2)} \delta_{ij} + 2\mu_2 \tilde{d}_{ij} + \mu_c \tilde{d}_{ik} \tilde{d}_{kj}, \tag{3}$$

where

$$\tilde{d}_{ij} = \frac{1}{2} \left(\tilde{\vec{q}}_{i,j}^{(2)} + \tilde{\vec{q}}_{j,i}^{(2)} \right),$$

$\tilde{\tau}_{ij}$ is the stress tensor, $\tilde{d}_{ij}$ is the rate of strain (deformation) tensor, μ_2 is the coefficient of viscosity and μ_c is the coefficient of cross-viscosity of Reiner-Rivlin liquid in region II. The velocity vector and the pressure for region II are respectively denoted by $\tilde{\vec{q}}^{(2)}$ and $\tilde{p}^{(2)}$.

Let (R, θ, ϕ) denote a spherical polar co-ordinate system with the origin at the center of the sphere of radius a. Since the flow is axially symmetric and the fluid flows in the meridian plane, all the physical quantities are independent of ϕ. Hence, we assume that the velocity vectors $\tilde{q}^{(1)}$ and $\tilde{q}^{(2)}$can be expressed as

$$\tilde{q}^{(i)} = \tilde{u}_R^{(i)}(R, \theta)\vec{e}_R + \tilde{u}_\theta^{(i)}(R, \theta)\vec{e}_\theta, \quad i = 1, 2. \tag{4}$$

In view of the incompressibility condition $\operatorname{div} \tilde{q}^{(1)} = 0$, we introduce the stream function $\tilde{\psi}^{(i)}(R, \theta) = 0, i = 1, 2$ through

$$\tilde{u}_R^{(i)} = \frac{-1}{R^2 \sin\theta} \frac{\partial \tilde{\psi}^{(i)}}{\partial \theta}, \quad \tilde{u}_\theta^{(i)} = \frac{1}{R \sin\theta} \frac{\partial \tilde{\psi}^{(i)}}{\partial R}. \tag{5}$$

In order to non-dimensionalize the values and operators appearing in the governing equations, we insert the following non-dimensional variables:

$$R = ar, \quad \tilde{u}_R = U u_r, \quad \tilde{u}_\theta = U u_\theta, \quad \tilde{\tau}_{ij} = \mu_i \tfrac{U}{a} \tau_{ij},$$
$$\tilde{d}_{ij} = \tfrac{U}{a} d_{ij}, \quad \tilde{p} = \mu_i \tfrac{U}{a} p, \quad \tilde{\psi} = U a^2 \psi,$$

where U and a represent some typical velocity and length of the flow field respectively.

We write the stream function and pressure in the form a power series in S for the internal flow within the liquid spheroid as follows

$$\psi^{(2)} = \psi_0 + \psi_1 S + \psi_2 S^2 + \dots,$$
$$p^{(2)} = p_0 + p_1 S + p_2 S^2 + \dots, \tag{6}$$

where $S = \frac{\mu_c U}{\mu_2 a}$ is the sufficiently small dimensionless number and therefore, the terms in (6) of $O(S > 2)$ can be neglected. The suffixes in (6) represent the zeroth, second and higher order approximation of the corresponding variables. Following [21], the stream functions ψ_0, ψ_1 and ψ_2 satisfy respectively the following differential equations

$$E^4\psi_0 = 0, \quad E^4\psi_1 = 8r\sin^2\theta\cos\theta, \quad E^4\psi_2 = \frac{32}{3}r^2\sin^2\theta. \tag{7}$$

Particular solutions of the equations mentioned in (7) are respectively given by

$$\psi_0 = (r^4 - r^2)\sin^2\theta, \quad \psi_1 = \frac{2}{21}r^5\sin^2\theta\cos\theta, \quad \psi_2 = \frac{2}{63}r^6\sin^2\theta. \tag{8}$$

Eliminating pressure from (1) and making use of (5) in the resulting equation, we get the following dimensionless equation for $\psi^{(1)}$

$$E^4\psi^{(1)} = 0, \tag{9}$$

where the second order differential operator E^2 is expressed as

$$E^2 = \frac{\partial^2}{\partial r^2} + \frac{1-\zeta^2}{r^2}\frac{\partial^2}{\partial\zeta^2}, \quad \zeta = \cos\theta. \tag{10}$$

The general solution of (9), which is non-singular everywhere in the flow field, is given as

$$\psi^{(1)} = [a_2r^2 + b_2r^{-1} + c_2r^4 + d_2r]G_2(\zeta) + [a_3r^3 + +b_3r^{-2} + c_3r^5 + d_3]G_3(\zeta) + \sum_{n=4}^{\infty}[A_nr^n + B_nr^{-n+1} + C_nr^{n+2} + D_nr^{-n+3}]G_n(\zeta), \tag{11}$$

where $1 + \alpha_kG_k \leq r \leq 1/\eta$.

While for the fluid flow within the Reiner-Rivlin liquid spheroid by [16], we may take

$$\psi^{(2)} = \psi_0 + \psi_1S + \psi_2S^2 + \sum_{n=4}^{\infty}[e_nr^n + f_nr^{n+2}]G_n(\zeta). \tag{12}$$

With the aid of (8), we can now write the relation (12) explicitly in the form:

$$\psi^{(2)}(r,\zeta) = [(e_2-2)r^2 + (f_2+2)r^4 + \frac{4}{63}S^2r^6]G_2(\zeta) + [e_3r^3 + (f_3 + \frac{4}{21}S)r^5]G_3(\zeta) + \sum_{n=4}^{\infty}[E_nr^n + F_nr^{n+2}]G_n(\zeta), \tag{13}$$

where $r \leq 1 + \alpha_kG_k$, $\zeta = \cos\theta$, and $G_n(\zeta)$ are the Gegenbauer functions, defined in the book by Abramowitz and Stegun [1], related to the Legendre functions $P_n(\zeta)$ by the relation

$$G_n(\zeta) = \frac{P_{n-1}(\zeta) - P_n(\zeta)}{2n-1}, \qquad n \geq 2.$$

Let the equation of the surface S of a spheroid approximating that of the sphere $R = a$, i.e., $r = 1$ in the polar form be $R = a[1 + f(\theta)]$ or $r = 1 + f(\theta)$. The orthogonality relations of Gegenbauer functions permit us, in general, to write $f(\theta)$ in the expanded form as $f(\theta) = \sum_{k=2}^{\infty}\alpha_kG_k(\zeta)$. Hence, we can take the surface S to be of the form

$$r = 1 + \alpha_kG_k(\zeta) \tag{14}$$

and neglect terms of $o(\alpha_k) \geq 2$. In case of Newtonian fluid past a perfect Reiner-Rivlin liquid sphere in spherical container the only coefficients that contribute to $\psi^{(1)}$ and $\psi^{(2)}$ are a_2, b_2, c_2, d_2, e_2, f_2 and f_3 and consequently we expect all other remaining coefficients appearing in the equations (11) and (13) to be of order α_k. Hence, except where the coefficients a_2, b_2, c_2, d_2, e_2, f_2 and f_3 appear, we may take the surface S to be a perfect sphere, i.e., $r = 1$ instead of its approximate form (14).

3. Boundary conditions and determination of arbitrary constants

The arbitrary constants appearing in (11) and (13) to be determined by applying the following the boundary conditions, which are physically realistic and mathematically consistent for the present proposed problem:

(a) The kinematic condition of mutual impenetrability at the interface implies that we may take

$$\psi^{(1)} = 0, \qquad \psi^{(2)} = 0 \qquad \text{on} \quad r = 1 + \alpha_k G_k(\zeta). \tag{15}$$

(b) We assume the continuity of tangential velocity across the interface, hence we have

$$\frac{\partial \psi^{(1)}}{\partial r} = \frac{\partial \psi^{(2)}}{\partial r} \qquad \text{on} \quad r = 1 + \alpha_k G_k(\zeta). \tag{16}$$

(c) We further assume that the presence of interfacial tension only produces a discontinuity in the normal stress τ_{rr} and does not in any way affect tangential stress $\tau_{r\theta}$, i.e., $\tau_{r\theta}^{(1)} = \tau_{r\theta}^{(2)}$ on the surface S of the spheroid which can be shown to be equivalent to

$$\lambda \frac{\partial}{\partial r}\left(\frac{1}{r^2}\frac{\partial \psi^{(1)}}{\partial r}\right) = \frac{\partial}{\partial r}\left(\frac{1}{r^2}\frac{\partial \psi^{(2)}}{\partial r}\right) \qquad \text{on} \quad r = 1 + \alpha_k G_k(\zeta). \tag{17}$$

(d) On the spherical container, the kinematic condition of impenetrability at the surface and the assumption of no slip there lead respectively to

$$\psi^{(1)} = \frac{1}{2} r^2 \sin^2\theta$$

and

$$\frac{\partial}{\partial r}\left(\psi^{(1)} - \frac{1}{2} r^2 \sin^2\theta\right) = 0 \quad \text{on} \quad r = \frac{1}{\eta}, \tag{18}$$

where $\lambda = \frac{\mu_1}{\mu_2}$ and $\eta = \frac{a}{b}$. The arbitrary constants encountering in (11) and (13) can be determined from the above boundary conditions, and the results are given in Appendix A. Thus, the stream functions for both the external and internal flow fields are now completely known and given respectively by

$$\begin{aligned}\psi^{(1)} = {} & [a_2 r^2 + b_2 r^{-1} + c_2 r^4 + d_2 r] G_2(\zeta) + (A_{k-2} r^{k-2} + B_{k-2} r^{-k+3} + C_{k-2} r^k + \\ & D_{k-2} r^{-k+5}) G_{k-2}(\zeta) + (A_k r^k + B_k r^{-k+1} + C_k r^{k+2} + D_k r^{-k+3}) G_k(\zeta) + \\ & (A_{k+2} r^{k+2} + B_{k+2} r^{-k-1} + C_{k+2} r^{k+4} + D_{k+2} r^{-k+1}) G_{k+2}(\zeta),\end{aligned} \tag{19}$$

$$\begin{aligned}\psi^{(2)} = {} & \left[(e_2 - 2) r^2 + (f_2 + 2) r^4 + \frac{4}{63} S^2 r^6\right] G_2(\zeta) + [E_{k-2} r^{k-2} + F_{k-2} r^k] G_{k-2}(\zeta) + \\ & (E_k r^k + F_k r^{k+1}) G_k(\zeta) + (E_{k+2} r^{k+2} + F_{k+2} r^{k+4}) G_{k+2}(\zeta).\end{aligned} \tag{20}$$

Therefore, the Stokes flow past a Reiner-Rivlin liquid spheroid in spherical container is duly determined.

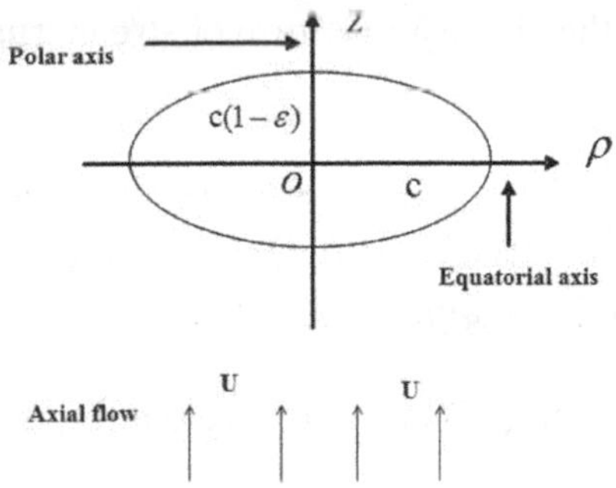

Fig. 2. Oblate spheroid in meridional two dimensional plane

4. Application to a Reiner-Rivlin liquid oblate spheroid

As an application of the foregoing discussion, we now address the particular case of a Reiner-Rivlin liquid oblate spheroid (see Fig. 2) at the instant it passes the centre a spherical container whose equation we take as

$$\frac{x^2+y^2}{c^2}+\frac{z^2}{c^2(1-\epsilon)^2}=1. \tag{21}$$

Therefore, the polar form of the equation (21) to the order $O(\epsilon)$ takes the form

$$R=a[1+\epsilon G_2(\zeta)] \text{ or } r=1+2\epsilon G_2(\zeta), \tag{22}$$

where $a = c(1-\epsilon)$. For an oblate fluid spheroid, the results of the previous section can be applied by taking $k = 2, \alpha_k = 2\epsilon$. Therefore, the expressions of stream functions for external and internal flows now get reduced respectively to

$$\begin{aligned}\psi^{(1)} &= [a_2r^2+b_2r^{-1}+c_2r^4+d_2r]G_2(\zeta)+[A_2r^2+B_2r^{-1}+C_2r^4+D_2r]G_2(\zeta)+\\ &\quad [A_4r^4+B_4r^{-3}+C_4r^6+D_4r^{-1}]G_4(\zeta), \qquad (23)\\ \psi^{(2)} &= \left[(e_2-2)r^2+(f_2+2)r^4+\frac{4}{63}S^2r^6\right]G_2(\zeta)+(E_2r^2+F_2r^4)G_2(\zeta)+\\ &\quad (E_4r^4+F_4r^6)G_4(\zeta). \qquad (24)\end{aligned}$$

5. Evaluation of drag on Reiner-Rivlin liquid oblate spheroid and associated wall effects

Now, for the fluid flow problems in spherical container, one of the most significant physical quantities is the drag force and associated wall effects experienced by the inner non-Newtonian liquid spheroid.

5.1. Drag force (F)

To evaluate the drag force experienced by the inner liquid spheroid in spherical container, we appeal to a simple elegant formula derived by Happel and Brenner [4]

$$F=\mu_1\pi\int_0^{\pi}\varpi^3\frac{\partial}{\partial R}\left(\frac{E^2\psi_d^{(1)}}{\varpi^2}\right)R\,\mathrm{d}\theta, \tag{25}$$

where $\varpi = R\sin\theta$ and $\psi_d^{(1)}$ is the dimensional form of stream function defined in (23) and after some working $E^2\psi_d^{(1)}$ simplifies to

$$\begin{aligned} E^2\psi_d^{(1)} &= Uc^2\sin^2\theta[40c^2R^5(1+2\epsilon)c_2 - 8c^5R^2(1-\epsilon)d_2 - 8c^5R^2(1-\epsilon)D_2 - \\ &\quad (15c^7 + 25c^7\cos 2\theta)D_4 + 40c^2R^5(1+2\epsilon)C_2 + \\ &\quad (27R^7 + 45R^7 \times \cos 2\theta)C_4]/8c^6R^3. \end{aligned} \tag{26}$$

Substituting the value of $E^2\psi_d^{(1)}$ in (25) and carrying out the integration with respect to θ, the dimensional drag simplifies to

$$F = 4c\pi U(1-\epsilon)(d_2 + D_2)\mu_1. \tag{27}$$

5.2. Wall correction factor (K):

The wall correction factor K is defined as the ratio of the actual drag experienced by the liquid spheroid in the enclosure and the drag on a liquid spheroid in an infinite expanse of fluid. With the aid of (27) and (36), this becomes

$$K = -\frac{(1+\lambda)(1-\epsilon)(d_2+D_2)}{\frac{1}{6}(1-\epsilon)\left(6\lambda + 9 + \frac{32}{63}S^2\right) + \frac{2\epsilon}{945}(567 + 378\lambda + 160S^2)}, \tag{28}$$

where

$$d_2 = [2(16S^2(2 - 5\eta^3 + 3\eta^5) + 189(3 + 3\eta^5(-1+\lambda) + 2\lambda))]\Delta, \tag{29}$$

$$\begin{aligned} D_2 &= (-1+\eta)^2[8\epsilon(16S^2(-1+\eta)^2(48\eta^7(-1+\lambda) - 40(1+\lambda) - 5\eta^4(-41+2\lambda) - \\ &\quad 40\eta^3(-1+4\lambda) + 20\eta^5(5+4\lambda) - 16\eta^2(7+13\lambda) - 8\eta(11+14\lambda) - \\ &\quad 8\eta(11+14\lambda) + 3\eta^6(-19+34\lambda)) - 189(2+4\eta+6\eta^2+3\eta^3)(6+10\lambda + \\ &\quad 4\lambda^2 + 3\eta^5(3 - 5\lambda + 2\lambda^2) + 5\eta^3(-3+\lambda+2\lambda^2))))]\Delta^2/945. \end{aligned} \tag{30}$$

and Δ is given by (A.8). These are respectively the new results for the drag and wall effects on a Reiner-Rivlin liquid spheroid at the instant it passes the center of a spherical envelope.

6. Results and discussion

It is important to consider some limiting situations of the drag force as discussed below:

(i) *A perfect Reiner-Rivlin liquid sphere in spherical container* ($\epsilon \to 0$):

We get

$$F = \frac{8c\pi U(16S^2(2 - 5\eta^3 + 3\eta^5) + 189(3 + 3\eta^5(-1+\lambda) + 2\lambda))\mu_1}{189(-1+\eta)^3(4\eta^3(-1+\lambda) + 4(1+\lambda+\eta^2(-3+6\lambda) + \eta(3+6\lambda)))}, \tag{31}$$

a result previously obtained by [16].

(ii) *A perfect Reiner-Rivlin liquid sphere in an unbounded medium* ($\eta \to 0$):

From (31), when $\eta \to 0$, we get

$$F = \frac{-2Uc\pi\mu_1(\frac{32}{63}s^2 + 6\lambda + 9)}{3(1+\lambda)}, \tag{32}$$

where $\lambda = \frac{\mu_1}{\mu_2}$. This result has been obtained previously by [21].

(iii) *Newtonian liquid oblate spheroid in spherical envelope* ($S \to 0$):

From (27), when $S \to 0$, we get

$$F = -8cU(-1+\epsilon)(-1+\eta)^2[4\epsilon(2+4\eta+6\eta^2+3\eta^3)(6+10\lambda+4\lambda^2+ 3\eta^5(3-5\lambda+2\lambda^2)+5\eta^3(-3+\lambda++2\lambda^2))+5(-1+\eta)(12\eta^8(-1+\lambda)^2+ 12\eta^5(-1+\lambda^2)+9\eta^7(1-3\lambda+2\lambda^2)+9\eta^6(-1-\lambda+2\lambda^2)+ 4\eta^3(-3+\lambda+2\lambda^2)+4(3+5\lambda+2\lambda^2)+3\eta^2(-3+4\lambda+4\lambda^2)+ 3\eta(3+8\lambda+4\lambda^2))]\Delta^2\mu_1/5. \tag{33}$$

This is the new result reported in this paper.

(iv) *Newtonian liquid sphere in spherical envelope* ($\epsilon \to 0$):

From (33), when $\epsilon \to 0$, we get

$$F = 1\,512c\pi\mu_1 U[3+3\eta^5(-1+\lambda)+2\lambda]\Delta, \tag{34}$$

a well known result reported earlier by [4] and later by [16].

(v) *Newtonian liquid oblate spheroid in unbounded medium* ($\eta \to 0$):

From (33), when $\eta \to 0$, we get

$$F = -\frac{6c\pi U\mu_1\left(1-\frac{1}{5}\epsilon\right)\left(1+\frac{2}{3}\lambda\right)}{1+\lambda}. \tag{35}$$

The result was earlier obtained by [22].

(vi) *Reiner-Rivlin liquid oblate spheroid in unbounded medium* ($\eta \to 0$):

From (27), we have

$$F = -\frac{4c\pi U\mu_1}{1+\lambda}\left[\frac{(1-\epsilon)\left(6\lambda+9+\frac{32}{63}S^2\right)}{6}+\frac{2\epsilon(567+378\lambda+160S^2)}{945}\right]. \tag{36}$$

The result was previously obtained by [19].

(vii) *Drag on a Reiner-Rivlin liquid sphere of volume equal to the volume of a Reiner-Rivlin liquid oblate spheroid:*

A perfect Reiner-Rivlin liquid sphere in spherical container of radius equal to $c(1-\epsilon/3)$ will have the volume as oblate spheroid. Therefore, the drag force experienced by a Reiner-Rivlin liquid sphere of radius $c(1-\epsilon/3)$ in spherical container is

$$F = 8c\pi\left(1-\frac{\epsilon}{3}\right)\mu_1 U[16S^2(2-5\eta^3+3\eta^5)+189(3+3\eta^5(-1+\lambda)+2\lambda)]\Delta. \tag{37}$$

On comparison with (27), we make out that a liquid sphere of equal volume experiences a smaller resistance than the liquid oblate spheroid(see Fig. 15).

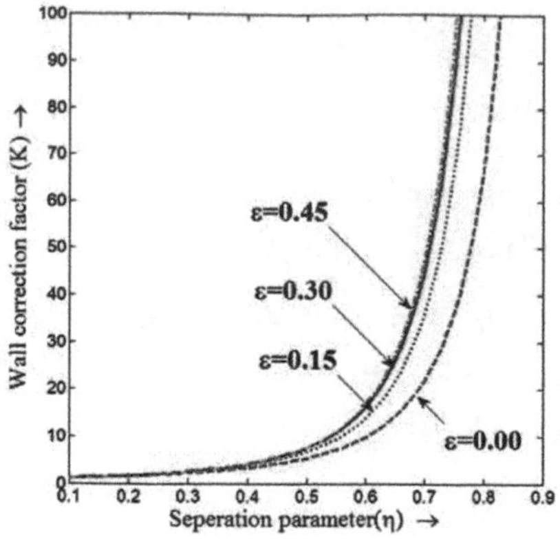

Fig. 3. Variation of the wall correction factor K with η for various values of deformation parameter ϵ at $S = 0.5$ and $\lambda = 0.5$

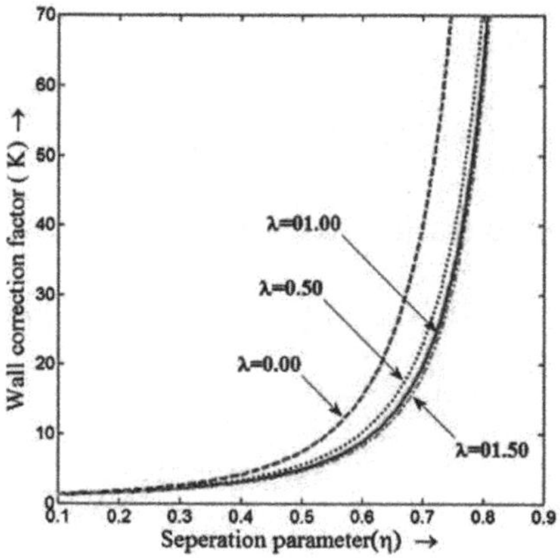

Fig. 4. Variation of the wall correction factor K with η for various values of viscosity ratio λ at $S = 0.5$ and $\epsilon = 0.25$

The effect of the deformation parameter ϵ and separation parameter η on the wall correction factor K has been plotted in Fig. 3 for a fixed value of λ and S. It can be observed from Fig. 3 that the wall correction factor K is increasing with the increasing deformity $(0 \leq \epsilon \leq 0.45)$ of the liquid spheroid and separation parameter η. The figure clearly indicates that the wall effects on Reiner-Rivlin liquid sphere is the lowest as compared with such effects on liquid spheroid. The figure also exhibits that the growth rate of the wall correction factors decreases as the deformation of the liquid sphere increases. For $(0.46 \leq \epsilon \leq 0.52)$, the wall correction factor K increases almost with the equal amounts for different values of ϵ in the range 0.46 to 0.52 when the separation parameter varies from 0.1 to 0.9. However, for the values of the deformation parameter ϵ greater than 0.5, there is reported a reversal in behavior of the wall correction factor K, i.e., the wall correction factor K decreases as the deformation parameter ϵ increases in the range 0.5 to 1. This bizarre and unexpected variation in wall correction factor K may be possibly due to the expansion and the contraction of the surface area and the volume of the oblate spheroid with decreasing and increasing of measure of the deformity respectively in contact with the fluid in spherical envelope as it is clear from Fig. 15.

The variation of the wall correction factor K against the separation parameter η with continuity of tangential stress for various values of viscosity ratio λ is illustrated in Fig. 4. From the figure, it can be observed that, as η increases, the wall effects on liquid spheroid increases. For

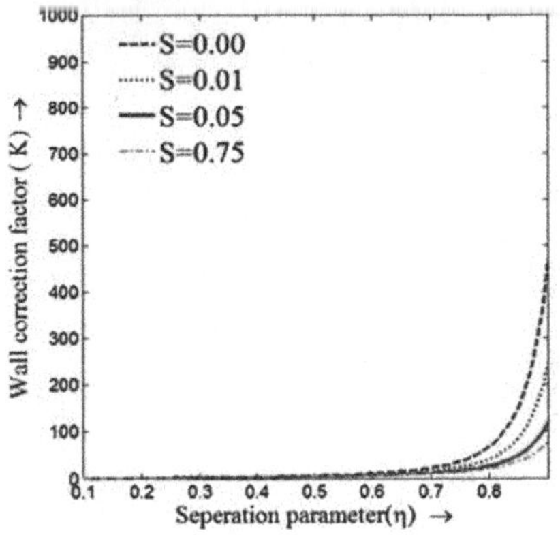

Fig. 5. Variation of the wall correction factor K with η for various values of S at $\lambda = 0.5$ and $\epsilon = 0$

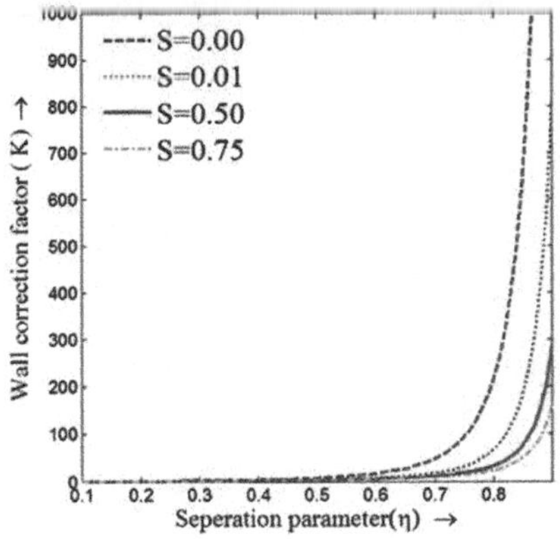

Fig. 6. Variation of the wall correction factor K with η for various values of S at $\lambda = 0.5$ and $\epsilon = 0.5$

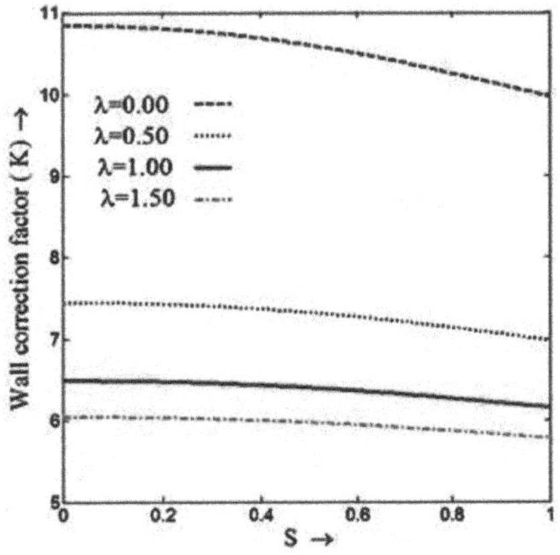

Fig. 7. Variation of the wall correction factor K with S for various values of λ at $\eta = 0.5$ and $\epsilon = 0.5$

smaller values of η, wall correction factor K increases slowly with approximately equal amount for all the values of viscosity ratio λ. But for comparatively larger values of η (> 0.3), K increases very rapidly and differently for different values of λ. Also, the wall correction factor K is decreasing with the increasing viscosity ratio λ. It is interesting to note that for $\lambda > 0.5$, the particle mobility varies slowly with the separation parameter η as compared with the case of lower viscosity ratio.

In Figs. 5 and 6, we have demonstrated the effects of S or the cross viscosity on the wall correction factor K with η for a fixed value of λ and ϵ. It is obvious from the figures that the wall correction factor K is increasing as the separation parameter η is increasing for any fixed S. The wall effects on liquid spheroid is decreasing with increasing S for any η, which is expected. It is also observed from Figs. 5 and 6 that the wall effects on Newtonian liquid spheroid is higher than such effects on Reiner-Rivlin liquid spheroid. In case when $\epsilon = 0$, the results of the present discussion coincide with the results obtained earlier by Ramkissoon and Rahaman [16]. Further, we also conclude from the figures that a perfect liquid sphere experiences less wall effects than a liquid spheroid when both are in spherical container.

We have plotted the wall correction factor K against S $(0 < S < 1)$, for various values of viscosity ratio λ in Fig. 7 and for a fixed value of separation parameter η and deformity of the

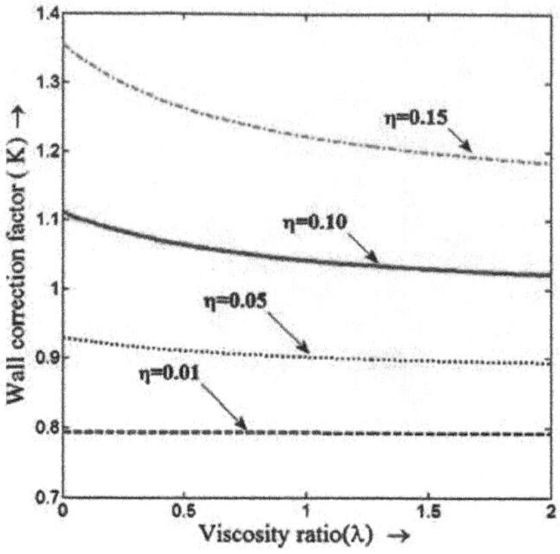

Fig. 8. Variation of the wall correction factor K with λ for various values of η at $S = 0.5$ and $\epsilon = 0.5$

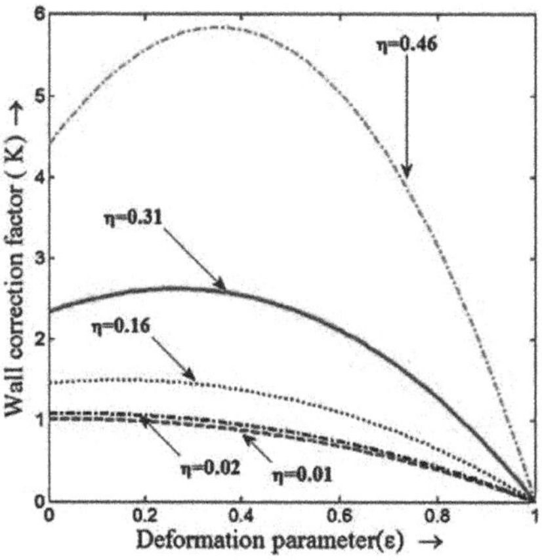

Fig. 9. Variation of the wall correction factor K with deformation parameter ϵ for various values of η at $S = 0.5$ and $\lambda = 0.5$

liquid spheroid ϵ. It can be visualized from the figure that the wall correction factor K, for a fixed λ, decreases with S. We also observe that, for a fixed S, the wall correction factor K shows the identical behavior with λ. It is interesting to note that the wall effects on rigid spheroid is higher than that of a liquid spheroid.

The variation of the wall effects with regard to viscosity ratio λ for various values of separation parameter η is depicted in Fig. 8 for any specified value of S and ϵ. The figure shows that, for relatively large values of separation parameter η, the wall effects on liquid spheroid after a slight decrease becomes almost constant with λ, whereas corresponding to very small values of η, the wall effects on it is almost constant in the entire range of viscosity ratio λ. The graph in Fig. 8 also shows that for a fixed value of λ, the wall effects on liquid spheroid increases with separation parameter η.

The variation of the wall effects on the surface of the Reiner-Rivlin liquid spheroid with deformation parameter ϵ is illustrated in Fig. 9 for numerous values of separation parameter η for a fixed value of S and λ. We can see that corresponding to very small values of η (≤ 0.02), the wall correction factor K is monotonically decreasing function of deformation parameter ϵ. In every other possibility of separation parameter η, wall effects on liquid spheroid first increases

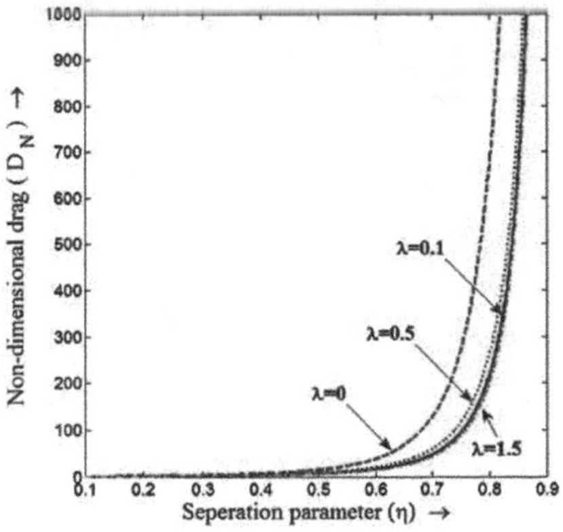

Fig. 10. Variation of the non-dimensional drag D_N with separation parameter η for various values of λ at $S = 0.5$ and $\epsilon = 0.5$

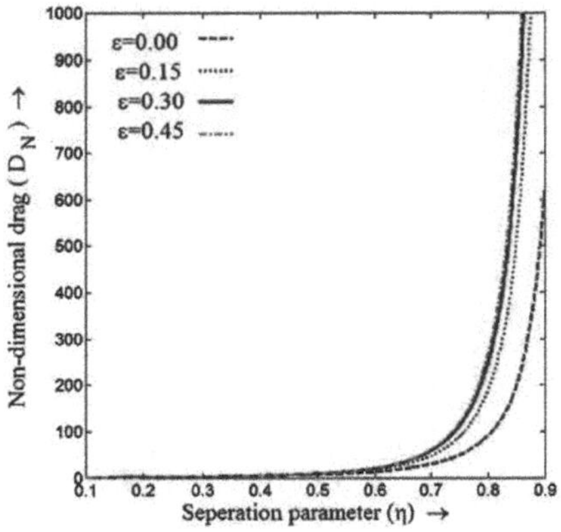

Fig. 11. Variation of the non-dimensional drag D_N with η for various values of deformation parameter ϵ at $S = 0.5$ and $\lambda = 0.5$

and attains to its maximum value at some certain value ϵ and then decreases significantly with the increasing ϵ and vanishes at $\epsilon = 1$. The graph in Fig. 9 also elucidates that the wall effects on liquid spheroid increases as separation parameter η increases for any fixed value of ϵ.

The non-dimensional drag D_N:

$$D_N = \frac{F}{4\pi U c \mu_1} = (1 - \epsilon)(d_2 + D_2). \tag{38}$$

The variation of D_N with respect to the separation parameter η is shown in Fig. 10 for various values of viscosity ratio λ. From the figure, it is observed that as the separation parameter η increases the drag coefficient D_N on Reiner-Rivlin liquid spheroid increases. Also D_N increases as viscosity ratio λ decreases. For relatively smaller values of η, D_N increases slowly with approximately equal amount for all the values of viscosity ratio λ. But for comparatively larger values of η (> 0.5), D_N increases very rapidly and differently with η for different values λ. It is interesting to note that for $\lambda > 1$ drag varies slowly with η as compared with the case of lower viscosity ratio $\lambda < 1$. Further, we also notice that the drag on a rigid spheroid is more than the drag experienced by a Reiner-Rivlin liquid spheroid which is expected.

Fig. 11 illustrates the effects of the deformation parameter ϵ and separation parameter η on the drag D_N for a fixed value of λ and S. It can be observed from Fig. 11 that D_N is increasing

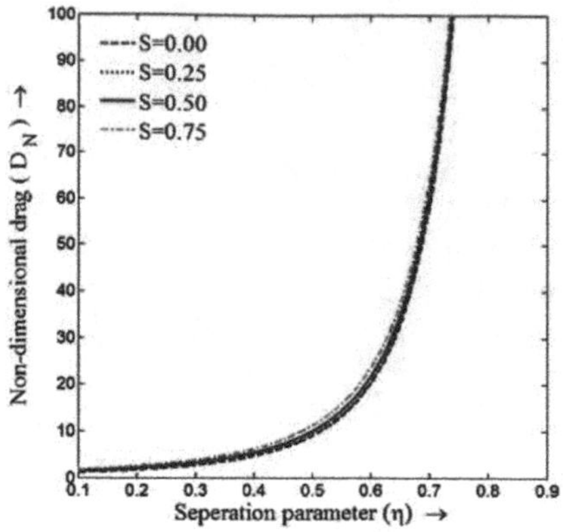

Fig. 12. Variation of the non-dimensional drag D_N with η for various values of S at $\epsilon = 0.5$ and $\lambda = 0.5$

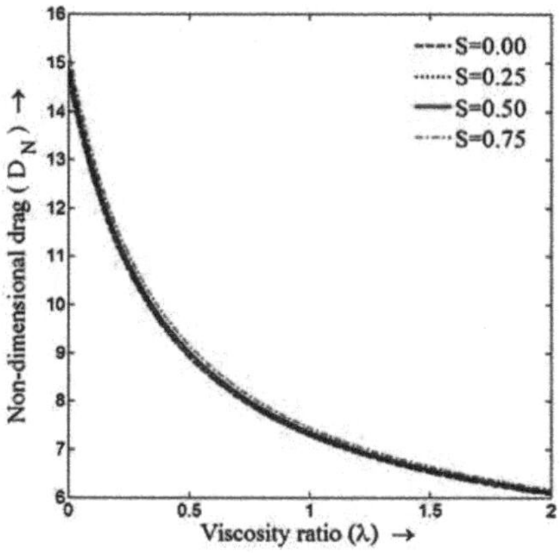

Fig. 13. Variation of the non-dimensional drag D_N with λ for various values of S at $\eta = 0.5$ and $\epsilon = 0.5$

as the separation parameter η is increasing. Also, the drag D_N is increasing with the increasing deformity of the liquid spheroid. The figure evidently indicates that the drag on Reiner-Rivlin liquid sphere is the lowest as compared with dag on the Reiner-Rivlin liquid spheroid. The figure also illustrates that the growth rate of drag D_N keeps on decreasing as the deformation of the liquid sphere is increasing.

We have plotted the drag D_N against η for various values of dimensionless parameter S in Fig. 12 and for a fixed value of deformation parameter ϵ and viscosity ratio λ. It can be seen from the figure that the drag D_N, for a fixed S, increases with η. We also observe that, for a fixed η, the drag D_N shows the identical behavior with S. It is interesting to note that the drag on Newtonian liquid spheroid is the lowest as compared with the drag experienced by a Reiner-Rivlin liquid spheroid.

Fig. 13 exhibits the variation of the drag D_N with λ for various values of dimensionless parameter S for a fixed value of deformation parameter ϵ and separation parameter η. It is observed from Fig. 13 that the drag D_N, for a fixed S, is decreasing with λ. We also observe that, for a fixed λ, the drag D_N is increasing with S. It is interesting to note that the drag on Newtonian liquid spheroid is the lowest as compared with the drag experienced by a Reiner-Rivlin liquid spheroid.

The variation in the drag with S is numerically studied for various values of separation parameter η for a fixed λ and ϵ and presented in Fig. 14 graphically. We observe from the figure that the drag D_N on the Reiner-Rivlin liquid spheroid is increasing with S. Here, the similar

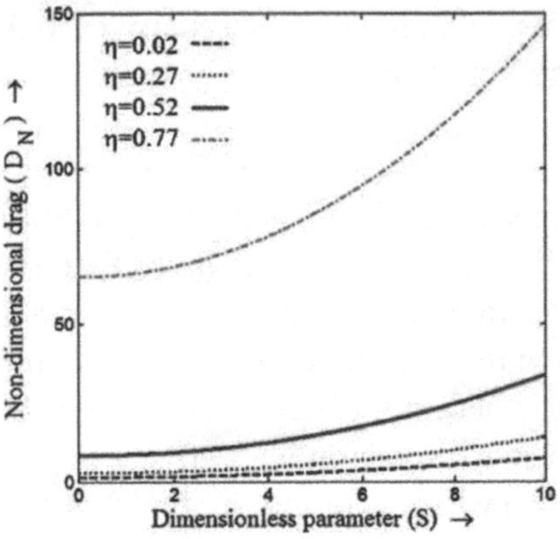

Fig. 14. Variation of the non-dimensional drag D_N with S for various values of separation parameter η at $\lambda = 0.5$ and $\epsilon = 0.01$

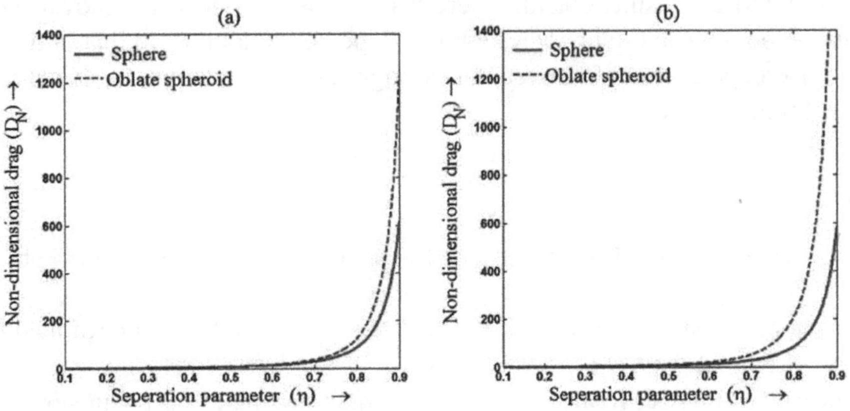

Fig. 15. Variation of the non-dimensional drag with separation parameter η over the sphere and spheroid of equal volumes at $\lambda = 0.5$ and $S = 0.5$ for (a) $\epsilon = 0.05$, (b) $\epsilon = 0.2$

behavior is observed in the variation of the drag on Reiner-Rivlin liquid spheroid with η for a fixed S and the only thing that draws our attention is the growth rate of drag on the body for comparatively lager values of separation parameter is higher than smaller values of η.

A comparative study of drag force experienced by a liquid sphere and a liquid spheroid having equal volumes is performed numerically and corresponding variations are shown graphically in Fig. 15. It can be easily observed that a liquid spheroid of equal volume experiences a larger drag than the liquid sphere. Also, the drag on spheroid increases whereas this drag decreases on sphere with the deformation parameter ϵ.

7. Conclusion

In this paper, an analytical solution to the problem of motion of a Reiner-Rivlin liquid spheroid at the instant it passes the centre of a spherical envelope filled with an incompressible Newtonian fluid is investigated by taking into account the Stokes equation for the external flow, whereas for internal flow expressing the stream function and pressure in a power series of S. An expression for the hydrodynamic drag force D_N on the Reiner-Rivlin liquid spheroid in a spherical shell and wall correction factor K are obtained in closed form in terms of separation parameter η, cross viscosity, i.e., dimensionless parameter S, viscosity ratio λ and deformation parameter ϵ. It is found that the drag on the liquid spheroid for smaller deformation is more than the drag on a liquid sphere in presence of spherical envelope. But for larger deformation, the behavior is reversed which is physically not possible. Therefore, our proposed model is valid only for small deformation. It is also observed that in case of equal volume and surface area, the drag force exerted on a Reiner-Rivlin liquid spheroid is more than the drag on a liquid sphere. The drag increases with separation parameter η, cross-viscosity μ_c and small deformation parameter ϵ and decreases with viscosity ratio λ. Although the very similar behavior in variation of wall effects on Reiner-Rivlin liquid spheroid is observed, but the cross-viscosity μ_c of Reiner-Rivlin liquid is reported to have a reverse impact on wall effects unlike drag force. Further, it is also observed that wall effects on Newtonian liquid spheroid is found to be more than the such effects on Reiner-Rivlin liquid spheroid and on the other hand the drag force on Newtonian liquid spheroid is found less than the drag on a Reiner-Rivlin liquid spheroid and this drag on liquid spheroid increases with increase in S, i.e., the cross-viscosity. By the analysis of the proposed model and the results obtained here can be easily applied to recover and to collect the information about those liquid bodies whose shape depart a bit from that of a sphere which are suitable for creeping or swift movement through the fluid medium enclosed in an infinite spherical envelope.

References

[1] Abramowitz, M., Stegun, I.A., Handbook of Mathematical Functions, Dover Publications, New York, 1970.

[2] Cunningham, E., On the velocity of steady fall of spherical particles through fluid medium, Proceedings of the Royal Society of London A 83 (1910) 357–365.

[3] Haberman, W.L., Sayre, R.M., Wall effects for rigid and fluid spheres in slow motion with a moving liquid, David Taylor model, Basin Report No. 1143, Washington DC, 1958.

[4] Happel, J., Brenner, H., Low Reynolds Number Hydrodynamics, Prentice-Hall, Engle-wood Cliffs, N. J., 1965.

[5] Iyengar, T.K.V., Srinivasacharya, D., Stokes flow of an incompressible micropolar fluid past an approximate sphere, International Journal of Engineering Science 31 (1) (1993) 115–123.

[6] Jaiswal, B.R., Gupta, B.R., Drag on Reiner-Rivlin liquid sphere placed in a micropolar fluid with non-zero boundary condition for microrotations, International Journal of Applied Mathematics and Mechanics 10 (7) (2014) 93–103.

[7] Jones, R.B., Dynamics of a colloid in a spherical cavity, Theoretical Methods for Micro Scale Viscous Flows, eds. Francois Feuillebois and Antoine Sellier, Transworld Research Network, 2009.

[8] Keh, H.J., Chou, J., Creeping motion of a composite sphere in a concentric spherical cavity, Chemical Engineering Science 59 (2) (2004) 407–415.

[9] Keh, H.J., Lu, Y.S., Creeping motions of a porous spherical shell in a concentric spherical cavity, Journal of Fluids and Structures 20 (5) (2005) 735–747.

[10] Kim, S., Karrila, S.J., Micro hydrodynamics: Principles and Selected Applications, Butterworth-Heinemann, Boston, MA, USA, 1991.

[11] Maul, C., Kim, S., Image of a point force in a spherical container and its connection to the Lorentz reflection formula, Journal of Engineering Mathematics 30 (1–2) (1996) 119–130.

[12] Maul, C., Kim, S., Image systems for a Stokeslet inside a rigid spherical container, Physics of Fluids 6 (6) (1994) 2221–2223.

[13] Oseen, C.W., Hydrodynamik, Akademische Verlagsgesellschaft, Leipzig, 1927.

[14] Palaniappan, D., Creeping flow about a slightly deformed sphere, Zeitschrift für Angewandte Mathematik und Physik 45 (1) (1994) 323–338.

[15] Ramkissoon, H., Majumadar, S.R., Micropolar fluid past a slightly deformed fluid sphere, Zeitschrift für Angewandte Mathematik und Mechanik 68 (3) (1988) 155–160.

[16] Ramkissoon, H., Rahaman, K., Non-Newtonian fluid sphere in a spherical container, Acta Mechanica 149 (1–44) (2001) 239–245.

[17] Ramkissoon, H., Rahaman, K., Wall effects on a spherical particle, International Journal of Engineering Science 41 (3–5) (2003) 283–290.

[18] Ramkissoon, H., Rahaman, K., Wall effects with slip, Zeitschrift für Angewandte Mathematik und Mechanik 83 (11) (2003) 773–778.

[19] Ramkissoon, H., Stokes flow past a non-Newtonian fluid spheroid, Zeitschrift für Angewandte Mathematik und Mechanik 78 (1) (1998) 61–66.

[20] Ramkissoon, H., Polar flow past a Reiner-Rivlin liquid sphere, Journal of Mathematical Sciences 10 (2) (1999) 63–68.

[21] Ramkissoon, H., Stokes flow past a Reiner-Rivlin fluid sphere, Zeitschrift für Angewandte Mathematik und Mechanik 69 (8) (1989) 259–261.

[22] Ramkissoon, H., Stokes flow past a slightly deformed fluid sphere, Journal of Applied Mathematics and Physics 37 (6) (1986) 859–866.

[23] Saad, E.I., Motion of a spheroidal particle in a micropolar fluid contained in a spherical envelope, Canadian Journal of Physics 86 (2008) 1039–1056.

[24] Saad, E.I., Translation and rotation of a porous spheroid in a spheroidal container, Canadian Journal of Physics 88 (2010) 689–700.

[25] Srinivasacharya, D., Krishna Prasad, M., Axi-symmetric motion of a porous approximate sphere in an approximate spherical container, Archives of Mechanics 65 (6) (2013) 485–509.

[26] Srinivasacharya, D., Krishna Prasad, M., Creeping motion of a porous approximate sphere with an impermeable core in spherical container, European Journal of Mechanics B-Fluids 36 (2012) 104–114.

[27] Srinivasacharya, D., Krishna Prasad, M., On the motion of a porous spherical shell in a bounded medium, Advances in Theoretical and Applied Mechanics 5 (6) (2012) 247–256.

[28] Srinivasacharya, D., Krishna Prasad, M., Slow steady rotation of a porous sphere in a spherical container, Journal of Porous Media 15 (12) (2012) 1 105–1 110.

[29] Srinivasacharya, D., Krishna Prasad, M., Steady rotation of a composite sphere in a concentric spherical cavity, Acta Mechanica Sinica 28 (3) (2012) 653–658.

[30] Srinivasacharya, D., Motion of a porous sphere in a spherical container, Compte Rendus Mécanique 333 (8) (2005) 612–616.

[31] Stokes, G. G., On the effect of the internal friction of fluid on the motion of pendulums, Transactions of the Cambridge Philosophical Society 9 (1851) 8–106.

[32] Williams, W.E., On the motion of a sphere in a viscous fluid, Philosophical Magazine 29 (1915) 526–550.

Appendix A

The unknowns appearing in (11) and (13) can be deduced by substituting these equations into the boundary conditions (15)–(18), which lead to the following system of algebraic equations:

$$(a_2 + b_2 + c_2 + d_2)G_2(\zeta) + (2a_2 - b_2 + 4c_2 + d_2)\alpha_k G_k(\zeta)G_2(\zeta)+$$
$$(a_3 + b_3 + c_3 + d_3)G_3(\zeta) + \sum_{n=4}^{\infty}(A_n + B_n + C_n + D_n)G_n(\zeta) = 0, \qquad \text{(A.1)}$$

$$\left(e_2+f_2+\frac{4}{63}S^2\right)G_2(\zeta)+\left(2e_2+4f_2+4+\frac{8}{21}S^2\right)\alpha_k G_k(\zeta)G_2(\zeta)+$$
$$(e_3+f_3+\frac{4}{21}S)G_3(\zeta)+5\left(f_3+\frac{4}{21}S\right)\alpha_k G_k(\zeta)G_3(\zeta)+\sum_{n=4}^{\infty}(E_n+F_n)G_n(\zeta)=0, \quad \text{(A.2)}$$

$$\left(2a_2-b_2+4c_2+d_2-2e_2-4f_2-4-\frac{8}{21}S^2\right)G_2(\zeta)+$$
$$(2a_2+2b_2+12c_2-2e_2-12f_2-20-\frac{40}{21}S^2)\alpha_k G_k(\zeta)G_2(\zeta)+$$
$$\left(3a_3-2b_3+5c_3-3e_3-5f_3-\frac{20}{21}S\right)G_3(\zeta)-$$
$$20\left(f_3+\frac{4}{21}S\right)\alpha_k G_k(\zeta)G_3(\zeta)+$$
$$\sum_{n=4}^{\infty}[nA_n-(n-1)B_n+(n+2)C_n+(-n+3)D_n-nE_n-(n+2)F_n]G_n(\zeta)=0, \quad \text{(A.3)}$$

$$[\lambda(-2a_2+4b_2+4c_2-2d_2)+2e_2-4f_2-12-\frac{8}{7}S^2]G_2(\zeta)+$$
$$[\lambda(4a_2-20b_2+6d_2)-4e_2+8-\frac{16}{7}S^2]\alpha_k G_k(\zeta)G_2(\zeta)+$$
$$\left[\lambda(10b_3-10c_3)-10\left(f_3+\frac{4}{21}S\right)\right]\times G_3(\zeta)-10\left(f_3+\frac{4}{21}S\right)\alpha_k G_k(\zeta)G_3(\zeta)+$$
$$\sum_{n=4}^{\infty}[\lambda(n(n-3)A_n+(n+2)(n-1)B_n+(n-1)(n+2)C_n+n(n-3)D_n)-$$
$$n(n-3)E_n-(n-1)(n+2)F_n]G_n(\zeta)=0, \quad \text{(A.4)}$$

$$\left(\frac{a_2-1}{\eta^2}+\eta b_2+\frac{c_2}{\eta^4}+\frac{d_2}{\eta}\right)G_2(\zeta)+\left(\frac{a_3}{\eta^3}+\eta^2 b_3+\frac{c_3}{\eta^5}+d_3\right)G_3(\zeta)+$$
$$\sum_{n=4}^{\infty}\left(\frac{A_n}{\eta^n}+\frac{B_n}{\eta^{-n+1}}+\frac{C_n}{\eta^{n+2}}+\frac{D_n}{\eta^{-n+3}}\right)G_n(\zeta)=0 \quad \text{(A.5)}$$

and

$$\left(\frac{2(a_2-1)}{\eta}-\eta^2 b_2+\frac{4c_2}{\eta^3}+d_2\right)G_2(\zeta)+\left(\frac{3a_3}{\eta^2}-2\eta^3 b_3+\frac{5c_3}{\eta^4}\right)G_3(\zeta)+$$
$$\sum_{n=4}^{\infty}\left(\frac{nA_n}{\eta^{n-1}}+\frac{(-n+1)B_n}{\eta^{-n}}+\frac{(n+2)C_n}{\eta^{n+1}}+\frac{(-n+3)D_n}{\eta^{-n+2}}\right)G_n(\zeta)=0. \quad \text{(A.6)}$$

The leading terms, i.e., the coefficients of $G_2(\zeta)$ and $G_3(\zeta)$ in the above system of equations (A.1)–(A.6) must vanish, i.e.,

$$a_2+b_2+c_2+d_2=0, \quad 3+b_3+c_3+d_3=0, \quad e_2+f_2+\frac{4}{63}S^2=0, \quad e_3+f_3+\frac{4}{21}S=0,$$

$$2a_2 - b_2 + 4c_2 + d_2 - 2e_2 - 4f_2 - 4 - \frac{8}{21}S^2 = 0, \quad 3a_3 - 2b_3 + 5c_3 - 3e_3 - 5f_3 - \frac{20}{21}S = 0,$$
$$\lambda(-2a_2 + 4b_2 + 4c_2 - 2d_2) + 2e_2 - 4f_2 - 12 - \frac{8}{7}S^2 = 0, \quad \lambda(b_3 - c_3) - f_3 - \frac{4}{21}S = 0,$$

$$\eta^2 a_2 + \eta^5 b_2 + c_2 + \eta^3 d_2 - \eta^2 = 0, \quad \eta^3 a_3 + \eta^7 b_3 + c_3 + \eta^5 d_3 = 0,$$
$$2\eta^2 a_2 - \eta^5 b_2 + 4c_2 + \eta^3 d_2 - 2\eta^2 = 0, \quad 3\eta^2 a_3 - 2\eta^7 b_3 + 5c_3 = 0. \tag{A.7}$$

Solving (A.7) yields

$$\begin{aligned}
&a_3 = 0, \quad b_3 = 0, \quad c_3 = 0, \quad d_3 = 0, \quad e_3 = 0, \quad f_3 = -\frac{4}{21}S,\\
e_2 &= [4S^2(42\eta^2 + \eta(7 - 6\lambda) + 4\eta^4(-7 + 3\lambda) - 4(7 + 3\lambda) + \eta^3(7 + 6\lambda)) +\\
&\quad 189(-8 + 2\eta + 8\eta^4(-1 + \lambda) - 6\lambda + 6\eta^2(2 + \lambda) + \eta^3(2 + 7\lambda))]\Delta,\\
f_2 &= -[8S^2(30\eta^2 + \eta(5 - 6\lambda) + 4\eta^4(-5 + 3\lambda) - 4(5 + 3\lambda) + \eta^3(5 + 6\lambda)) +\\
&\quad 189(-8 + 2\eta + 8\eta^4(-1 + \lambda) - 6\lambda + 6\eta^2(2 + \lambda) + \eta^3(2 + 7\lambda))]\Delta,\\
a_2 &= -[32S^2(3\eta - 5\eta^3 + 2\eta^6) + 189(5\eta^3 + 4(1 + \lambda) + \eta^5(-9 + 6\lambda))]\Delta,\\
b_2 &= -[32S^2(2 - 3\eta + \eta^3) + 378(1 + \eta^3(-1 + \lambda))]\Delta,\\
c_2 &= [\eta^3(32S^2(-1 + \eta)^2(1 + 2\eta) + 189(3 - 3\eta^2 + 2\lambda))]\Delta,\\
d_2 &= [2(16S^2(2 - 5\eta^3 + 3\eta^5) + 189(3 + 3\eta^5(-1 + \lambda) + 2\lambda))]\Delta,
\end{aligned} \tag{A.8}$$

where

$$\frac{1}{\Delta} = 189(-1 + \eta)^3[4\eta^3(-1 + \lambda) + 4(1 + \lambda) + \eta^2(-3 + 6\lambda) + \eta(3 + 6\lambda)].$$

By using these values, the equations (A.1)–(A.6) now get reduced respectively to

$$\xi_1 \alpha_k G_k(\zeta) G_2(\zeta) + \sum_{n=4}^{\infty} (A_n + B_n + C_n + D_n) G_n(\zeta) = 0, \tag{A.9}$$

$$\xi_2 \alpha_k G_k(\zeta) G_2(\zeta) + \sum_{n=4}^{\infty} (E_n + F_n) G_n(\zeta) = 0, \tag{A.10}$$

$$\begin{aligned}\xi_4 \alpha_k G_k(\zeta) G_2(\zeta) + \sum_{n=4}^{\infty} [nA_n + (-n + 1)B_n + (n + 2)C_n +\\
(-n + 3)D_n - nE_n - (n + 2)F_n] G_n(\zeta) = 0,\end{aligned} \tag{A.11}$$

$$\begin{aligned}\xi_6 \alpha_k G_k(\zeta) G_2(\zeta) + \sum_{n=4}^{\infty} [\lambda(n(n - 3)A_n + (n + 2)(n - 1)B_n +\\
(n - 1)(n + +2)C_n + n(n - 3)D_n) - n(n - 3)E_n - (n - 1)(n + 2)F_n] G_n(\zeta) = 0,\end{aligned} \tag{A.12}$$

$$\sum_{n=4}^{\infty} \left(\frac{A_n}{\eta^n} + \frac{B_n}{\eta^{-n+1}} + \frac{C_n}{\eta^{n+2}} + \frac{D_n}{\eta^{-n+3}} \right) G_n(\zeta) = 0, \tag{A.13}$$

$$\sum_{n=4}^{\infty} \left(\frac{nA_n}{\eta^{n-1}} + \frac{(-n + 1)B_n}{\eta^{-n}} + \frac{(n + 2)C_n}{\eta^{n+1}} + \frac{(-n + 3)D_n}{\eta^{-n+2}} \right) G_n(\zeta) = 0, \tag{A.14}$$

where

$$\begin{aligned}\xi_1 = \xi_2 &= [2(16S^2(-1+\eta)^2(4+7\eta+4\eta^2) - 189(2+4\eta+6\eta^2+3\eta^3)\lambda)]\Lambda,\\ \xi_4 &= -3[2(-189(2+4\eta+6\eta^2+3\eta^3)+32S^2(-2-\eta+\eta^3+2\eta^4))(-1+\lambda)]\Lambda,\\ \xi_6 &= -[4(-189\lambda(13+10\eta0^3(-1+\lambda)+3\eta^5(-1+\lambda)+2\lambda)+8S^2(68\eta^6(-1+\lambda)-\\ &\quad 10\eta^3(17+\lambda)-9\eta^5(-17+12\lambda)+9\eta(17+18\lambda)-4(17+28\lambda)))]\Lambda\end{aligned}$$

and

$$\frac{1}{\Lambda} = 189(-1+\eta)[4\eta^3(-1+\lambda)+4(1+\lambda)+\eta^2(-3+6\lambda)+\eta(3+6\lambda).$$

To determine the remaining arbitrary constants in equations (A.9)–(A.14), we use the following identity

$$\begin{aligned}G_kG_2 = &-\frac{(k-2)(k-3)}{2(2k-1)(2k-3)}G_{k-2} + \frac{k(k-1)}{(2k+1)(2k-3)}G_k - \\ &\frac{(k+1)(k+2)}{2(2k-1)(2k+1)}G_{k+2}, \qquad k \geq 2.\end{aligned} \tag{A.15}$$

In solving the system of equations (A.9)–(A.14), we glimpse that $A_n = B_n = C_n = D_n = E_n = F_n = 0$, if $n \neq k-2, k, k+2$ and when $n = k-2, k, k+2$, we have the following system of equations:

$$\xi_1\alpha_k\phi_n + A_n + B_n + C_n + D_n = 0, \tag{A.16}$$

$$\xi_2\alpha_k\phi_n + E_n + F_n = 0, \tag{A.17}$$

$$\begin{aligned}\xi_4\alpha_k\phi_n + nA_n - (n-1)B_n + (n+2)C_n - (n-3)D_n - nE_n - \\ (n+2)F_n = 0,\end{aligned} \tag{A.18}$$

$$\begin{aligned}\xi_6\alpha_k\phi_n + \lambda[n(n-3)A_n + (n+2)(n-1)B_n + (n-1)(n+2)C_n + \\ n(n-3)D_n] - n(n-3)E_n - (n-1)(n+2)F_n = 0,\end{aligned} \tag{A.19}$$

$$\frac{A_n}{\eta^n} + \frac{B_n}{\eta^{-n+1}} + \frac{C_n}{\eta^{n+2}} + \frac{D_n}{\eta^{-n+3}} = 0, \tag{A.20}$$

$$\frac{nA_n}{\eta^{n-1}} + \frac{(-n+1)B_n}{\eta^{-n}} + \frac{(n+2)C_n}{\eta^{n+1}} + \frac{(-n+3)D_n}{\eta^{-n+2}} = 0, \tag{A.21}$$

where

$$\phi_{k-2} = -\frac{(k-2)(k-3)}{2(2k-1)(2k-3)}, \quad \phi_k = \frac{k(k-1)}{(2k+1)(2k-3)}, \quad \phi_{k+2} = -\frac{(k+1)(k+2)}{2(2k-1)(2k+1)}.$$

The solutions of the equations (A.16)–(A.21) yield the expressions for A_n, B_n, C_n, D_n, E_n, and F_n when $n = k-2, k, k+2$.

16

Influence of the mass of the weight on the dynamic response of the asymmetric laboratory fibre-driven mechanical system

P. Polach[a,*], M. Hajžman[a], Z. Šika[b], O. Červená[a], P. Svatoš[b]

[a] *Section of Materials and Mechanical Engineering Research, Výzkumný a zkušební ústav Plzeň s.r.o., Tylova 1581/46, 301 00 Plzeň, Czech Republic*

[b] *Department of Mechanics, Biomechanics and Mechatronics, Faculty of Mechanical Engineering, Czech Technical University in Prague, Technická 4, 166 07 Praha, Czech Republic*

Abstract

Experimental measurements focused on the investigation of a fibre behaviour are performed on an assembled weigh-fibre-pulley-drive mechanical system. The fibre is driven with one drive and it is led over a pulley. On its other end there is a prism-shaped steel weight, which moves in a prismatic linkage on an inclined plane. The position of the weight is asymmetric with respect to the vertical plane of drive-pulley symmetry. Drive exciting signals can be of a rectangular, a trapezoidal and a quasi-sinusoidal shape and there is a possibility of variation of a signal rate. Dynamic responses of the weight and the fibre are measured. The same system is numerically investigated by means of a multibody model. The influence of the mass of the weight and the influence of the weight asymmetry on the coincidence of results of experimental measurements and the simulations results are evaluated. The simulations aim is to create a phenomenological model of a fibre, which will be utilizable in fibre modelling in the case of more complicated mechanical or mechatronic systems.

Keywords: fibre, mechanical system, dynamic response, phenomenological model, experiment, computer simulations

---

1. Introduction

The replacement of the chosen rigid elements of manipulators or mechanisms by fibres or cables [3] is advantageous due to the achievement of a lower moving inertia, which can lead to a higher machine speed, and lower production costs. Drawbacks of using the flexible elements like that can be associated with the fact that cables should be only in tension (e.g. [7,30]) in the course of a motion.

Experimental measurements focused on the investigation of the fibre behaviour were performed on an assembled weigh-fibre-pulley-drive system [20,21,24]. A fibre is driven with one drive, it is led over a pulley and on its other end there is a prism-shaped steel weight, which moves on an inclined plane. The position of the weight can be symmetric or asymmetric with respect to the plane of drive-pulley symmetry (in presented case asymmetric position is considered, see Fig. 1). It is possible to add an extra mass to the weight (in presented case the added mass is considered). The same system is numerically investigated using a multibody model created in the *alaska* simulation tool [14]. The influence of the model parameters on the coincidence of the results of experimental measurements and the simulations results is evaluated. The simulation aim is to create a phenomenological model of a fibre, which will be utilizable in fibre

*Corresponding author. e-mail: polach@vzuplzen.cz.

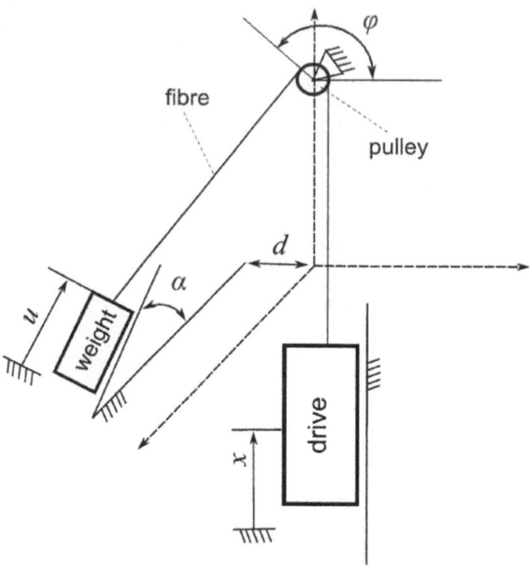

Fig. 1. Scheme and a real weight-fibre-pulley-drive mechanical system with asymmetric position of the weight

modelling in the case of more complicated mechanical or mechatronic systems. A spherical tilting mechanism named QuadroSphere (see Fig. 2) is an example of such a mechatronic system, at which the phenomenological computational model of fibre will be utilized. The QuadroSphere is a mechanism with a spherical motion of a platform and accurate measurement of its position. The platform position is controlled by four fibres; each fibre is guided by a pulley from linear guidance to the platform (see Fig. 3).

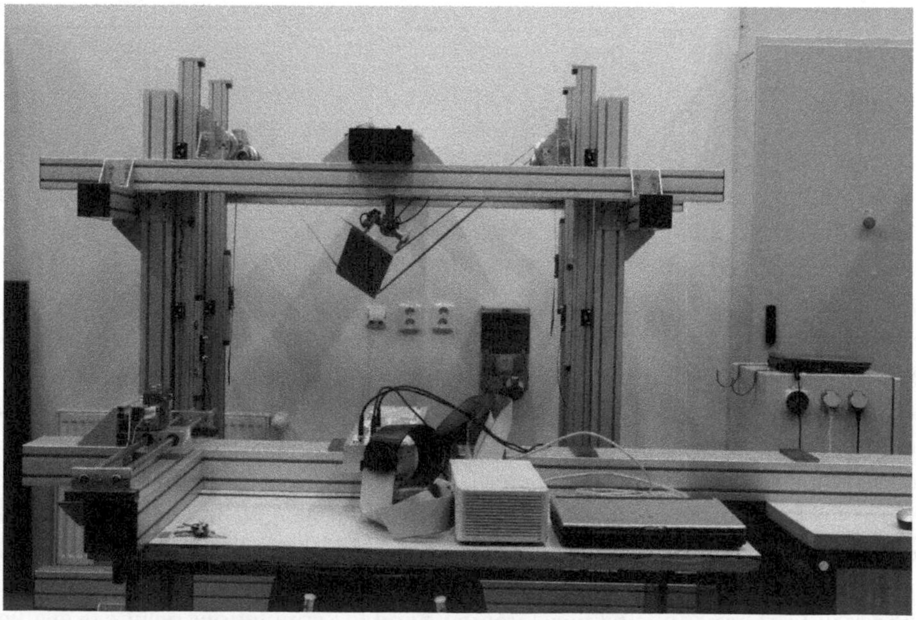

Fig. 2. The QuadroSphere spherical tilting mechanism

Fig. 3. Detail of the QuadroSphere platform

The first pieces of knowledge concerning the phenomenological model of a simple fibre-mass system (the system consists of a moving weight coupled with a frame by a fibre) creation are given in [18] and [19]. The paper continues investigating the weight-fibre-pulley-drive system given in [20] and [24] where the position of the weight was symmetric with respect to the plane of a drive-pulley symmetry, and in [21], where the position of the weight was asymmetric and the weight was without added mass.

2. Experimental stand

Originally it was supposed that for the experimental measurement focused on determining the phenomenological model of the fibre an inverted pendulum driven by two fibres attached to a frame would be used. Its properties were investigated very thoroughly applying calculation models (see e.g. [17]). But strength calculation results drew attention to a high loading of fibres which were to be used in experimental measurements (carbon or wattled steel wire) and to the possibility of their breaking [22].

Due to those reasons a different mechanical system was chosen for the experimental measurements (its geometrical arrangement was changed several times on the basis of various pieces of knowledge). Experimental measurements focused on the investigation of the fibre behaviour are performed on an assembled weigh-fibre-pulley-drive mechanical system (see Fig. 1). A carbon fibre with a silicone coating (see e.g. [22]) is driven with one drive and it is led over a pulley. The fibre length is 1.82 meters (fibre weight is 4.95 grams), the pulley diameter is 80 millimetres. The weight position can be symmetric [20,24] or asymmetric [21] with respect to the vertical plane of drive-pulley symmetry (as it has been already mentioned asymmetric position is considered in this paper). Distance of the weight from the vertical plane of drive-pulley symmetry is $d = 280$ mm in the case of the asymmetric weight position (see Fig. 1). At the drive the fibre is fixed on a force gauge. On the other end of the fibre there is a prism-shaped steel weight (weight 3.096 kilograms), which moves in a prismatic linkage on an inclined plane. It is possible to add an extra mass (of the weight 5.035 kilograms) to the weight (as it has been already mentioned the added mass is considered in this paper). The angle of inclination of the inclined plane could be changed (in this case the angle is $\alpha = 30.6$ degrees and the pulley-fibre angle is $\varphi = 124$ degrees). Drive exciting signals can be of a rectangular, a trapezoidal and a quasi-sinusoidal shape and there is a possibility of variation of a signal rate [15]. The amplitudes of the drive displacements are up to 90 millimetres. Time histories of the weight position u (in direction of the inclined plane; measured by means of a dial gauge), of the drive position x (in vertical direction) and of the force acting in the fibre (measured on a force gauge at drive) were recorded using sample rate of 2 kHz.

3. Possibilities of the fibre modelling

The fibre (cable, wire etc.) modelling [9] should be based on considering the fibre flexibility and suitable approaches can be based on the flexible multibody dynamics (see e.g. [6,28]). Flexible multibody dynamics is a rapidly growing branch of computational mechanics and many industrial applications can be solved using newly proposed flexible multibody dynamics approaches. Studied problems are characterized by a general large motion of interconnected rigid and flexible bodies with the possible presence of various nonlinear forces and torques. There are many approaches to the modelling of flexible bodies in the framework of multibody systems [8]. Comprehensive reviews of these approaches can be found in [28] or in [31]. Further development together with other multibody dynamics trends was introduced in [27]. Details of multibody formalisms and means of the creation of equations of motion can be found e.g. in [1,29].

The simplest way how to incorporate fibres in equations of motion of a mechanism is the force representation of a fibre (e.g. [4]; the massless fibre model). It is assumed that the mass of fibres is low to such an extent comparing to the other moving parts that the inertia of fibres is negligible with respect to the other parts. The fibre is represented by the force dependent on the fibre deformation and its stiffness and damping properties. This way of the fibre modelling is probably the most frequently used one in the cable-driven robot dynamics and control (e.g. [11, 33]). The fibre-mass system fulfils all requirements for modelling the fibre using the force representation of a fibre. A more precise approach is based on the representation of the fibre by means of a point-mass model (e.g. [12]). It has the advantage of a lumped point-mass model. The point masses can be connected by forces or constraints. In order to represent bending behaviour of fibres their discretization using the finite segment method [28] or so called rigid finite elements [32] is possible. Standard multibody codes (SIMPACK, MSC.ADAMS, alaska etc.) can be used for this purpose. Other more complex approaches can utilize nonlinear three-dimensional finite elements [5] or can employ the absolute nodal coordinate formulation (ANCF) elements [6,10,13,28].

The massless fibre model is considered in this phase of investigation of the weight-fibre-pulley-drive system. The fibre model is considered to be phenomenological and it is modelled by the forces which comprise e.g. influences of fibre transversal vibration, "jumping" from the pulley etc. The weight (with added mass), the pulley, the cradle of pulley and the drive are considered to be rigid bodies [21]. The number of degrees of freedom in kinematic joints is 6. A planar joint between the weight and the base (prismatic linkage), a revolute joint between the cradle of pulley and the base, a revolute joint between the pulley and the cradle of pulley and a prismatic joint between the drive and the base (the movement of the drive is kinematically prescribed) are considered. Behaviour of this nonlinear system is investigated using the *alaska* simulation tool [14].

4. Simulation and experimental results

As it has already been stated the simulations aim was to create a phenomenological model of a fibre. When looking for compliance of the results of experimental measurement with the simulation results influences of the following system parameters are considered: the fibre stiffness, the fibre damping coefficient and the friction force acting between the weight and the prismatic linkage in which the weight moves.

Investigation of the (carbon) fibre properties eliminating the influence of the drive and of the pulley was an intermediate stage before the measurement on the stand [18,19]. A phenomenological model dependent on the fibre stiffness, on the fibre damping coefficient and on the friction force acting between the weight and the prismatic linkage was the result of this investigation. When looking for the fibre model [18] that would ensure the similarity of time histories of the weight displacement and time histories of the dynamic force acting in a fibre as high as possible fibre stiffness and fibre damping coefficient were considered to be constant in this phase of the fibre behaviour research. The friction force course (in dependence on the weight velocity) was considered nonlinear (basis for the determination of the friction force course was especially [2,25]). A general phenomenological model of the fibre (at "quicker" tested situation [20,21,24]) was not determined, but general influences of individual parameters on the system behaviour, which are usable for all systems containing fibre-mass subsystem(s), were assessed. A suitable fibre model, but only in dependence on the definite simulated test situation, was determined.

"Starting" values at the phenomenological model creating are, identically with [19], fibre stiffness measured on a tensile testing machine [22] ($94 \cdot 10^3$ N/m) and fibre damping coefficient derived on the basis of experience (46.9 N $\cdot$ s/m). The "starting" friction force between the weight and the prismatic linkage is considered to be zero [20].

Final values were calculated on the basis of the final values determined in [20] and are the same as in [20,21] and [24] (stiffness $= 34 \cdot 10^3$ N/m, damping coefficient $= 27.5$ N $\cdot$ s/m). Friction force course determined at investigating the weight-fibre mechanical system [19] with the angle of inclination of the inclined plane $\alpha = 30$ degrees (see Fig. 4) was applied in the model of the weight-fibre-pulley-drive mechanical system [20,21,24].

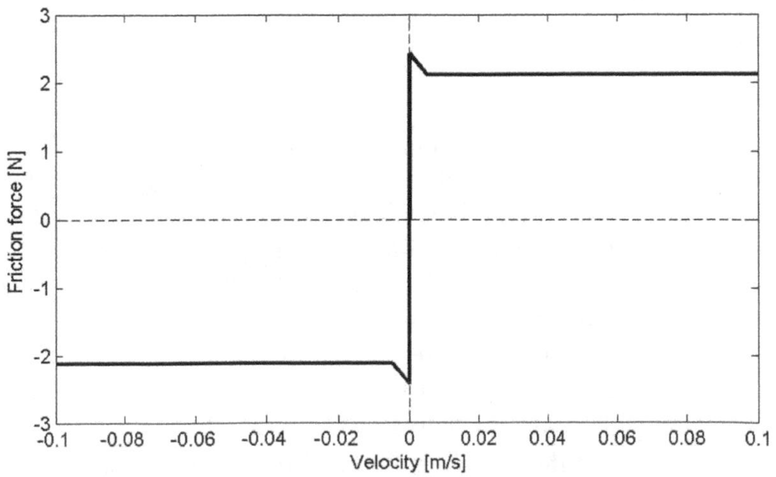

Fig. 4. Friction force acting between the weight and the prismatic linkage

Results of experimental measurements and simulations of four tested situations are presented (altogether twelve situations were tested). Two tested situations at a "quicker" drive motion (situations 10 and 11) and two situations at a "slower" drive motion (situations 12a and 14) are presented in this paper (see time histories of drive motion in Figs. 5, 8b and 11b). Frequencies of drive motion (i.e. frequencies of input signal) higher than 1 Hz are designated as "quicker" drive motions, frequency of drive motion lower than 1 Hz are designated as "slower" drive motions.

The influence of the fibre stiffness, the fibre damping coefficient and the friction force acting between the weight and the prismatic linkage on time histories of the weight displacement and also on time histories of the dynamic force acting in the fibre was evaluated partly visually and partly on the basis of the value of the correlation coefficient between the records of the experimental measurements and the simulation results. Application of the approach based on the calculation of the statistical quantities that enables to express directly the relation between two time series has appeared to be suitable for comparing two time series in various cases, e.g. [16,23].

Correlation coefficient $R(p)$ [26] defined for two discrete time series $x^{(1)}$ (the time history recorded at experimental measurement) and $x^{(2)}(p)$ (the time history determined at simulation with the multibody model; function of investigated parameters p) was calculated

$$R(p) = \frac{\sum\limits_{i=1}^{n} \left(x_i^{(1)} - \mu_1\right) \cdot \left[x_i^{(2)}(p) - \mu_2(p)\right]}{\sqrt{\sum\limits_{i=1}^{n} \left(x_i^{(1)} - \mu_1\right)^2 \cdot \sum\limits_{i=1}^{n} \left[x_i^{(2)}(p) - \mu_2(p)\right]^2}}, \tag{1}$$

where μ_1 and $\mu_2(p)$ are mean values of the appropriate time series. The maximum value of the correlation coefficient is one. The more the compared time series are similar to each other the more the correlation coefficient tends to one. The advantage of the correlation coefficient is that it quantifies very well the similarity of two time series by scalar value, which is obtained using a simple calculation.

The problem is possible to be put as the problem of the minimization of the objective function in the form

$$\psi(p) = (1 - R(p))^2. \tag{2}$$

In case of the computer simulations in the *alaska 2.3* simulation tool, the whole process of the optimization was limited by the impossibility of executing the analysis from the statement line and evaluating the results of numerical simulations without the necessary human intervention. The whole process could not be automated. "Manual" change in the parameters on the basis of the chosen optimization method was the only solution. Comparing to automated optimization process it is not possible to perform so many iteration cycles in a short time. But the advantage is that during the evaluation it is possible to respect criteria that do not have to be strictly mathematically formulated (the coefficient of correlation given by relation (1) enables to imagine coincidence of (time) series, but it is not "universal").

The monitored quantities at the experimental measurements and the computer simulations are presented in Figs. 5 to 16. In Table 1 there are values of correlation coefficient $R(p)$ for the "starting" values of parameters and for the values of parameters of the mechanical fibre-mass system model taken from [20].

Table 1. Values of correlation coefficient $R(p)$ [–]

Tested situation	Shape of an input signal	Comparison of time histories of the weight displacement		Comparison of time histories of dynamic force acting in a fibre	
		"Starting" value	Final value	"Starting" value	Final value
10	trapezoidal	0.999 300	0.991 8	0.054 69	0.546 9
11	trapezoidal	0.006 408	0.746 5	0.121 60	0.285 3
12a	trapezoidal	0.999 900	0.999 4	0.010 47	0.499 0
14	quasi-sinusoidal	1.000 000	1.000 0	0.032 60	0.830 6

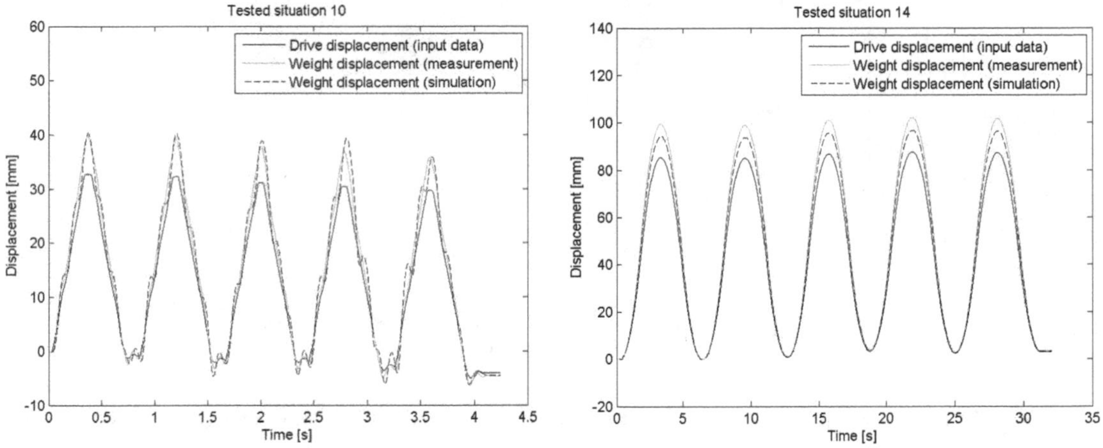

a) b)

Fig. 5. Time histories of the weight displacement, a) at "quicker" tested situation 10, b) at "slower" tested situation 14

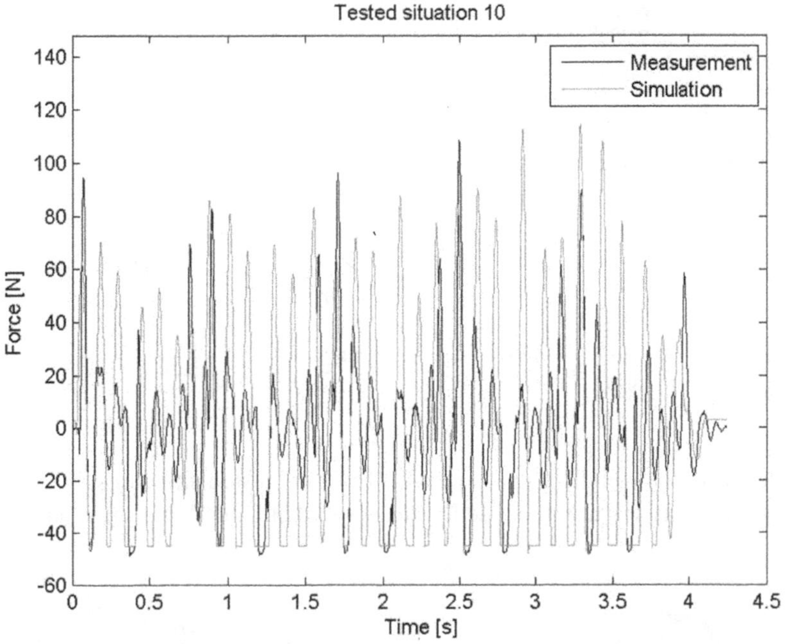

Fig. 6. Time histories of the dynamic force acting in a fibre at "quicker" tested situation 10

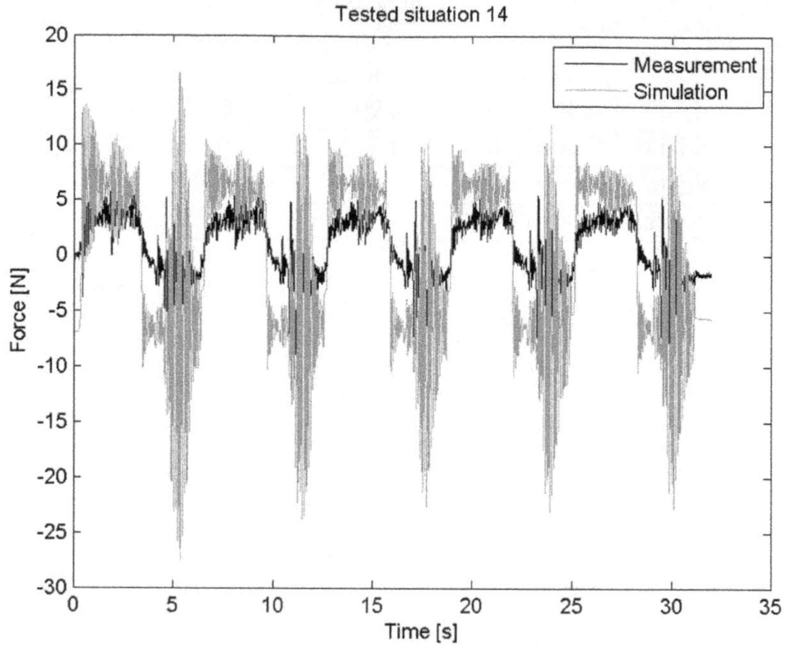

Fig. 7. Time histories of the dynamic force acting in a fibre at "slower" tested situation 14

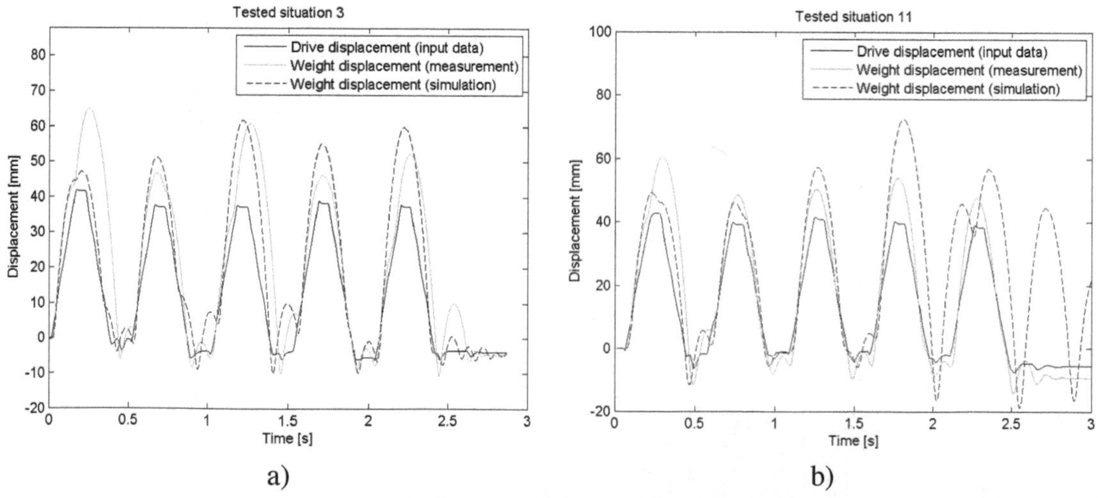

Fig. 8. Time histories of the weight displacement at "quicker" tested situations, influence of the mass of the weight, a) situation 3 (weight without added mass — taken from [21]), b) situation 11 (weight with added mass)

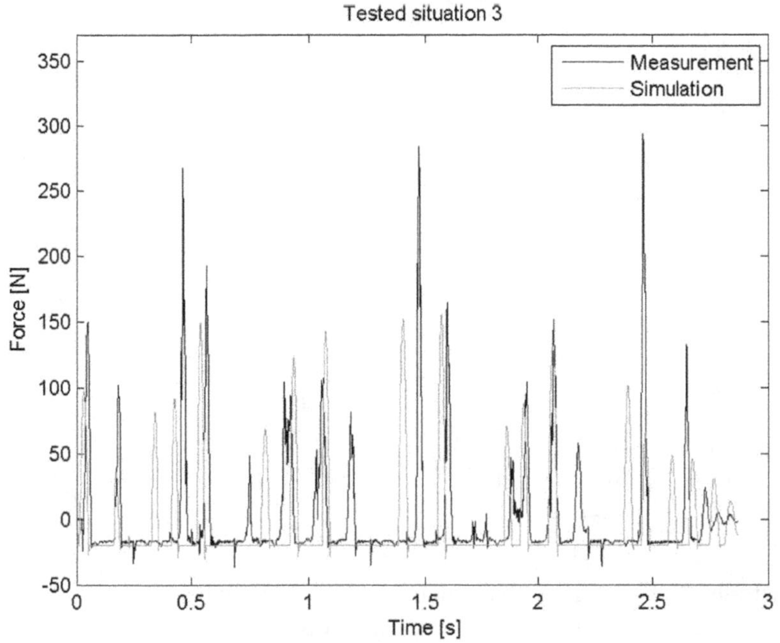

Fig. 9. Time histories of the dynamic force acting in a fibre at "quicker" tested situation 3, weight without added mass (taken from [21]), influence of the mass of the weight

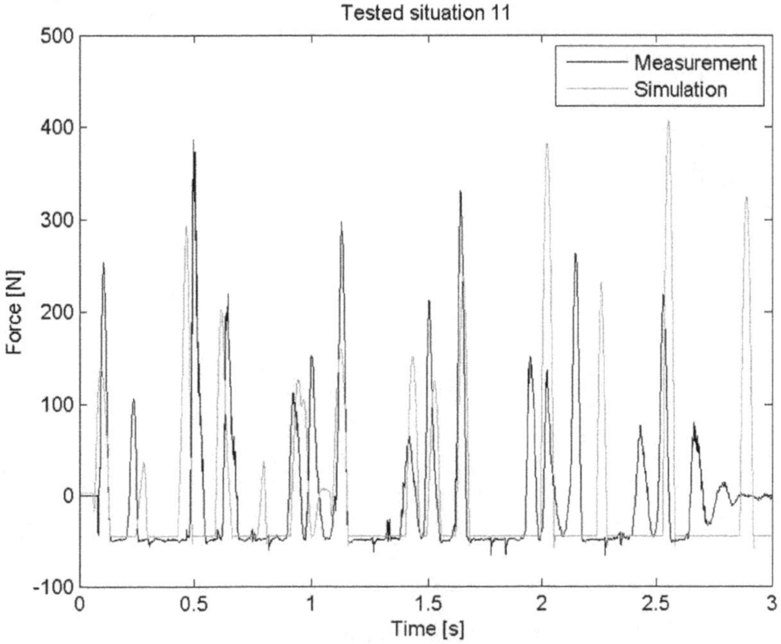

Fig. 10. Time histories of the dynamic force acting in a fibre at "quicker" tested situation 11, weight with added mass, influence of the mass of the weight

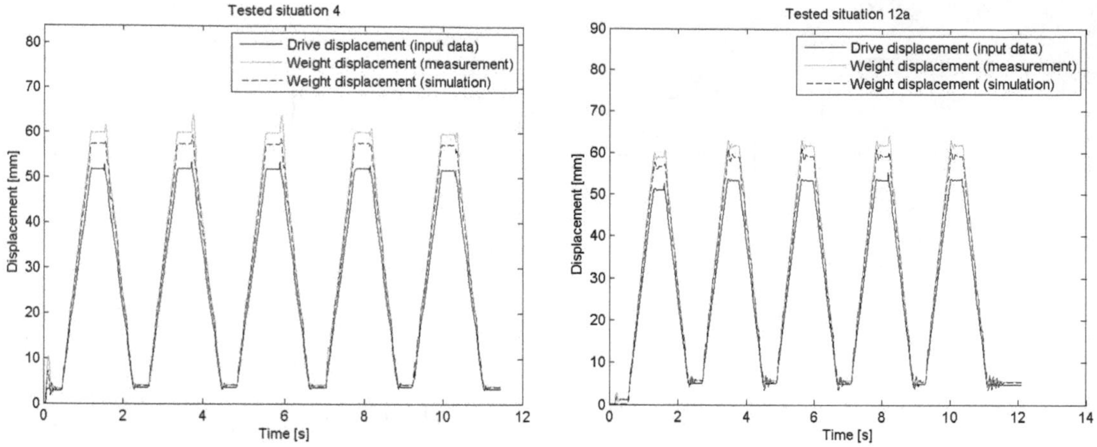

a) b)

Fig. 11. Time histories of the weight displacement at "slower" tested situations, influence of the mass of the weight, a) situation 4 (weight without added mass — taken from [21]), b) situation 12a (weight with added mass)

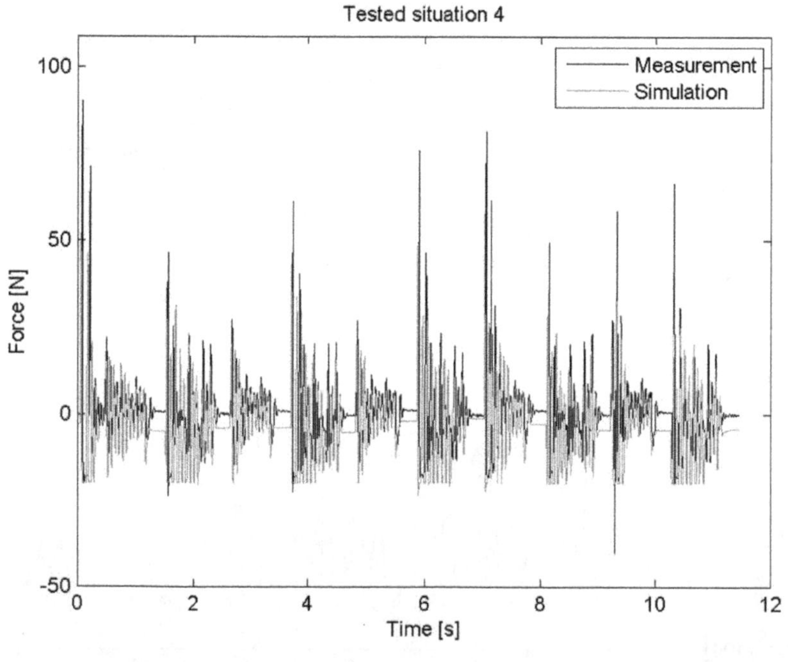

Fig. 12. Time histories of the dynamic force acting in a fibre at "slower" tested situation 4, weight without added mass (taken from [21]), influence of the mass of the weight

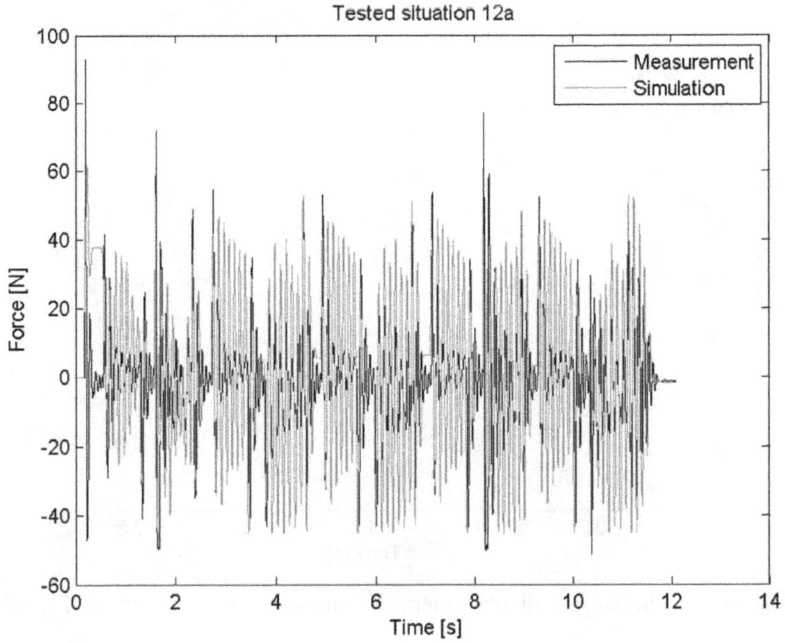

Fig. 13. Time histories of the dynamic force acting in a fibre at "slower" tested situation 12a, weight with added mass, influence of the mass of the weight

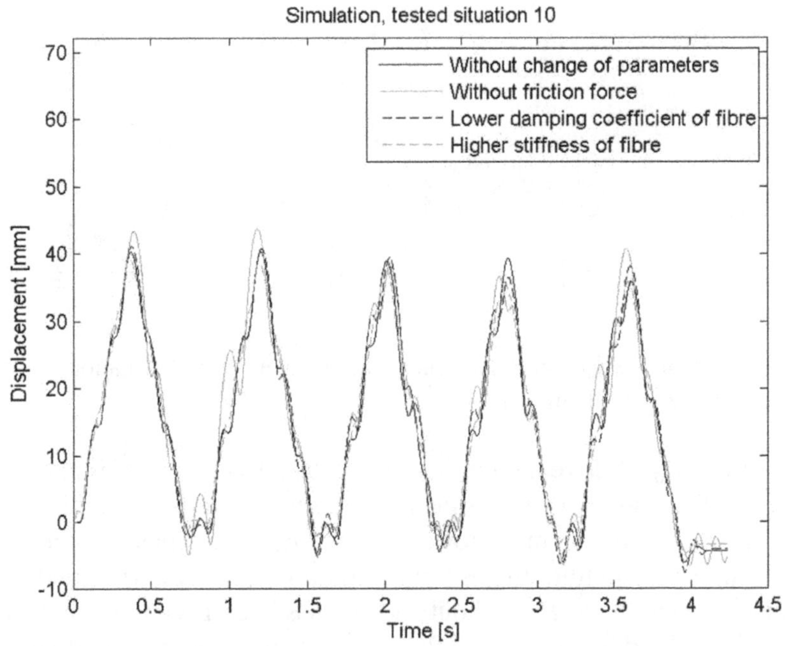

Fig. 14. Time histories of the weight displacement at "quicker" tested situation 10, influence of model parameters

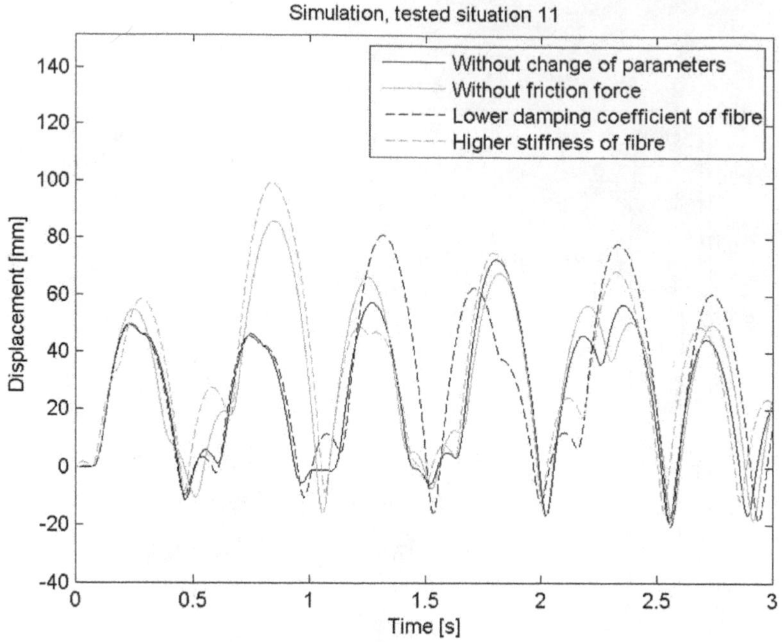

Fig. 15. Time histories of the weight displacement at "quicker" tested situation 11, influence of model parameters

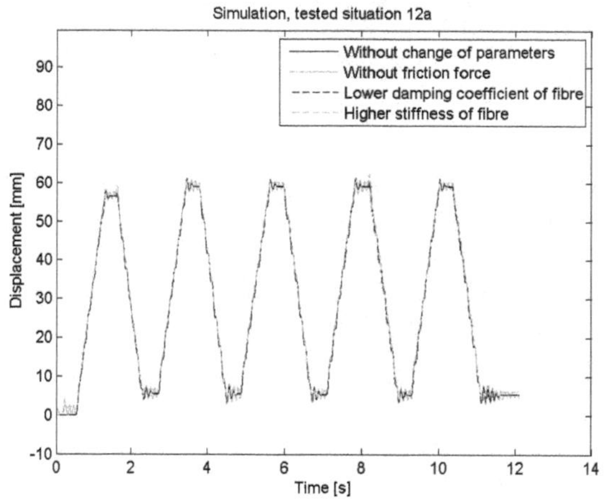

a)

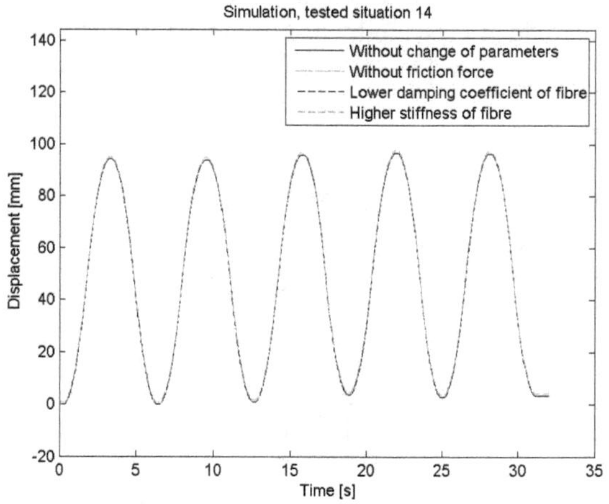

b)

Fig. 16. Time histories of the weight displacement at "slower" tested situations, influence of model parameters, a) situation 12a, b) situation 14

The highest frequency of drive motion (i.e. the highest frequency of input signal) at investigation of the weight-fibre-pulley-drive system is 2 Hz, see situations 3 and 11 (see Figs. 8 to 10). This frequency of drive motion is much lower than natural frequencies of the computer model of linearized system in an equilibrium position. Natural frequency corresponding to the weight vibrations of the system with weight without added mass is 25 Hz and natural frequency of the system with weight with added mass is 15.25 Hz. It means that in case of weigh vibration at "quicker" tested situations the excitation of resonant vibrations is not concerned, but vibrations that are given by strongly nonlinear behaviour of a fibre (as it has been already stated, fibres are able to transfer only tensile force, in "compression" they are not able to transfer any force), which can even have the character of chaos, are involved.

In Figs. 5a and 6 there are given time histories of monitored quantities at tested situation 10. Though this situation is specified as a "quicker" one, it still is not a typical "quicker" situation. In the record of time histories of the weight displacement in Fig. 5a the results of the measurement and that of the simulation seem to be identical, but in reality they are influenced by the parameters of the phenomenological model, with which the simulation had been performed (see the following paragraph). Time histories of the dynamic force acting in a fibre in Fig. 6, determined at measurement and using the simulation, are of the same character. The phenomenological model of fibre is to cover, as it has been stated, e.g. influences of fibre transversal vibration, "jumping" from pulley etc. As it does not include those phenomena physically (but by the change in the already introduced model parameters), it is evident, that it is not possible to expect that the introduced time histories of dynamic force acting in fibre will be of the same course.

Time histories of the monitored quantities at tested situation 14 are shown in Figs. 5b and 7. A typical "slower" situation is concerned. In the record of time histories of the weight displacement in Fig. 5b measurement and simulation results seem to be more different than at simulating "quicker" situation 10 mentioned in the previous paragraph. The cause of differences in local extremes of deflections was (probably) an incorrect calibration of a dial gauge used for the measurement of weight displacement (the differences occurred in all cases at the investigation of the weight-fibre-pulley-drive system, at which the position of the weight was asymmetric with respect to the plane of a drive-pulley symmetry [21]; at measurements at which the position of the weight was symmetric this problem has not occurred [20,24]). But this fact cannot be 100 % verified because immediately after the measurement the experimental stand was dismounted. Time histories of the dynamic force acting in the fibre in Fig. 7, determined at measurement and using the simulation, are of the same character. The reason of different course of time histories of the dynamic force acting in the fibre is the same as in the case of tested situation 10. The character of time histories of the dynamic force acting in the fibre shown in Fig. 7 is seemingly visually more different than the character of time histories of the dynamic force acting in the fibre shown in Fig. 6. But it is necessary to realize that lower magnitudes of the dynamic force by orders are concerned.

Time histories of the weight displacement at "quicker" tested situation 3, at which the weight was without added mass [21], and at tested situation 11, at which the weight was with added mass are shown in Fig. 8. An identical drive motion was used at those tested situations. Similarly, time histories of the weight displacement at "slower" tested situation 4, at which the weight was without added mass [21], and at tested situation 12a, at which the weight was with added mass, are shown in Fig. 11. At those tested situations an identical drive motion was also used. Figs. 9, 10, 12 and 13 show time histories of dynamic force acting in fibre for those identical tested situations. As to the influence of the added mass to the weight on the experimental measurements and computer simulations results, a higher mass is shown in higher magnitudes of time histories of the weight displacement at the "quicker" situations (see Fig. 8). At the "slower" situations a higher mass of the weight does not influence the magnitudes of time histories of the weight displacement in any way (see Fig. 11). The added mass of the weight is shown in higher magnitudes of dynamic forces acting in the fibre independently of the input signal rate. The higher the mass of the weight the higher the influence of the input signal rate (see Figs. 8 to 10, 12 and 13).

Time histories of the weight displacement given in Figs. 14 to 16 inform about the influence of the change in the phenomenological model parameters (i.e. the fibre stiffness, the fibre damping coefficient and the friction force) on the course of these time histories. When finding out the influence of individual parameters value of only one parameter was changed, other ones remained unchanged. "Starting" value (already mentioned in this paper) was used as the parameter changed value. From Fig. 16 it is evident that at simulating the experimental measurements at the "slower" drive motion the monitored time histories of the weight displacement are identical independently of the fibre stiffness, the fibre damping coefficient and the friction force (between the weight and the prismatic linkage). At the "quicker" tested situations (10 and 11) the measured and computed time histories of the weight displacement are of the same character (see Figs. 5a and 8b). At simulating the experimental measurements at "quicker" drive motion (see Figs. 14 and 15) the local extremes of the monitored time histories of the weight displacement are dependent on all the phenomenological model parameters (i.e. on the fibre stiffness, the fibre damping coefficient and the friction force). All findings given in this paragraph correspond to the results obtained in previous investigation of the weight-fibre-pulley-drive [20,21,24].

At all the simulations when changing the computational model the time histories of dynamic force acting in the fibre are different (more or less) but their character remains the same (the same finding as in [20,21] again). From Figs. 6, 7, 10, and 13 and Table 1 it is evident that time histories of dynamic force acting in the fibre are less suitable for searching for the parameters of the fibre phenomenological model.

From the obtained results it is evident that parameters of the fibre phenomenological model must be, in addition, considered dependent on the speed of the weight motion (i.e. on the input signal rate).

For searching for the parameters of the fibre phenomenological model it would be useful to perform more experimental measurements with the "quicker" drive motion. In reality the "quicker" drive motion is limited on the one hand by its parameters (limited amplitudes of the drive displacements and limited drive speed) and on the other hand by the danger of the weight "flinging out" of the prismatic linkage (during one test measurement at a higher drive speed it really occurred). The possibility of performing experimental measurements with other time histories of drive motion or with a different geometrical arrangement of the experimental stand will be analysed.

5. Conclusion

The approach to the fibre modelling based on the force representations was utilised for the investigation of the motion of the weight in the weigh-fibre-pulley-drive mechanical system. The simulation aim is to create a phenomenological model of a fibre, which will be utilizable in fibre modelling in the case of more complicated mechanical or mechatronic systems. The created phenomenological model is assumed to be dependent on the fibre stiffness, on the fibre damping coefficient and on the friction force acting between the weight and the prismatic linkage in which the weight moves.

Development of the fibre phenomenological model will continue. From the obtained results it is evident that parameters of the fibre phenomenological model must be, in addition, considered dependent on the speed of the weight motion. The question is if it is possible to create the phenomenological model like that.

In addition it must be stated that the model of the fibre-pulley contact appears to be problematic in the computational model.

Acknowledgements

The article has originated in the framework of solving No. P101/11/1627 project of the Czech Science Foundation entitled "Tilting Mechanisms Based on Fibre Parallel Kinematical Structure with Antibacklash Control".

References

[1] Awrejcewicz, J., Classical mechanics: Dynamics, Springer, New York, 2012.

[2] Awrejcewicz, J., Olejnik, P., Analysis of dynamic systems with various friction laws, Applied Mechanics Review 58 (6) (2005) 389–411.

[3] Chan, E. H. M., Design and implementation of a high-speed cable-based parallel manipulator, PhD Thesis, University of Waterloo, Waterloo, 2005.

[4] Diao, X., Ma, O., Vibration analysis of cable-driven parallel manipulators, Multibody System Dynamics 21 (4) (2009) 347–360.

[5] Freire, A., Negrão, J., Nonlinear dynamics of highly flexible partially collapsed structures, Proceedings of the III European Conference on Computational Mechanics, Solids, Structures and Coupled Problems in Engineering, Lisbon, Springer, 2006.

[6] Gerstmayr, J., Sugiyama, H., Mikkola, A., Developments and future outlook of the absolute nodal coordinate formulation, Proceedings of the 2nd Joint International Conference on Multibody System Dynamics, Stuttgart, University of Stuttgart, Institute of Engineering and Computational Mechanics, 2012.

[7] Gosselin, C., Grenier, M., On the determination of the force distribution in overconstrained cable-driven parallel mechanisms, Meccanica 46 (1) (2011) 3–15.

[8] Hajžman, M., Polach, P., Modelling of flexible bodies in the framework of multibody systems, Proceedings of the 6th International Conference Dynamics of Rigid and Deformable Bodies 2008, Ústí nad Labem, Faculty of Production Technology and Management, Jan Evangelista Purkyně University in Ústí nad Labem, 2008, pp. 33–42.

[9] Hajžman, M., Polach, P., Modelling of cables for application in cable-based manipulators design, Proceedings of the ECCOMAS Thematic Conference Multibody Dynamics 2011, Université catholique de Louvain, Brussels, 2011.

[10] Hajžman, M., Polach, P., Nonlinear finite element formulation and its simple application for cables, Computational and Experimental Methods in Applied Mechanics I, Faculty of Production Technology and Management, Jan Evangelista Purkyně University in Ústí nad Labem, Ústí nad Labem, 2013, pp. 187–196.

[11] Heyden, T., Woernle, C., Dynamics and flatness-based control of a kinematically undetermined cable suspension manipulator, Multibody System Dynamics 16 (2) (2006) 155–177.

[12] Kamman, J. W., Huston, R. L., Multibody dynamics modeling of variable length cable systems, Multibody System Dynamic 5 (3) (2001) 211–221.

[13] Liu, Ch., Tian, Q., Hu, H., New spatial curved beam and cylindrical shell elements of gradient-deficient Absolute Nodal Coordinate Formulation, Nonlinear Dynamics 70 (3) (2012) 1 903–1 918.

[14] Maißer, P., Wolf, C.-D., Keil, A., Hendel, K., Jungnickel, U., Hermsdorf, H., Tuan, P. A., Kielau, G., Enge, O., Parsche, U., Härtel, T., Freudenberg, H., *alaska*, User manual, Version 2.3, Institute of Mechatronics, Chemnitz, 1998.

[15] Michalík, J., Janík, D., Development of software control system in LabView for support of experiment with linear drive controlled by Emerson Unidrive SP converter, University of West Bohemia, Plzeň, 2012. (in Czech)

[16] Polach, P., Hajžman, M., Design of characteristics of air-pressure-controlled hydraulic shock absorbers in an intercity bus, Multibody System Dynamics 19 (1–2) (2008) 73–90.

[17] Polach, P., Hajžman, M., Šika, Z., Influence of crucial parameters of the system of an inverted pendulum driven by fibres on its dynamic behaviour, Applied and Computational Mechanics 6 (2) (2012) 173–184.

[18] Polach, P., Hajžman, M., Tuček, O., Validation of the point-mass modelling approach for fibres in the inverted pendulum model, Proceedings of the 18th International Conference Engineering Mechanics 2012, Svratka, Institute of Theoretical and Applied Mechanics Academy of Sciences of the Czech Republic, 2012, pp. 443–452.

[19] Polach, P., Hajžman, M., Václavík, J., Simple fibre-mass model and experimental investigation, Proceedings of the National Colloquium with International Participation Dynamics of Machines 2013, Prague, Institute of Thermomechanics Academy of Sciences of the Czech Republic, 2013, pp. 79–84.

[20] Polach, P., Hajžman, M., Václavík, J., Šika, Z., Svatoš, P., Model parameters influence of a simple mechanical system with fibre and pulley with respect to experimental measurements, Proceedings of the ECCOMAS Thematic Conference Multibody Dynamics 2013, Zagreb, University of Zagreb, Faculty of Mechanical Engineering and Naval Architecture, 2013, pp. 473–482.

[21] Polach, P., Hajžman, M., Václavík, J., Šika, Z., Valášek, M., Experimental and computational investigation of a simple mechanical system with fibre and pulley, Proceedings of the 12th Conference on Dynamical Systems — Theory and Applications, Dynamical Systems — Applications, Łódź, Department of Automation, Biomechanics and Mechatronics, Łódź University of Technology, 2013, pp. 717–728.

[22] Polach, P., Václavík, J., Hajžman, M., Load of fibres driving an inverted pendulum system, Proceedings of the 50th Annual International Conference on Experimental Stress Analysis, Tábor, Czech Technical University in Prague, Faculty of Mechanical Engineering, 2012, pp. 337–344.

[23] Polach, P., Václavík, J., Hajžman, M., Verification of the multibody models of the TriHyBus on the basis of experimental measurements, Proceedings of the 6th Asian Conference on Multibody Dynamics ACMD2012, Shanghai, Shanghai Jiao Tong University, 2012.

[24] Polach, P., Václavík, J., Hajžman, M., Šika, Z., Valášek, M., Influence of the mass of the weight on the dynamic response of the simple mechanical system with fibre, Book of Extended Abstracts of the 29th Conference with International Participation Computational Mechanics 2013,Špičák, University of West Bohemia in Plzeň, 2013, pp. 97–98.

[25] Půst, L., Pešek, L., Radolfová, A., Various types of dry friction characteristics for vibration damping, Engineering Mechanics 18 (3–4) (2011) 203–224.

[26] Rektorys, K., et al., Survey of applicable mathematics, Vol. II, Kluwer Academic Publishers, Dordrecht, 1994.

[27] Schiehlen, W., Research trends in multibody system dynamics, Multibody System Dynamics 18 (1) (2007) 3–13.

[28] Shabana, A. A., Flexible multibody dynamics: Review of past and recent developments, Multibody System Dynamics 1 (2) (1997) 189–222.

[29] Stejskal, V., Valášek, M., Kinematics and dynamics of machinery, Marcel Dekker, Inc., New York, 1996.

[30] Valášek, M., Karásek, M., HexaSphere with cable actuation, recent advances in mechatronics: 2008–2009, Springer-Verlag, Berlin, 2009, pp. 239–244.

[31] Wasfy, T. M., Noor, A. K., Computational strategies for flexible multibody systems, Applied Mechanics Review 56 (6) (2003) 553–613.

[32] Wittbrodt, E., Adamiec-Wójcik, I., Wojciech, S., Dynamics of flexible multibody systems – Rigid finite element method, Springer, Berlin, 2006.

[33] Zi, B., Duan, B. Y., Du, J. L., Bao, H., Dynamic modeling and active control of a cable-suspended parallel robot, Mechatronics 18 (1) (2008) 1–12.

17

The role of vortices in animal locomotion in fluids

R. Dvořák[a,*]

[a] *Institute of Thermomechanics v.v.i., Academy of Sciences of the Czech Republic, Dolejškova 1402/5, 182 00 Prague, Czech Republic*

Abstract

The aim of this paper is to show the significance of vortices in animal locomotion in fluids on two deliberately chosen examples. The first example concerns lift generation by bird and insect wings, the second example briefly mentiones swimming and walking on water. In all the examples, the vortices generated by the moving animal impart the necessary momentum to the surrounding fluid, the reaction to which is the force moving or lifting the animal.

Keywords: animal locomotion, bird flight, flapping wings, insect flight, swimming, walking on water, vortices in animal propulsion

1. Introduction

The topic is far too diverse to cover all aspects of animal locomotion both in (or on) water and air. All animals in water and air have inhabitated this planet for 300 million years (fishes and insects, twice as long as birds who are here about 150 million years), and they have had well enough time to develop their skill of swimming and flying. From the whole number of existing animal species almost 80 % have the capability of flying, out of which almost 99 % are insects (about 10^6 species). It is only less than half a century when people have begun to uncover the mechanism of their — often uncomprehensible — way of locomotion.

In order for any animal to move, it must apply a force to its environment, the reaction force is then propelling the animal forward. The driving force is by most animals transmitted entirely to the environment by vortices.

In this article we will only deal with two examples – both incomplete and highly simplified – the *leading edge vortices on insect wings* and a *couple of two counterrotating vortices* (like, e.g., the *fish tail vortices* used in swimming and in *walking on water*). Most of the results relevant to the first example have been obtained while investigating aerodynamics of micro-air-vehicles [6].

2. Vortices in nature

Vortex is the most frequent phenomenon in fluid dynamics. In nature, it appears at dimensions ranging from 10^{-10} m in liquid helium, or 10^{-5} m in turbulent flows, up to 10^5 m in oceans and hurricanes, or even more in planetary atmosphere (e.g., orders of light years in galaxies).

There are many definitions of a vortex. Not in all cases we can see vortices directly or by visualizing the flow field investigated. We have to identify them even in cases where we can

*Corresponding author. e-mail: dvorak@it.cas.cz.

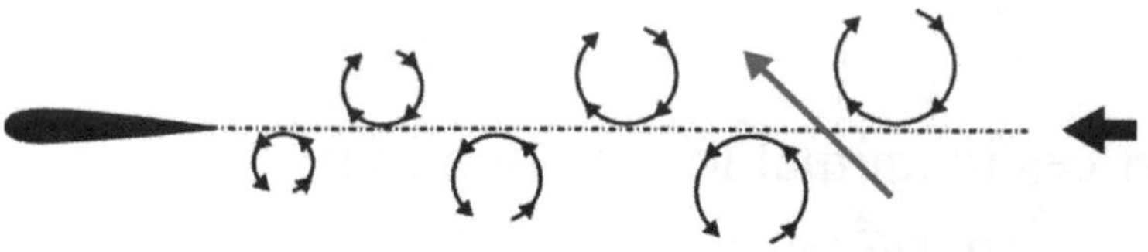

Fig. 1. Vortices in a von Kármán vortex street

Fig. 2. Vortices formed on leading edge (leading edge vortex) and trailing edge (starting vortex)

only deduce on their existence from measuring the whole flow field, or, by calculating the whole flow field with a CFD code. For more details see [5,9] and [11].

There are no vortices in uniform homogeneous flows. Vortices are generated by nonuniformity of the flow field, or by inserting any foreign body that will generate them. The behaviour of vortices is often described by applying the laws of vortex motion as described already by Helmholtz in 1858 (see, e.g., [5] and [11]).

For the vortices to survive in the flow field, they have to draw energy from the surrounding flow field, thus representing always a loss of energy. The same holds true for vortices used by animals to move in fluids. Even these vortices need an energy supply from the animal's musculoskeletal system, and it is amazing how efficiently can the animals control their motion to minimize the energy consumption.

3. Preliminaries

To understand why are the vortices so important in animal locomotion, it is useful to look first at the origin and behaviour of vortices in *steady uniform fluid flows*, and in flows past bodies immersed in fluids.

Vortices are generated on sharp edges of plates placed perpendicular to the fluid motion, or behind blunt trailing edges of various bodies and profiles. In steady flows they stick to these edges and their development depends mainly on velocity of the oncoming flow and — in real fluids — on the viscosity of the fluid. They depend also on the body dimensions, or strictly speaking, on the *Reynolds number* (a ratio of inertial and viscous forces), $Re = U \cdot L/\nu$. At small Re (typically order of unity) are the viscous forces so dominant that these vortices do not appear at all, i.e., there is not enough energy to establish these vortices.

In real fluids the wake behind a 2D profile or bodies consists of a double row of alternating vortices, known as a *von Kármán vortex street* (Fig. 1). Any of the two neighbouring vortices form a couple driving the fluid against the direction the profile is moving, contributing thus to the loss of momentum and representing drag of the body.

If the profile is inclined against the oncoming stream under a certain angle of attack the starting vortex is formed behind the trailing edge as a result of different lengths of fluid paths on the upper and lower surfaces (Fig. 2). According to the *Kelvin's law*, circulation of this starting vortex generates in turn circulation of the same intensity, but opposite sence of rotation on the profile. This circulation multiplied by the mass flow density and the profile area determines the lift force (N. J. Žukovskij, 1902). Vortex at the edge remains there for the whole time the profile is moving. The starting vortex is blown off the trailing edge with the stream velocity.

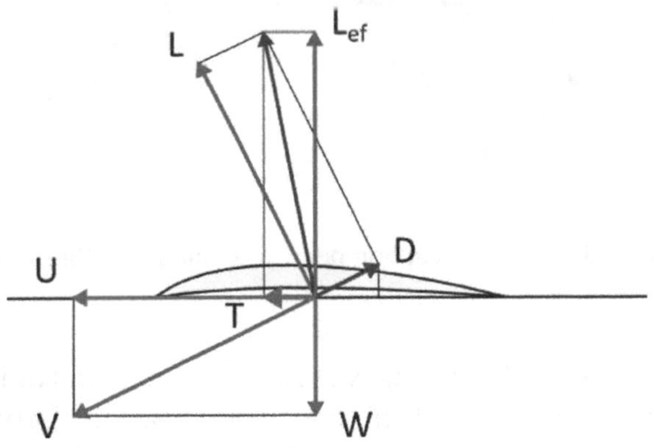

Fig. 3. Diagramme of forces acting on a moving wing, U – flight velocity, W – velocity of the moving wing, V – actual velocity of the moving wing, L – lift, D – drag, L_{ef} – effective lift (vertical component of the resultant aerodynamic force on the airfoil), T – thrust (horizontal component of the resultant aerodynamic force)

One of the already mentioned Helmholtz laws states that the vortex cannot have a free end inside the fluid. It can end on the free surface (if it is available), or it must form a closed loop (e.g. a ring). A good example is a wing of finite aspect ratio, as in the well known *Prandtl's wing theory*. The starting vortex and the circulation on the profile (the bound vortex) are joint together via tip vortices.

However, any animal locomotion in fluids (swimming, flying, hovering) requires a different approach. To generate lift and thrust the animals need either *moving wings* (birds, insects, bats), tails, fins, or *movement of the whole body or its part* (fishes). Flow past these moving parts is *unsteady* and *threedimensional*, i.e., much different from most aircraft wings.

A simple diagramme of forces on the profile during its movement is in Fig. 3. The actual velocity of the oncoming flow is now a vector sum of the flight velocity and the moving wing velocity. Lift is perpendicular to the actual velocity, however, to compensate the weight of the flyer, the vertical projection of the lift force (the effective lift) has to be considered. From the same diagramme we can infer that even a much higher drag and much smaller lift can result in an acceptable effective lift. The projection of the resulting aerodynamic force on the horizontal direction represents thrust of the flyer [4].

At this point we reveal the importance of considering the *Reynolds number*. At higher Re inertia dominates over viscosity and once the body (e.g., ship) is set into motion it has to generate a considerable force to stop it. This is not the case with small animals (fish) where also the Re is small (typically 1 or even less). Viscosity is in this case so dominant that the animal stops as soon as it ceases to propel itself. The same holds true for small insects, generating the leading edge vortex which is indispensable for lift production.

If the same profile as in Fig. 1 is set into motion perpendicular to the velocity of the oncoming stream, at certain values of the perpendicular velocity and frequency of this motion we arive at a situation, where the vortex street behind the profile have opposite sense of the vortex circulation (Fig. 4), see, e.g., [10]. Any of the two neighbouring vortices now form a couple driving the fluid in the direction the profile is moving. Clearly, this vortex street generates thrust (this phenomenon is often called *Katzmayr effect*).

Despite the apparent similarity of both cases (Fig. 1 and Fig. 4), the mechanism generating the vortex streets is in both cases different. In the first case the boundary layers from either sides

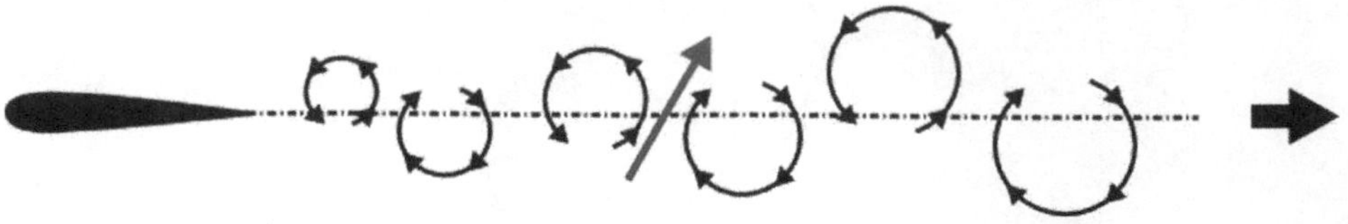

Fig. 4. “Reversed von Kármán vortex street” on perpendicularly oscillating wing (Katzmayer effect)

of the profile turn into vortices at the trailing edge, and a certain kind of instability transfer them into the vortex street as described in Fig. 1. In case of the perpendicularly moving profile (Fig. 4) vortices appear at the trailing edge during the profile movement. At the point where the profile begins to return the vortex is blown down with the stream velocity. The phase lag between the vortices blown off the trailing edge at the points where the profile stops to return is proportional to the velocity, frequency and amplitude of the profile motion, and it determines the position of the vortex in the vortex street (Fig. 5). Evidently, the situation is not so univocal as in case of the Kármán vortex street.

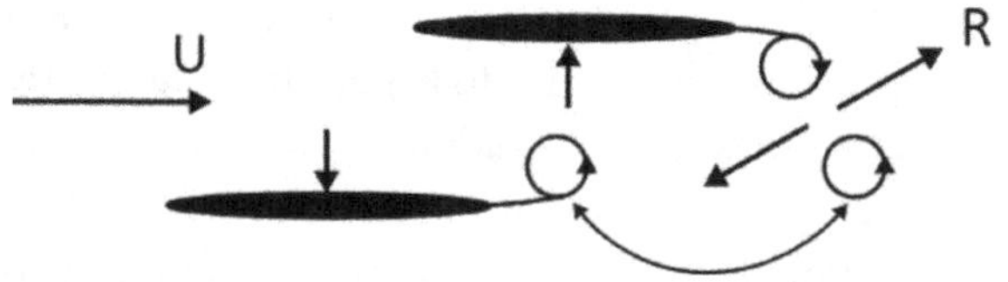

Fig. 5. Mechanism explaining the Katzmayer effect, R is the reaction of the fluid

Bird’s wing during the working phase (downstroke) moves almost perpendicularly to the flight velocity. It generates vortices on both edges — on the leading edge and the trailing edge — rotating towards the wing (Fig. 6). They thus form a vortex couple driving the fluid downwards and giving it a considerable momentum. The reaction to this momentum is the lift, generated by the bird’s wing.

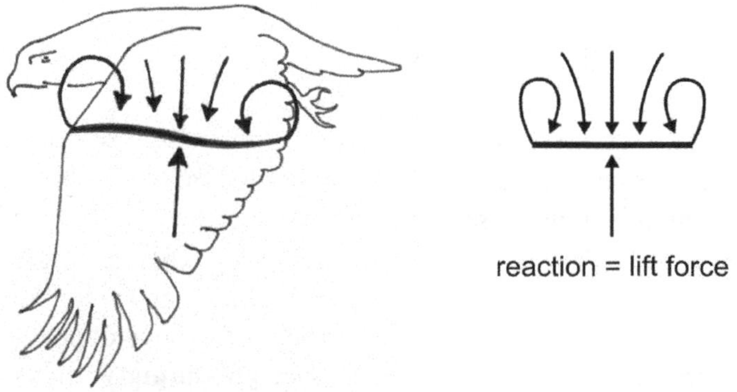

Fig. 6. Leading edge and trailing edge vortices impart downward momentum to the fluid they drive in from the surrounding atmosphere; reaction to this momentum is the lift force

4. Leading edge vortices on insect wings

To keep the insect flying the insect has to generate sufficient lift to compensate the weight (body + payload). The lift force is a reaction to the downward velocity and momentum imparted by the wing to the air. There are two basic forms of insect wing movements — the *"bird-like" form* — wings are moving up and down as in Fig. 6 — and the actual *"insect-like" form* — wings are oscillating almost in one plane. Lift generation in both these cases aptly uses *leading edge vortices*, as apparent from Fig. 7.

When the wings move down from the initial dorsal position, a vortex is formed at the leading edge, entraining the surrounding fluid into the gap between the wings and thus generating lift. The leading edge vortex left at the wing initial position from the wing's returning motion, together with the leading edge vortex of the downward moving wing, form a vortex couple which helps to open the wings and start the already described process of lift generation. This process has been called *"clap and fling"* and was described by Thorsten Weis-Fogh in 1976 (see, e.g., [13]). It is used by butterflies, moths, and many other insects who use even the other mode (the oscillating wing).

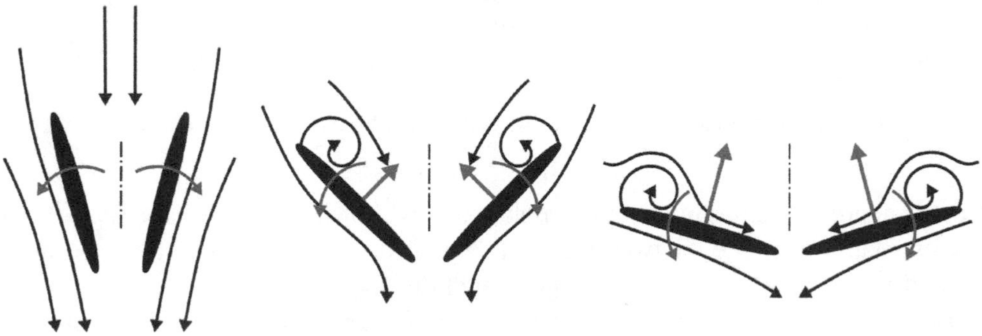

Fig. 7. Mechanism of the "clap and fling" insect wing motion

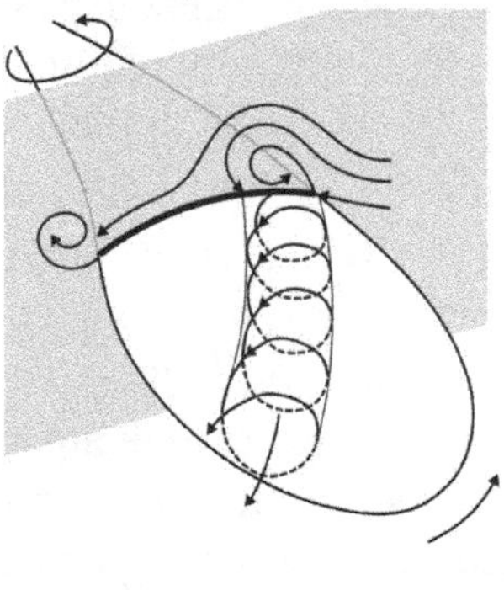

Fig. 8. Leading edge vortex on an oscillating wing

The whole process of lift generation on *oscillating insect wings* is, however, much more sophisticated. The leading edge vortices, or mainly their displacement effect, create a high-lift-profile by enlarging effectively the wing upper side (Fig. 8). The wing is moving on semicircular paths with very high frequency within about 160° and at each end of the wing amplitude it has to revert (rotate) to move always with the leading edge ahead. This movement generates intensive centrifugal flow inside the leading edge vortex, which stabilizes the vortex.

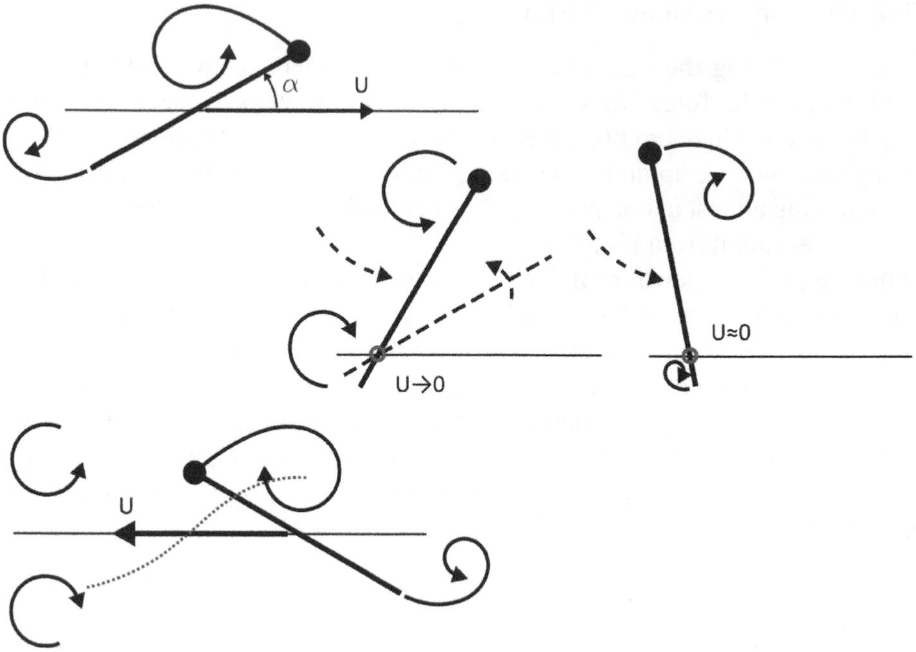

Fig. 9. Mechanism of increased lift generation by oscillating insect wings

Measurements, as well as numerical simulation, has shown that lift is not only generated by the wing oscillation between the two end points, but also by rotating (reverting) the wing (see, e.g., [3] and [13]). Leading edge vortex generated during rotation can contribute up to 35 % of total lift, depending on the position of the center of rotation and on the moment (phase) of the rotation beginning. Even more, when the wing returns it may meet the starting vortex from the preceding cycle, and — both vortices having the same sense of rotation — increase thus intensity of the leading edge vortex. The whole process is schematized in Fig. 9. This process of lift generation is quite unique for insects, and is made possible only by a special wing base — a pivotal joint with the capability to set almost any required wing motion, and a system of the flight muscles, controlling all the degrees of freedom the wing has.

The aerodynamics of insect wings is strongly affected by small values of the Reynolds number, i.e., by low ratio of inertial to viscous forces. Reynolds number for large insects is of the order 10^3 to 10^4, however, for smallest insects it may be 10 or even less. At these Reynolds numbers the flow past insect wings is even in air highly "viscous".

Only at these Reynolds numbers the wings can operate at high angles of attack with a strong leading edge vortex, which almost does not move during one stroke. The strong centrifugal flow inside the vortex stabilizes it during wing oscillations. Nothing similar can happen at high Reynolds numbers — as, e.g., on aircraft wings where the flow at these angles of attack will separate, leading ultimately to complete break-down of the lift.

Also only at these Reynolds numbers the wings can have very unusual surface (from the classical aerodynamics point of view), and nevertheless achieve very good lift and drag. In Fig. 10 is an example of the cross section of a corrugated dragon-fly wing. Wind tunnel experiments, as well as numerical simulation has proved that at very small Re the corrugation grooves are filled in with highly viscous vortices, so that the effective final shape of the wing profile acquires good aerodynamic quality (see, e.g., [8,12]).

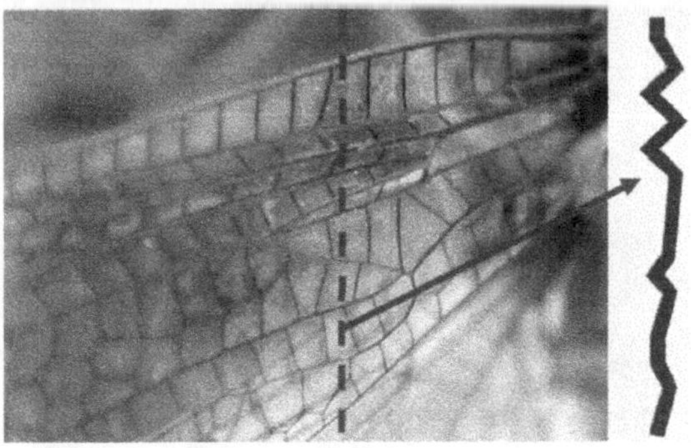

Fig. 10. Corrugated dragon-fly wing (adapted from [8])

There is still another possible view on the high lift generation by insect wings, similar to that we have already used in describing the bird's wing-lift-generation and to that we shall use in explaining the driving mechanism in aquatic locomotion. The leading edge vortex together with the starting vortex (see Fig. 2) represent a couple of counter-rotating vortices, which impart downward oriented momentum to the entrained air. Reaction of the surrounding stagnant air to this momentum is the insect-wing-generated lift. Explaining the lift of the oscillating insect wing using this model seems much more relevant than referring to the analogy with an "aircraft-like-generated-lift". Everybody can easily feel the vortex-generated momentum when putting his hand under the flying or hovering insect.

5. Vortices in animal locomotion in water

It is not easy to understand and describe how movements initiated by the muscoskeletal and nervous system result in producing thrust and in generating forces to maneuvre the body. In aquatic locomotion forces exerted by the body and fins against the surrounding water (or stream) yield in response to this action forces needed for propulsion and maneuvring. However, if we concentrate only on the most simple way of animal locomotion in water, we will see again, that the propelling force is a reaction of water to the action of vortices.

The basic fluid dynamic mechanism is astonishingly simple. A couple of two counterrotating vortices entrains fluid from their environment and drives it against the mass of stagnant water. Reaction against the momentum of the vortex driven fluid imparts a momentum to the vortices and sets them into motion off the stagnant water (Fig. 11) The most simple generator of a vortex couple is a caudal fin, or the whole fish tail (like at dolphins), or the leg of a water strider, etc. The pectoral fins are used mainly in maneuvering. However, the most important vortex couple that forms in rapidly turning fish is generated by bending the whole body (see, e.g., [14]).

Fig. 11. Forward momentum generation in water — a reaction of the surrounding mass of fluid against a 2D vortex couple (adapted from [14])

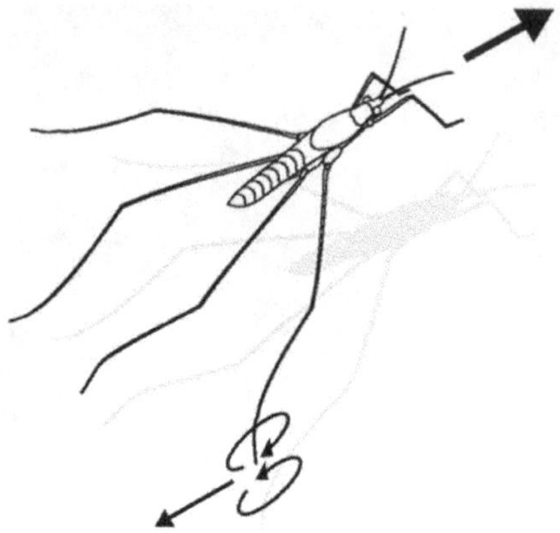

Fig. 12. Subsurface vortices generated by a water strider's leg penetrating the water level (from [2])

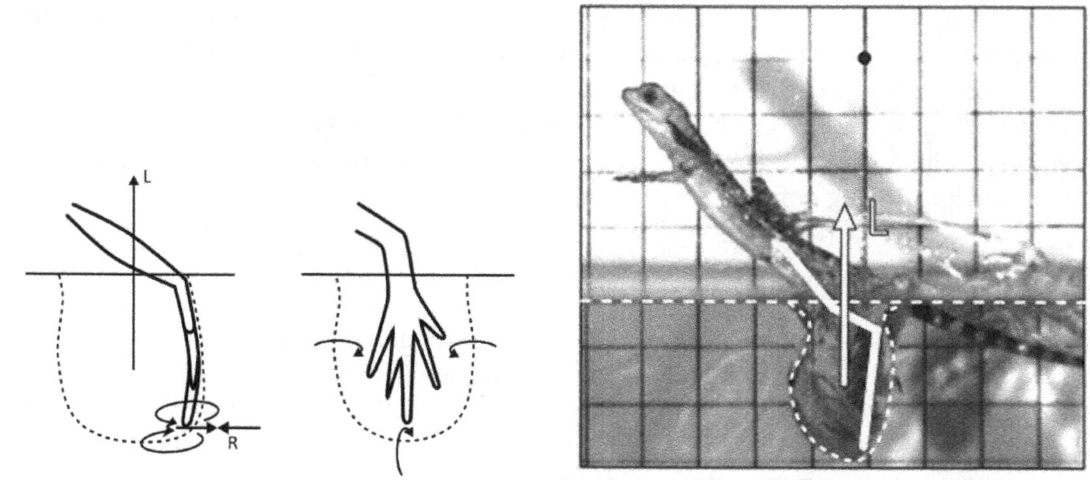

Fig. 13. Subsurface vortices generated by a basilisk lizard's leg penetrating the water level. Note the great cavity behind the lizard's leg, L is the Archimedes lift of the cavity, R is the reaction of the mass of water (adapted from [7])

Similar vortices are generated in swimming frogs. Each leg generates a separate vortex ring and no interference by the other leg has been observed. As intensity of the generated vortices is much smaller then in fishes, even the propulsion effectivity is smaller. This experience has also influenced the swimming style in humans.

Vortices are used for propulsion even in water-walking arthropods (Fig. 12). Flow visualization revealed that the wakes of water walkers are formed by a series of subsurface vortices shed by the driving stroke. The water strider when not moving rests on the water surface due to surface tension, however, even they use the vortex reaction to water walking [1].

There is, however, a much heavier creature capable of water-walking (or better "running") — the basilisk lizard — who can sprint across the water surface at speeds about $1.6\,\mathrm{ms}^{-1}$. To run on water he does not use only the vortices, but makes a very clever use of the Archimedes lift. Fig. 13 explains the way he does it. After he strikes the water level with one of his legs the leg gradually penetrates into water forming a bubble in its wake. It is this bubble which receives the

Archimedes lift and supports the lizard. The bubble closes very quickly, so the lizard does not have much time to drive the leg back with minimum friction, and to continue the same action with the other leg. The reaction of the water which represents here the actual lift force has in this case two parts — the reaction to the momentum imparted to the stagnant water by the lizard's leg during the stroke, and the Archimedes lift of the bubble and the volume of the leg. The whole trick is in repeating the strokes with such a frequency that the described model can be applied. The frequency depends on the bubble volume and its closing time, and, through it on the water density and viscosity.

The reaction force which develops in response to the stroke of the lizard's leg is generated by the same mechanism as in water-walking arthropods (see Fig. 12 and Fig. 13). The leg penetrating the water surface generates vortices — two counter-rotating vortices in a 2D case, or a vortex- ring- like structure in a 3D case. These vortical structures remain where they were generated even when the leg is driven off this place. They entrain the water from their neighbourhood and drive it in the direction of the stroke. The reaction has then the same momentum but opposite direction.

6. Closing remarks

It was only as late as the second half of the last century that scientists were successful in unveiling the secrets of animal locomotion in fluids — capabilities that the animals have been developing for millions of years.

Investigation into physics of animal locomotion has revealed that it depends largely on vortices. A couple of two counterrotating vortices or a vortex ring entrain fluid from their environment and drive it against the mass of stagnant fluid. Reaction of the momentum of the vortex driven fluid imparts a momentum to the vortices (vortex ring) and sets them into motion.

Investigation into birds' and insects' flight has also discovered the incomprehensible perfection of the neuroskeletal system controlling and moving the wings. This represents another large category of problems still open to research. Only after such research has been completed, can we think of successfully building and exploiting the micro-air-vehicles.

References

[1] Bush, J. W. M., Hu, D. L., Walking on water: Biolocomotion at the interface, Annual Review of Fluid Mechanics 38 (1) (2006) 339–369.

[2] Denny, M. W., Paradox lost: answers and questions about walking on water, Journal of Experimental Biology 207 (2004) 1 601–1 606.

[3] Dickinson, M. H., Lehmann, F.-O., Sane, S. P., Wing rotation and the aerodynamic basis of insect flight, Science 284 (5422) (1999) 1 954–1 960.

[4] Dvořák, R., How they fly, submitted to Academia (The Publishing House of the Academy of Sciences of the Czech Republic), 2013 (in Czech).

[5] Dvořák, R., Vortices in moving fluids, Proceedings of Engineering Mechanics 1997, Svratka, 1997 (in Czech).

[6] Ellington, C. P., The novel aerodynamics of insect flight: Application to micro-air-vehicles, Journal of Experimental Biology 202 (1999) 3 439–3 448.

[7] Hsieh, S. T, Lauder, G. V., Running on water: Three-dimensional force generation by basilisk lizards, Proceedings of the National Academy of Sciences 101 (48) (2004) 16 784–16 788.

[8] Hu, H., Tamai, M., Bioinspired corrugated airfoil at low Reynolds numbers, Journal of Aircraft 45 (6) (2008) 2 068–2 077.

[9] Jeong, J., Hussain, F., On the identification of a vortex, Journal of Fluid Mechanics 285 (1995) 69–94.

[10] Koochesfahani, M. M., Vortical patterns in the wake of an oscillating airfoil, AIAA Journal 27 (9) (1989) 1 200–1 205.

[11] Lugt, H. J., Vortex flow in nature and technology, John Wiley & Sons, New York, 1983.

[12] Platzer, M. F., Jones, K. D., Young, J., Lai, J. C. S., Flapping wing aerodynamics: Progress and challenges, AIAA Journal 46 (9) (2008) 2 136–2 149.

[13] Sane, S. P., The aerodynamics of insect flight, The Journal of Experimental Biology 206 (2003) 4 191–4 208.

[14] Triantafyllou, M. S., Triantafyllou, G. S., Yue, D. K. P., Hydrodynamics of fishlike swimming, Annual Review of Fluid Mechanics 32 (2000) 33–53.

18

Analytic solution of simplified Cardan's shaft model

M. Zajíček[a,*], J. Dupal[a]

[a] *Faculty of Applied Sciences, University of West Bohemia, Univerzitní 22, 306 14 Plzeň, Czech Republic*

Abstract

Torsional oscillations and stability assessment of the homokinetic Cardan shaft with a small misalignment angle is described in this paper. The simplified mathematical model of this system leads to the linearized equation of the Mathieu's type. This equation with and without a stationary damping parameter is considered. The solution of the original differential equation is identical with those one of the Fredholm's integral equation with degenerated kernel assembled by means of a periodic Green's function. The conditions of solvability of such problem enable the identification of the borders between stability and instability regions. These results are presented in the form of stability charts and they are verified using the Floquet theory. The correctness of oscillation results for the system with periodic stiffness is then validated by means of the Runge-Kutta integration method.

Keywords: Cardan shaft, analytic periodic solution, stability assessment, linearized equation, Mathieu's equation

1. Introduction

The Cardan's shaft is used as a component of many mechanisms for the transmission of torque through angularly misaligned rotating shafts. Typical applications can be found, e.g., in automobile industry or shipbuilding. It could be said that the problems relating to angularly misaligned shafts are so far at the subject of research. This can be illustrated by the development of various types of joints for transfer constant velocity, see, e.g., [16]. The typical design of the Cardan's shaft is composed of two Hooke's joints in series and mutually connected with a shaft that usually enables an axial dilatation. A major problem with the use of Hooke's joint, also known as the Cardan's joint or the universal joint, is that it transforms a constant input speed to a periodically fluctuating one. This means that the dynamic system with these joints is parametrically excited. It introduces a number of specific resonance conditions or dynamic instability of the system. The main goal of analyzing such systems is assessment of the instability conditions as well as the periodic solution of the steady state motion.

It seems Porter [13] to be the first who predicted the critical speed ranges associated with such a system. A linearized one-degree-of-freedom (1 DOF) model was considered to predict the two primary parametric resonance zones of a stability chart obtained using the Floquet theory [9]. Investigation was concerned, among others, with the effects of parameters such as stiffness ratio and joint angle on the critical speed ranges. These analyses were later extended to a nonlinear model given in [14]. The first approximation of the Krylov-Bogoliubov method was used to get the system behavior in the two primary parametric resonance zones. Porter and Gregory [14] showed that the system ultimately executes a limit cycle of oscillation and that (in certain critical speed ranges) the amplitude of this oscillation is large if the system is lightly

*Corresponding author. e-mail: zajicek@kme.zcu.cz.

damped. The stability problem was studied through Poincaré-Lyapunov method described, e.g., in [10]. Zahradka [17] brought the solution for a 1 DOF driving system incorporating the Cardan shafts by means of the approximate Van der Pol's method of slowly-varying coefficients. The resulting nonlinear equation of motion with periodic coefficients was investigated in the region of subharmonic and subultraharmonic resonance. Zahradka showed that a number of solutions exist for one tuning coefficient of a system and the width of attraction domains depends mainly on the system damping and the angular misalignment of the Hooke's joints. Chang [5] revisited nonlinear equation presented in [13] and obtained the higher order stability map for the damped system by using a perturbation technique. Moreover, the application of higher order averaging (see [12]) to the equation of motion leads to amplitude equations consisting of a finite number of terms including quadratic and cubic nonlinearities.

Porter [15] investigated the problem in [14] by considering a 2 DOF model and established the instability conditions as well as the amplitudes of the steady state motion. Asokanthan and Hwang [1] also solved the system with a 2 DOF. They used the method of averaging and established the closed form instability conditions associated with combined resonance. The same problem of instabilities was investigated through calculations of the Lyapunov exponent in [3]. In both cases mentioned above, only linearized models were employed, and as a result the critical speed ranges due to quasi-periodic and chaotic motion were not predicted. Later, Asokanthan and Methan [2] considered a nonlinear model and showed via numerical simulations that the system may exhibit chaotic behavior under certain circumstances. The dynamic stability problem was solved in [4] by means of a monodromy matrix method. Linearized n DOF system was solved by Zeman in [18,19] and Kotera in [11].

The goal of the present study is to apply to the Cardan's shaft problem solving methodology developed by the authors in [8]. The use of this solution is suitable for a linear system with periodically varying stiffness and excitation, see [7]. Therefore, the governing linearized equation representing the torsional motion of the system is derived. The model with and without a stationary damping parameter is considered. The periodic analytic solutions in a steady state including the boundary curves of (in)stability regions are determined using the presented method.

2. Governing equation

For the purposes of the present study, the simplified model of the homokinetic Cardan's shaft which is depicted in Fig. 1 is derived. Let us assume that the torsional stiffness of shafts 1 and 3 is closing to infinity, and so $\varphi_2 \equiv \varphi_1$ and $\varphi_5 \equiv \varphi_6$. Furthermore, the rotary inertia of disks 1 and 6 is dominant and the others are neglected. The transformation relations between φ_3, φ_2 and φ_4, φ_5 have the form (see, e.g. [6])

$$\tan\varphi_3 = \tan\varphi_2 \cos\delta = \tan\varphi_1 \cos\delta \qquad \text{and} \qquad \tan\varphi_4 = \tan\varphi_5 \cos\delta = \tan\varphi_6 \cos\delta\,, \quad (1)$$

where the initial configuration of the Hooke's joints is shown in position on Fig. 1. However, if the initial position of joints is turned at 90°, the relations have the form

$$\tan\varphi_3 = \frac{\tan\varphi_2}{\cos\delta} = \frac{\tan\varphi_1}{\cos\delta} \qquad \text{and} \qquad \tan\varphi_4 = \frac{\tan\varphi_5}{\cos\delta} = \frac{\tan\varphi_6}{\cos\delta}\,. \quad (2)$$

Parameter δ denotes an angular misalignment. Let us assume that the angle δ is small. For this reason, the approximations are used as follows:

$$\cos\delta \approx 1 - \delta^2/2\,, \qquad 1/\cos\delta \approx 1 + \delta^2/2\,. \quad (3)$$

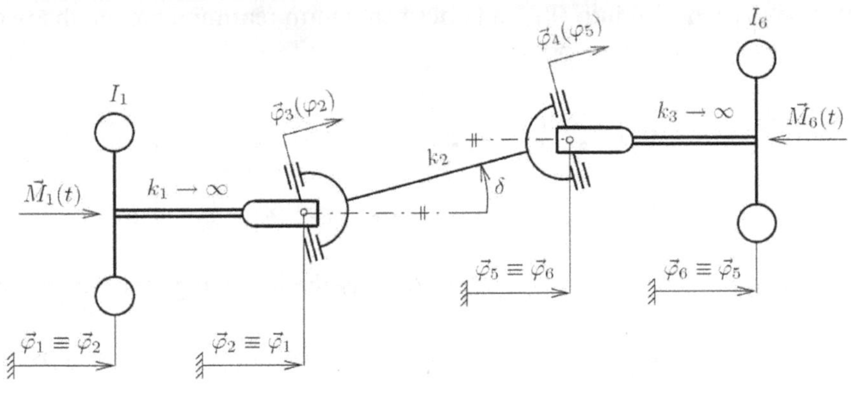

Fig. 1. Simplified scheme of Cardan shaft

Then equations (1) and (2) can be rewritten by means of (3) as

$$\tan\varphi_3 \approx (1 \mp \frac{\delta^2}{2})\tan\varphi_1 \qquad \text{and} \qquad \tan\varphi_4 \approx (1 \mp \frac{\delta^2}{2})\tan\varphi_6 \tag{4}$$

with respect to both initial configurations of the Hooke's joints. When the functions $\tan\varphi_3$ and $\tan\varphi_4$ are expanded to the Taylor series with regard to the first two terms,

$$\begin{aligned} \tan\varphi_3 &= \tan(\varphi_1 + \Delta\varphi_1) \approx \tan\varphi_1 + \frac{1}{\cos^2\varphi_1}\Delta\varphi_1\,, \\ \tan\varphi_4 &= \tan(\varphi_6 + \Delta\varphi_6) \approx \tan\varphi_6 + \frac{1}{\cos^2\varphi_6}\Delta\varphi_6\,, \end{aligned} \tag{5}$$

it is simply proved with the aid of (4) that the increments $\Delta\varphi_1$ and $\Delta\varphi_6$ are as follows:

$$\begin{aligned} \Delta\varphi_1 &\approx \mp\frac{\delta^2}{2}\tan\varphi_1\cos^2\varphi_1 = \mp\frac{\delta^2}{4}\sin 2\varphi_1\,, \\ \Delta\varphi_6 &\approx \mp\frac{\delta^2}{2}\tan\varphi_6\cos^2\varphi_6 = \mp\frac{\delta^2}{4}\sin 2\varphi_6\,. \end{aligned} \tag{6}$$

Therefore, the angles φ_3 and φ_4 are approximated such as

$$\varphi_3 = \varphi_1 + \Delta\varphi_1 \approx \varphi_1 \mp \frac{\delta^2}{4}\sin 2\varphi_1 \qquad \text{and} \qquad \varphi_4 = \varphi_6 + \Delta\varphi_6 \approx \varphi_6 \mp \frac{\delta^2}{4}\sin 2\varphi_6\,. \tag{7}$$

The equations of motion in terms of the twist variables $\boldsymbol{\varphi} = [\varphi_1, \varphi_6]^{\mathrm{T}}$ may be derived using Lagrange's equations

$$\frac{\mathrm{d}}{\mathrm{d}t}\left(\frac{\partial E_{\mathrm{k}}}{\partial\dot{\boldsymbol{\varphi}}}\right) - \frac{\partial E_{\mathrm{k}}}{\partial\boldsymbol{\varphi}} = \frac{\delta W}{\delta\boldsymbol{\varphi}} - \frac{\partial E_{\mathrm{p}}}{\partial\boldsymbol{\varphi}}\,, \tag{8}$$

where

$$E_{\mathrm{k}} = \frac{1}{2}I_1\dot{\varphi}_1^{\,2} + \frac{1}{2}I_6\dot{\varphi}_6^{\,2} \qquad \text{and} \qquad E_{\mathrm{p}} = \frac{1}{2}k_2\left[\varphi_4(\varphi_6) - \varphi_3(\varphi_1)\right]^2 \tag{9}$$

is kinetic and potential energy of the system, respectively. The virtual work δW of non-conservative forces takes the form

$$\delta W = M_1(t)\delta\varphi_1 - M_6(t)\delta\varphi_6\,. \tag{10}$$

After substitution (9) and (10) into (8), and after some term rearrangements, the two equations of motion can be expressed in the form

$$I_1\ddot{\varphi}_1 + k_2(\varphi_1 - \varphi_6) \mp k_2(\varphi_1 - \varphi_6)\frac{\delta^2}{2}\cos 2\varphi_1 = M_1(t)\,,$$
$$I_6\ddot{\varphi}_6 - k_2(\varphi_1 - \varphi_6) \pm k_2(\varphi_1 - \varphi_6)\frac{\delta^2}{2}\cos 2\varphi_6 = -M_6(t)\,. \tag{11}$$

The system of equations (11) is further linearized using the following simplifications:

$$\cos 2\varphi_1 = \cos 2(\omega t + \Delta\varphi_1) \doteq \cos 2\omega t\,, \quad \cos 2\varphi_6 = \cos 2(\omega t + \Delta\varphi_6) \doteq \cos 2\omega t\,. \tag{12}$$

Multiplying the first equation in (11) by $1/I_1$ and the second one (11) by $1/I_6$, and subtracting the two resulting equations lead to

$$\ddot{\varphi} + \Omega^2(1 \mp \varepsilon\cos 2\omega t)\varphi = \frac{M_1(t)}{I_1} + \frac{M_6(t)}{I_6}\,, \tag{13}$$

which is the equation of the relative torsional vibrations of the Cardan's shaft shown in Fig. 1. In the above equation,

$$\varepsilon = \frac{\delta^2}{2}\,, \qquad \Omega^2 = \frac{k_2}{I_1} + \frac{k_2}{I_6} \qquad \text{and} \qquad \varphi = \varphi_1 - \varphi_6\,, \tag{14}$$

where ε is a measure of stiffness modulation. The signs $\mp$ in (13) indicate what the initial configuration of the Hooke's joints is considered. After dimensionless transformation

$$\tau = \Omega t \qquad \text{and} \qquad \eta = \omega/\Omega \tag{15}$$

and some arrangements

$$\ddot{\varphi}(t) = \Omega^2\varphi''(\tau) \qquad \text{and} \qquad \omega t = \eta\tau\,, \tag{16}$$

the equation of motion (13) is further written as

$$\varphi'' + (1 \mp \varepsilon\cos 2\eta\tau)\varphi = \frac{1}{\Omega^2}\left(\frac{M_1(\tau)}{I_1} + \frac{M_6(\tau)}{I_6}\right)\,. \tag{17}$$

Let us consider that the system is damped. When only a stationary damping is taken into account, the equations of motion (13) and (17) can be extended and rewritten in the form

$$\ddot{\varphi} + 2D\Omega\dot{\varphi} + \Omega^2(1 \mp \varepsilon\cos 2\omega t)\varphi = \frac{M_1(t)}{I_1} + \frac{M_6(t)}{I_6}\,, \tag{18}$$
$$\varphi'' + 2D\varphi' + (1 \mp \varepsilon\cos 2\eta\tau)\varphi = \frac{1}{\Omega^2}\left(\frac{M_1(\tau)}{I_1} + \frac{M_6(\tau)}{I_6}\right)\,, \tag{19}$$

respectively, where D is a damping ratio. If the functions $M_1(t)$ and $M_6(t)$ are periodic with period $T = 2\pi/\omega$, the steady state solution of the derived equation of motion with and/or without damping can be found analytically.

3. Analytic solution in steady state and stability assessment

The presented method of solution is based on the knowledge of a periodic Green's function $H(t)$ which is constructed as a response of the stationary part of the equation of motion to the T-periodic Dirac chain, see [8]. While respecting (18) where the effect of damping is considered, equation for finding $H(t)$ can be written as

$$\ddot{H}(t) + 2D\Omega\dot{H}(t) + \Omega^2 H(t) = \frac{1}{T}\sum_{n=-\infty}^{\infty} e_n(t)\,, \tag{20}$$

where

$$e_n(t) = \mathrm{e}^{in\omega t} \qquad \text{and} \qquad \mathrm{i}^2 = -1\,. \tag{21}$$

Superposition principle in solving (20) leads to finding the periodic Green's function in the form

$$H(t) = \frac{1}{T}\sum_{n=-\infty}^{\infty} L_n e_n(t)\,, \tag{22}$$

where

$$L_n = \frac{1}{\Omega^2}\frac{1}{1 + 2\mathrm{i}Dn\eta - n^2\eta^2}\,. \tag{23}$$

Including the parametric term $(\Omega^2\varepsilon\cos 2\omega t)\varphi$ into the exitation, the steady state response corresponding to equation (18) may then be expressed as a sum of convolution integrals

$$\varphi(t) = \pm\varepsilon\int_0^T H(t-s)k(s)\varphi(s)\,\mathrm{d}s + \int_0^T H(t-s)f(s)\,\mathrm{d}s\,, \tag{24}$$

where

$$k(s) = \Omega^2\cos 2\omega s \qquad \text{and} \qquad f(s) = \frac{M_1(s)}{I_1} + \frac{M_6(s)}{I_6}\,. \tag{25}$$

With regard to the solution mentioned in [8], the function $\varphi(t)$ can then be written in the form

$$\varphi(t) = \mathbf{e}^{\mathrm{T}}(t)\left[\mathbf{I} \pm \varepsilon\left(\mathbf{I} \mp \varepsilon\mathbf{LH}\right)^{-1}\mathbf{LH}\right]\mathbf{Lf}\,, \tag{26}$$

where $\mathbf{I}$ is the infinity identity matrix, $\mathbf{L}$ is the infinity diagonal matrix

$$\mathbf{L} = \mathrm{diag}\left\{L_n\right\} \tag{27}$$

and $\mathbf{H}$ is a symmetric matrix defined as

$$\mathbf{H} = \frac{1}{2}\Omega^2\begin{bmatrix} \ddots & \vdots & \vdots & \vdots & \vdots & \vdots & \iddots \\ \cdots & 0 & 0 & 1 & 0 & 0 & \cdots \\ \cdots & 0 & 0 & 0 & 1 & 0 & \cdots \\ \cdots & 1 & 0 & 0 & 0 & 1 & \cdots \\ \cdots & 0 & 1 & 0 & 0 & 0 & \cdots \\ \cdots & 0 & 0 & 1 & 0 & 0 & \cdots \\ \iddots & \vdots & \vdots & \vdots & \vdots & \vdots & \ddots \end{bmatrix}. \tag{28}$$

Further, the vectors in equation (26) are given as

$$\mathbf{e}(t) = \left[\ldots, e_{-n}(t), \ldots, e_{-1}(t), 1, e_1(t), \ldots, e_n(t), \ldots\right]^{\mathrm{T}}, \tag{29}$$

$$\mathbf{f} = \left[\ldots, f_{-n}, \ldots, f_{-1}, f_0, f_1, \ldots, f_n, \ldots\right]^{\mathrm{T}} \qquad \text{with} \qquad f(t) = \mathbf{f}^{\mathrm{T}}\mathbf{e}(t)\,. \tag{30}$$

It is obvious that solving $\varphi(t)$ exists only if the matrix $\mathbf{I} \mp \varepsilon\mathbf{LH}$ is invertible. Then, the real (for a real system) eigenvalues $1/\varepsilon$ of the matrix $\mathbf{LH}$ determine the borders of (in)stability. However, the boundaries between the stable and unstable regions at combinations of parameters η and ε may occur not only for the ratio $\varphi(t+T)/\varphi(t) = 1$ but also for the ratio -1. While in the first case, the solution $\varphi(t)$ has a period T, in the second case it has a period $2T$. Hence, it follows that it is necessary to take into account the resonant stage with frequency $\omega/2$. The eigenvalues are then determined for a matrix $\mathbf{L^*H^*}$, where

$$\mathbf{L}^* = \operatorname{diag}\left\{\ldots, L_{-n}, L_{-n+\frac{1}{2}}, L_{-n+1}, \ldots, L_0, \ldots, L_{n-1}, L_{n-\frac{1}{2}}, L_n, \ldots\right\} \tag{31}$$

and

$$\mathbf{H}^* = \frac{1}{2}\Omega^2 \begin{bmatrix} \ddots & \vdots & \vdots & \vdots & \vdots & \vdots & \vdots & \iddots \\ \cdots & 0 & 0 & 0 & 0 & 1 & 0 & \cdots \\ \cdots & 0 & 0 & 0 & 0 & 0 & 1 & \cdots \\ \cdots & 0 & 0 & 0 & 0 & 0 & 0 & \cdots \\ \cdots & 0 & 0 & 0 & 0 & 0 & 0 & \cdots \\ \cdots & 1 & 0 & 0 & 0 & 0 & 0 & \cdots \\ \cdots & 0 & 1 & 0 & 0 & 0 & 0 & \cdots \\ \iddots & \vdots & \vdots & \vdots & \vdots & \vdots & \vdots & \ddots \end{bmatrix}. \tag{32}$$

It has been proved in [8] that the spectra of matrices $\mathbf{LH}$ and $\mathbf{L^*H^*}$ satisfy the condition

$$\mathbf{\Sigma}(\mathbf{LH}) \subset \mathbf{\Sigma}(\mathbf{L^*H^*}). \tag{33}$$

The real calculations of $\varphi(t)$ or ε are not possible for an infinite system. Therefore, the finite systems of equations have to be solved. Let us denote the solutions $\varphi_N(t)$ and ε_N approximated by the system of $2N+1$ and $4N+1$ equations, respectively. The solutions are assumed to be correct, i.e. $\varphi_N(t) = \varphi(t)$ and $\varepsilon_N = \varepsilon$, if the conditions

$$\|\varphi_{N+1}(t) - \varphi_N(t)\| < \varepsilon_\varphi \quad \text{for} \quad \varphi_{N+1}, \varphi_N \in L_2(0,T) \qquad \text{and} \qquad \left(\frac{\varepsilon_{N+1}}{\varepsilon_N} - 1\right)^2 < \varepsilon_\lambda \tag{34}$$

are satisfied while the parameters ε_φ and ε_λ are small positive numbers.

4. Numerical results and discussion

As further shown in this section, the existence of the searched solution $\varphi(t)$ is closely related to the investigation of stability regions. The equation of motion derived for the Cardan's shaft has limitations in use only for small values of the parameter δ, and therefore it makes sense to analyze a limited zones of stability with respect to this parameter. As known, the problem of stability is not dependent on the excitation. Identified regions of stability have so general validity for the analyzed equation of motion that is of Mathieu's type. The following numerical computation of stability problems are performed for the number $N = 18$.

The stability chart of an undamped system is shown in Fig. 2. It is evident that the (in)stability borders are symmetric about the axis $\varepsilon = 0$ and that the system may become unstable for values $|\delta| < 30°$ only in small regions. It is also interesting that the stable regions are repeatedly divided by the (in)stability borders. This fact is well seen in Fig. 3 showing the values of the determinant $d = \det(\mathbf{I} - \varepsilon\mathbf{L^*H^*})$. This one takes only the non-negative values in the places

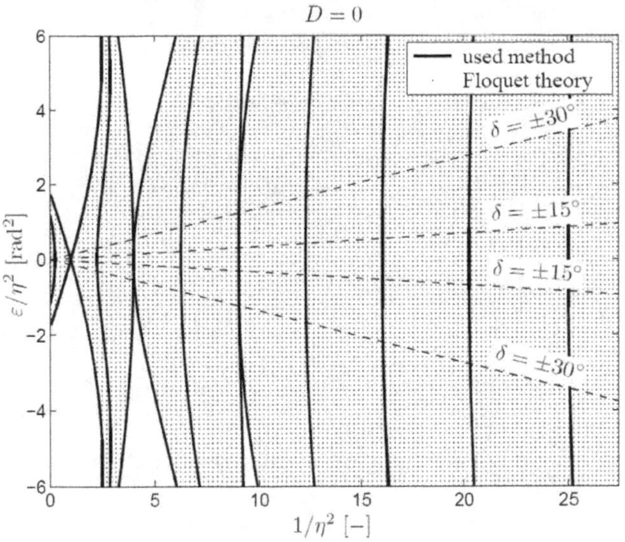

Fig. 2. Stable and unstable regions, problem without damping

of interest. Numerical experiments have demonstrated that there are double eigenvalues (points where $d = 0$). Fig. 3 also shows one important observation that is proved to be valid in all solved cases. The investigated systems with and without damping are stable when the determinant takes positive values.

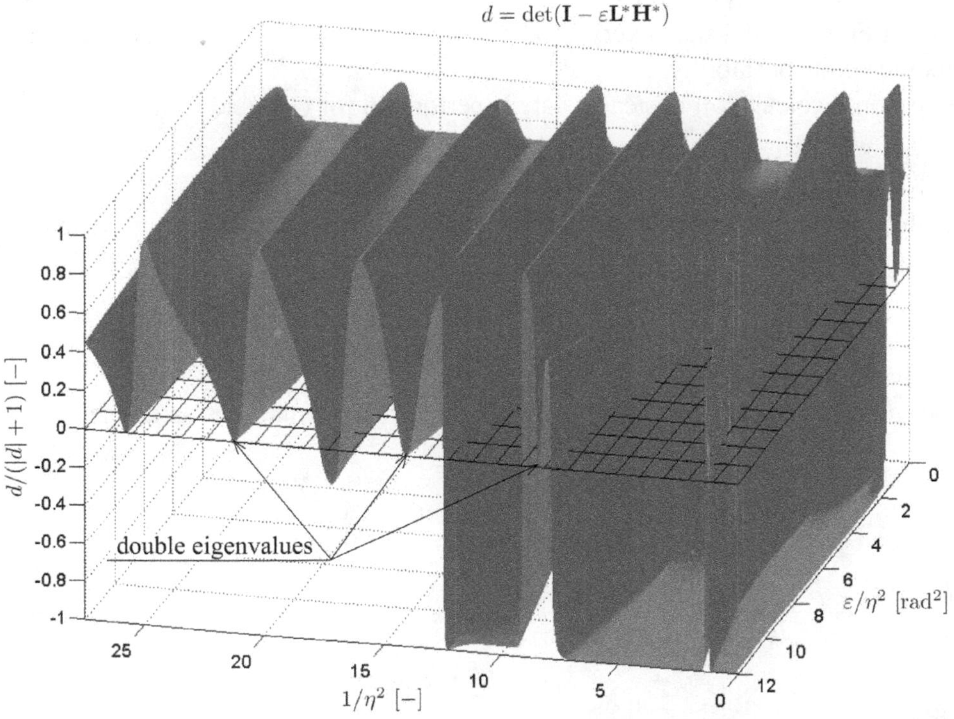

Fig. 3. Calculated values of determinant d in stable and unstable regions, problem without damping

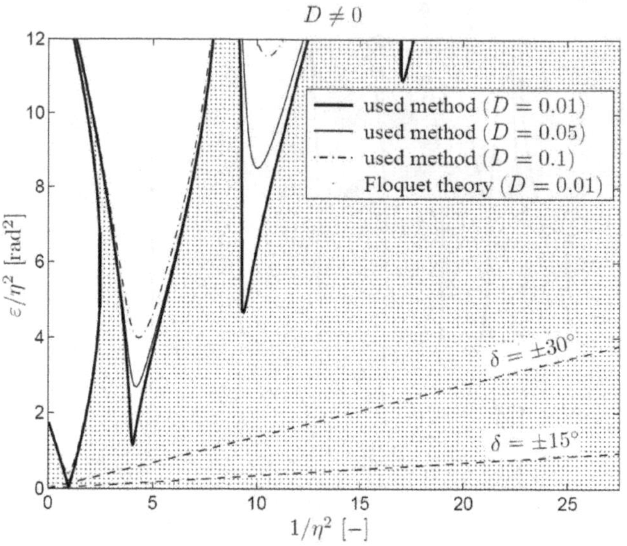

Fig. 4. Stable and unstable regions, problem with damping

The shape of the stable and unstable regions for different values of a damping ratio D is depicted in Fig. 4. The (in)stability borders are also symmetric about the axis $\varepsilon = 0$ as in the case $D = 0$, and therefore only the upper half ($\varepsilon \geq 0$) of the figure is presented. It is obvious from Fig. 4 that the system is stable for all values of η if $|\delta| < 15°$ and $D \geq 0.01$. In contrast, if the angle $\delta = \pm 30°$ and $D = 0.01$, the system is unstable for values close to $\eta = 1$. Furthermore, it is found that the unstable region of a set of zero measure do not occur in cases with damping. The correctness of detected boundaries is verified using Floquet theory [9]. It is clear from Figs. 2 and 4 that a very good agreement is found. The dot marks in both figures represent the points of stability.

The response calculation in steady state is performed for (17) and (19). It is supposed that the driving torque is

$$M_1(t) = M_1 = \text{constant}, \tag{35}$$

which is the typical characteristic of DC motors. The powered torque is then considered in the form of a Fourier series

$$M_6(t) = m_0 + \sum_{j=1}^{\infty} m_j \cos(j\omega t + \phi_j) \approx M_1 \left[r_0 + r_1 \cos(\omega t + \phi) + r_2 \cos 2\omega t\right], \tag{36}$$

while the moment is approximated by only the first three terms. Let us define the following ratios:

$$r_j = \frac{m_j}{M_1} \qquad \text{for} \qquad j = 0, 1, 2\,. \tag{37}$$

The right-hand side of equations (17) and (19) then takes the form

$$\begin{aligned} \frac{1}{\Omega^2}\left(\frac{M_1(\tau)}{I_1} + \frac{M_6(\tau)}{I_6}\right) &= \frac{M_1}{k_2}\frac{I_1 I_6}{I_1 + I_6}\left\{\frac{1}{I_1} + \frac{1}{I_6}\left[r_0 + r_1\cos(\eta\tau + \phi) + r_2\cos 2\eta\tau\right]\right\} = \\ &= \frac{\varphi_\mathrm{m}}{1 + \lambda_\mathrm{I}}\left[\lambda_\mathrm{I} + r_0 + r_1\cos(\eta\tau + \phi) + r_2\cos 2\eta\tau\right], \end{aligned} \tag{38}$$

where

$$\varphi_\mathrm{m} = \frac{M_1}{k_2} \qquad \text{and} \qquad \lambda_\mathrm{I} = \frac{I_6}{I_1}\,. \tag{39}$$

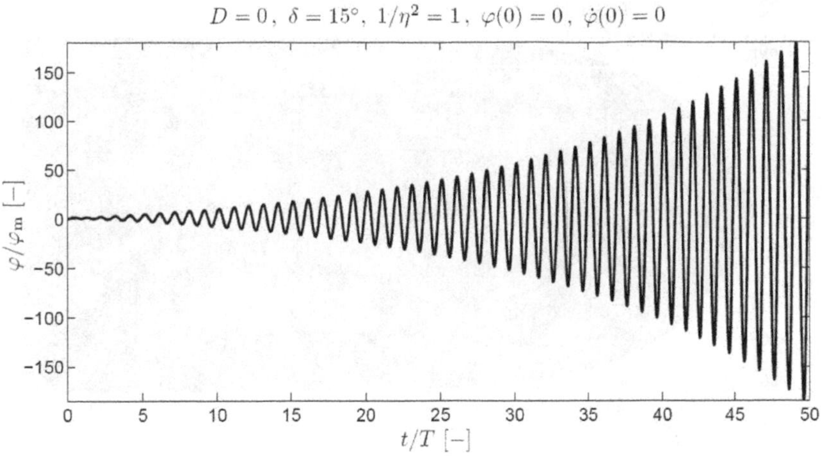

(a)

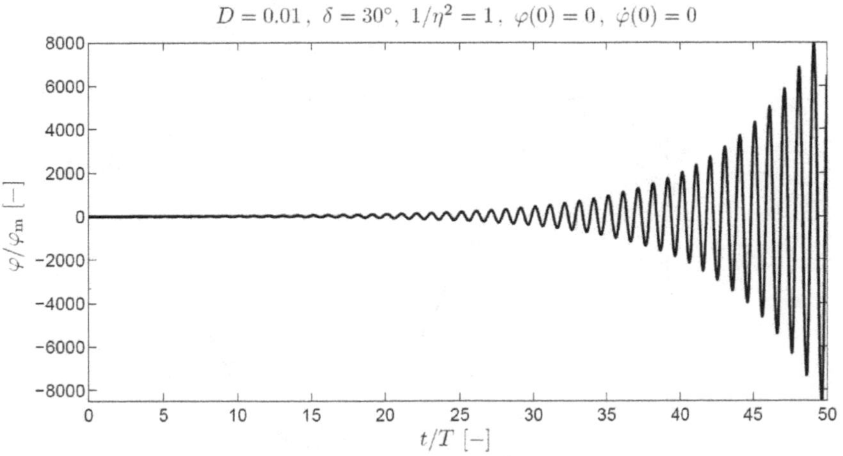

(b)

Fig. 5. Examples of calculation of response by the R-KI method in an unstable region

The above mentioned parameters are chosen for the numerical simulations as follows: $r_0 = 1.0, r_1 = 0.5, r_2 = 0.25, \phi = \pi/4$ and $\lambda_I = 1$. The work also includes analyses of results for two values of an angular misalignment $\delta = \{15°, 30°\}$. The parameter N specifies the number of Fourier series terms used for the description of the function $\varphi(\tau)$ and is set with respect to the numerical experiments. The value $N = 14$ is used for all subsequent calculations.

Typical characteristics of the system without and with damping in the unstable region is shown in Fig. 5(a) and (b), respectively. The system without damping is unstable respecting both $\delta = 15°$ and $\delta = 30°$ while the system with damping is unstable only for $\delta = 30°$. It can be seen in Figs. 2 and 4. Because the analytic solution given in equation (26) is valid only in the stable regions, it is necessary to find solutions through other methods. The Runge-Kutta integration (R-KI) method implemented in MATLAB function `ode45` is used. It is clear that the system is unstable with respect to an increasing amplitude in both cases, see Fig. 5. Computations were done for equations (17) and (19) having on the left side the periodic term "$-\varepsilon\varphi\cos 2\eta\tau$".

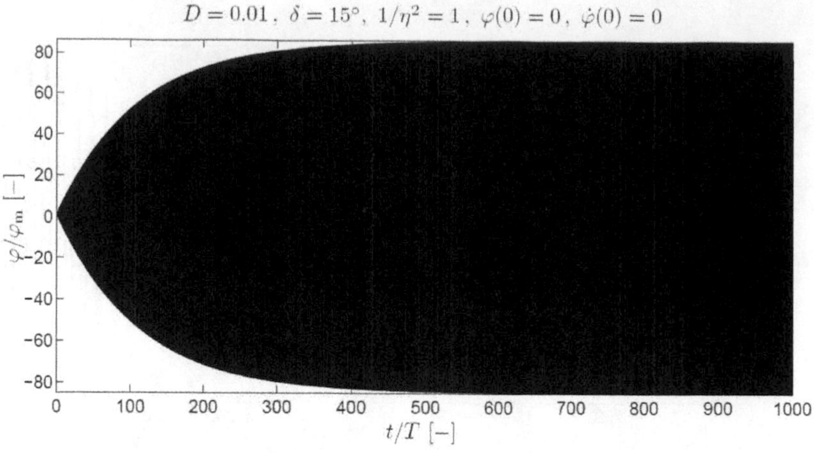

(a)

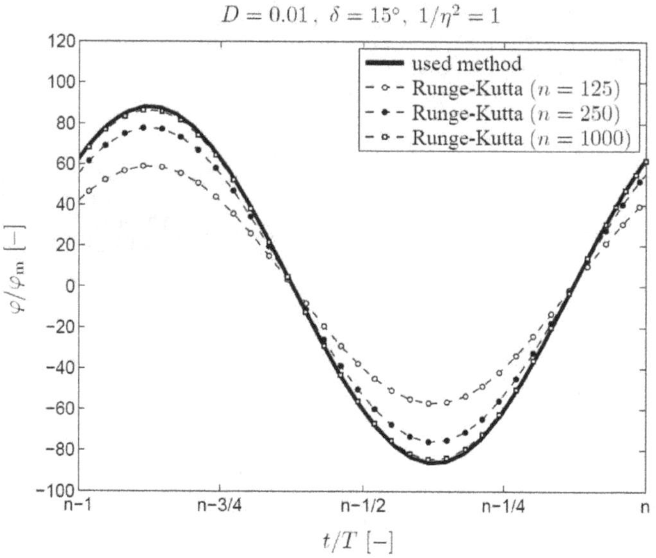

(b)

Fig. 6. Response in a stable region calculated by the R-KI method (a)–(b), and by the used method (b); considered periodic term "$-\varepsilon\varphi \cos 2\eta\tau$"

Response φ obtained for the similar parameters ($D = 0.01$, $\delta = 15°$, $\eta = 1$) but in the stable region is shown in Figs. 6 and 7. In the first case (Fig. 6), the equation of motion (19) with the periodic term "$-\varepsilon\varphi \cos 2\eta\tau$" is considered. Subsequently, the curves depicted in Fig. 7 correspond to the case where the periodic term is "$+\varepsilon\varphi \cos 2\eta\tau$". While the sign of this term has no influence on the stability charts because the (in)stability borders are symmetric about axis $\varepsilon = 0$, the effect on the response φ is apparent and it is caused by a different phase delay of excitation and stiffness modulation. Although the curves have similar shape, the ratio φ/φ_m gives significantly different values, see Figs. 6(b) and 7(b). It is also obvious in Figs. 6(a) and 7(a) that it is necessary to consider a different number of cycles t/T using R-KI method when comparable results with the analytic solution be provided.

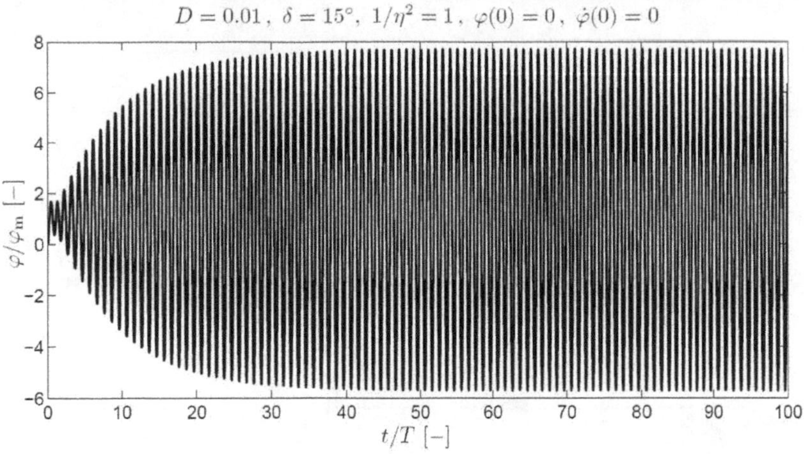

(a)

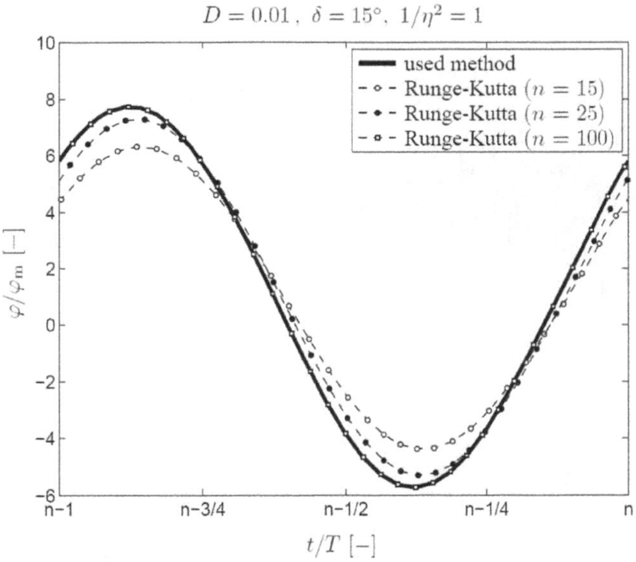

(b)

Fig. 7. Response in a stable region calculated by the R-KI method (a)–(b), and by the used method (b); considered periodic term "$+\varepsilon\varphi\cos 2\eta\tau$"

Previous examples of calculations made using the R-KI method show only the problems with homogeneous initial conditions. As known from the Floquet theory, the eigenvalues ρ of a monodromy matrix called the characteristic multipliers decide about the system stability. If all $\rho_i \in \mathbb{C}$, $\forall i$, satisfy the condition $|\rho_i| < 1$, then the system is stable and the solution φ corresponds to the stable limit cycle. This fact is demonstrated in Fig. 8. The same results using the R-KI method are obtained for the system with homogeneous and inhomogeneous ($\varphi(0) = -1$, $\dot{\varphi}(0) = 1$) initial conditions after a number of cycles $t/T = 25$. A very good agreement of the analytic and numeric results is shown in Fig. 8(c) and (d).

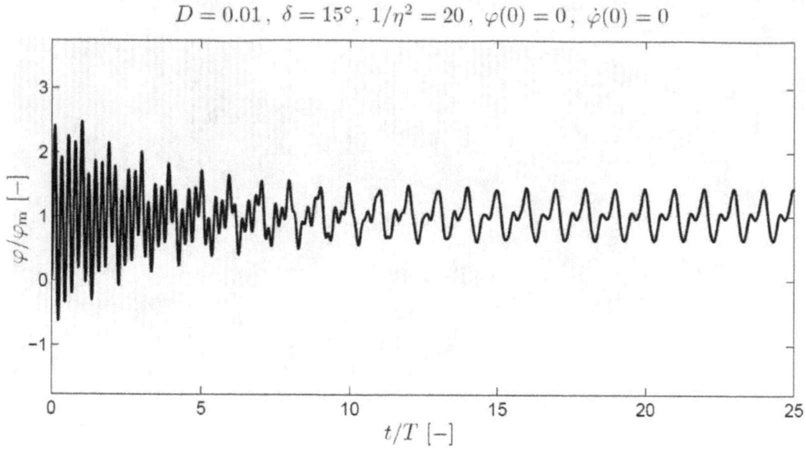

(a)

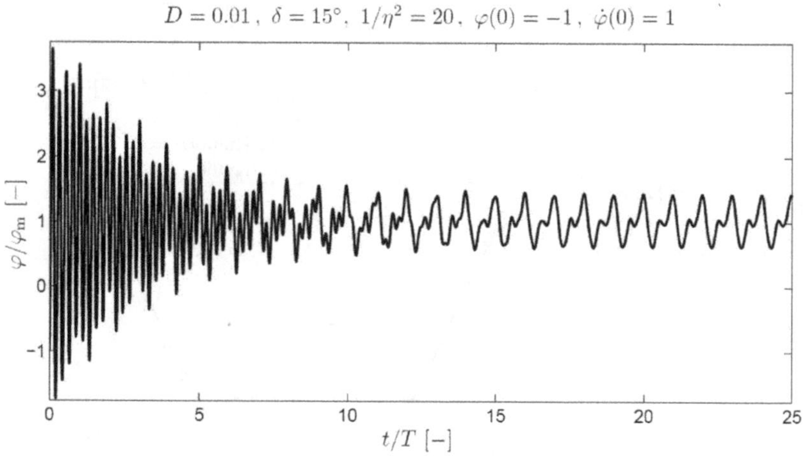

(b)

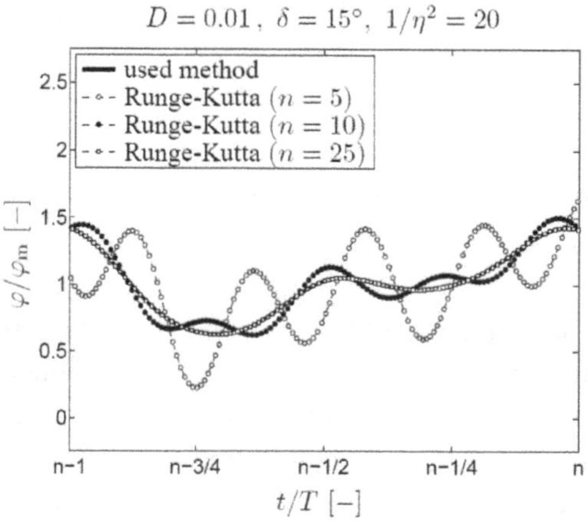

(c)

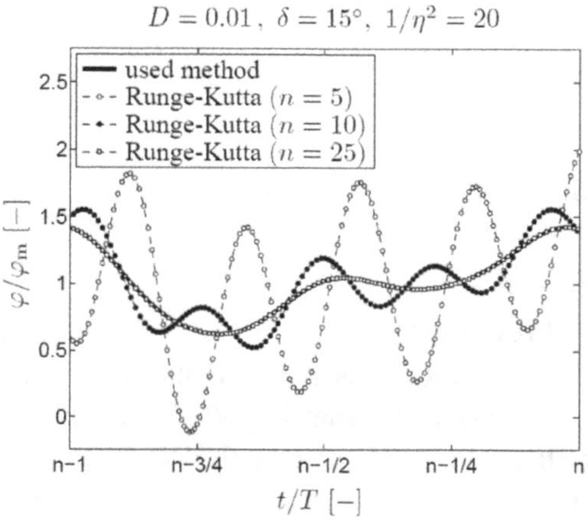

(d)

Fig. 8. Response calculated by the used method (c), (d), and the R-KI method (a)–(d) in a stable region: the R-KI method with homogeneous (a), (c), and inhomogeneous (b), (d) initial conditions; considered periodic term "$-\varepsilon\varphi\cos 2\eta\tau$"

5. Conclusion

The presented paper brings the analytic solution of the Cardan's shaft with a small misalignment angle. With regard to this assumption, the torsional motion equation of Mathieu's type has been found. If this system is stable and is excited by a periodic function, the analytic solution is given in the form of a particular solution of investigated differential equation. The results obtained by the Runge-Kutta integration method are identical with the analytic solution in spite of different initial conditions. Then the presented method is much more efficient for finding of the steady state solution. It has been shown in a number of numerical experiments.

Moreover, the analytic solution enables to find the boundaries of (in)stability. The real eigenvalues of the matrix $(\mathbf{L}^*\mathbf{H}^*)^{-1}$ determine these borders. Furthermore, it was numerically demonstrated that the determinant of the matrix $(\mathbf{I} - \varepsilon\mathbf{L}^*\mathbf{H}^*)$ is positive only if the system is stable. The Floquet theory results were used for verification. Based on the calculations, it can be stated that the damped system with the damping ratio $D \geq 0.01$ and with the misalignment angle $\delta \leq 15^\circ$ is always stable.

Acknowledgements

This work was supported by the European Regional Development Fund (ERDF), project "NTIS — New Technologies for the Information Society", European Centre of Excellence, CZ.1.05/1.1.00/02.0090 and by the project TE01020068 "Centre of research and experimental development of reliable energy production" of the Technology Agency of the Czech Republic.

References

[1] Asokanthan, S. F., Hwang, M. C., Torsional instabilities in a system incorporating a Hooke's joint, Journal of Vibration and Acoustics 118 (3) (1996) 368–374.

[2] Asokanthan, S. F., Methan, P. A., Non-linear vibration of torsional system driven by a Hooke's joint, Journal of Sound and Vibration 233 (2) (2000) 297–310.

[3] Asokanthan, S. F., Wang, X. H., Characterization of torsional instabilities in a Hooke's joint driven system via maximal Lyapunov exponents, Journal of Sound and Vibration 194 (1) (1996) 83–91.

[4] Bulut, G., Parlar, Z., Dynamic stability of a shaft system connected through a Hooke's point, Mechanism and Machine Theory 46 (2011) 1 689–1 695.

[5] Chang, S. I., Torsional instabilities and non-linear oscillation of a system incorporating a Hooke's joint, Journal of Sound and Vibration 229 (4) (2000) 993–1 002.

[6] Duditza, F., Cardan transmissions, Editions Eyrolles, Paris, 1971. (in French)

[7] Dupal, J., Zajíček, M., Analytical solution of the drive vibration with time varying parameters, Proceedings of the ASME 2011 International Design Engineering Technical Conferences & Computers and Information in Engineering Conference, Washington DC, USA, 2011, pp. 1 365–1 370.

[8] Dupal, J., Zajíček, M., Analytical periodic solution and stability assessment of 1 DOF parametric systems with time varying stiffness, Applied Mathematics and Computation 243 (2014) 138–151.

[9] Floquet, G., Sur les équations différentielles linéaires à coefficients périodiques, Annales scientifiques de l'École Normale Supérieure 12 (1883) 47–88. (in French)

[10] Jordan, D. W., Smith, P., Nonlinear ordinary differential equations: An introduction for scientists and engineers, Oxford University Press, Oxford, 2007.

[11] Kotera, T., Instability of torsional vibrations of a system with a cardan joint, Memoirs of the Faculty of Engineering, 26, Kobe University, 1980, pp. 19–30.

[12] Murdock, J. A., Perturbations theory and methods, New York, Wiley, 1991.

[13] Porter, B., A theoretical analysis of the torsional oscillation of a system incorporating a Hooke's joint, The Journal of Mechanical Engineering Science 3 (4) (1961) 324–329.

[14] Porter, B., Gregory, R. W., Non-linear torsional oscillation of a system incorporating a Hooke's joint, The Journal of Mechanical Engineering Science 5 (2) (1963) 191–200.

[15] Porter, B., Non-linear torsional oscillation of a two-degree-of-freedom system incorporating a Hooke's point, Proceedings of the Royal Society of London Series A 277, 1964, pp. 92–106.

[16] Watson, I., Prusty, B. G., Olsen, J., Conceptual design optimisation of a constant-velocity coupling, Mechanism and Machine Theory 68 (2013) 18–34.

[17] Zahradka, J., Torsional vibrations of a non-linear driving system with cardan shafts, Journal of Sound and Vibration 26 (4) (1973) 533–550.

[18] Zeman, V., Stability of motion of mechanical systems with joints, Acta Technica CSAV 1 (1977) 52–62.

[19] Zeman, V., Dynamik der Drehsysteme mit Kardangelenken, Mechanism and Machine Theory 13 (2) (1978) 107–118. (in German)

Permissions

All chapters in this book were first published in ACM, by University of West Bohemia; hereby published with permission under the Creative Commons Attribution License or equivalent. Every chapter published in this book has been scrutinized by our experts. Their significance has been extensively debated. The topics covered herein carry significant findings which will fuel the growth of the discipline. They may even be implemented as practical applications or may be referred to as a beginning point for another development.

The contributors of this book come from diverse backgrounds, making this book a truly international effort. This book will bring forth new frontiers with its revolutionizing research information and detailed analysis of the nascent developments around the world.

We would like to thank all the contributing authors for lending their expertise to make the book truly unique. They have played a crucial role in the development of this book. Without their invaluable contributions this book wouldn't have been possible. They have made vital efforts to compile up to date information on the varied aspects of this subject to make this book a valuable addition to the collection of many professionals and students.

This book was conceptualized with the vision of imparting up-to-date information and advanced data in this field. To ensure the same, a matchless editorial board was set up. Every individual on the board went through rigorous rounds of assessment to prove their worth. After which they invested a large part of their time researching and compiling the most relevant data for our readers.

The editorial board has been involved in producing this book since its inception. They have spent rigorous hours researching and exploring the diverse topics which have resulted in the successful publishing of this book. They have passed on their knowledge of decades through this book. To expedite this challenging task, the publisher supported the team at every step. A small team of assistant editors was also appointed to further simplify the editing procedure and attain best results for the readers.

Apart from the editorial board, the designing team has also invested a significant amount of their time in understanding the subject and creating the most relevant covers. They scrutinized every image to scout for the most suitable representation of the subject and create an appropriate cover for the book.

The publishing team has been an ardent support to the editorial, designing and production team. Their endless efforts to recruit the best for this project, has resulted in the accomplishment of this book. They are a veteran in the field of academics and their pool of knowledge is as vast as their experience in printing. Their expertise and guidance has proved useful at every step. Their uncompromising quality standards have made this book an exceptional effort. Their encouragement from time to time has been an inspiration for everyone.

The publisher and the editorial board hope that this book will prove to be a valuable piece of knowledge for researchers, students, practitioners and scholars across the globe.

List of Contributors

A. A. Gholampour, M. Ghassemieh and H. Razavi
Department of Civil Engineering, University of Tehran, Tehran, Iran

L. Půst, L. Pešek, J.Košina and A. Radolfováa
Institute of Thermomechanics, AS CR, v.v.i., Dolejškova 5, 182 00 Prague, Czech Republic

A. Belhocine
Faculty of Mechanical Engineering, USTO Oran University 31000 Oran, Algeria

C.-D. Cho
Department of Mechanical Engineering, Inha University, Incheon, 402-751, Republic of Korea

M. Nouby
Department of Mechanical Engineering, Faculty of Engineering, South Valley University, Qena-83523, Egypt

Y. B.Yi
Department of Mechanical and Materials Engineering, University of Denver, 2390 S York St Denver, CO 80208, USA

A. R. Abu Bakare
Department of Automotive Engineering, Universiti Teknologi Malaysia, 81310 UTM Skudai, Malaysia

J. Špička, L. Hynčík and M. Hajžman
Faculty of Applied Sciences, University of West Bohemia in Pilsen, Univerzitní 8, 306 14 Plzeň, Czech Republic

J. Turjanicová and E. Rohan
Faculty of Applied Sciences, University of West Bohemia, Univerzitní 22, 306 14 Plzeň, Czech Republic

S. Naili
Laboratoir modelisation et simulation multiéchelle, Université Paris-est, 61 avenue

F. Benkhaldoun
LAGA, University Paris 13, 99 Av. J. B. Clement, 93430 Villetaneuse, France

J. Karel
LAGA, University Paris 13, 99 Av. J. B. Clement, 93430 Villetaneuse, France
Faculty of Mechanical Engineering, CTU in Prague, Karlovo namesti 13, 121 35 Prague, Czech Republic

D. Trdlička and J. Fořt
Faculty of Mechanical Engineering, CTU in Prague, Karlovo namesti 13, 121 35 Prague, Czech Republic

K. Hassouni
LSPM, University Paris 13, 99 Av. J. B. Clement, 93430 Villetaneuse, France

R. Dvořák
Institute of Thermomechanics v.v.i., Academy of Sciences of the Czech Republic, Dolejškova 1402/5, 182 00 Prague, Czech Republic

J. Zapoměl, P. Ferfecki and J.Kozánek
Department of Dynamics and Vibrations, Institute of Thermomechanics, Department of Mechanics, Dolejškova 1402/5, 182 00 Praha 8, Czech Republic

B. R. Jaiswal and B. R. Gupta
Department of Mathematics, Jaypee University of Engineering and Technology, 473226 Guna, M. P., India

N. Khader
Department of Mechanical Engineering, Jordan University of Science & Technology (JUST), P.O. Box 3030, Irbid 22110, Jordan

Z. Hlaváč and V. Zeman
Faculty of Applied Sciences, University of West Bohemia, Univerzitní 22, 306 14 Plzeň, Czech Republic

M. Jansová, L. Hynčík and H. Čechová
New Technologies - Research Centre, University of West Bohemia, Univerzitní 8, 306 14 Plzeň, Czech Republic

J. Toczyski and D. Gierczycka-Zbrozek
Institute of Aeronautics and Applied Mechanics, Warsaw University of Technology, ul Nowowiejska 29, 00-665 Warsaw, Poland

P. Baudrit
CEESAR - European Centre of Studies on Şafety and Risk Analysis, 132, Rue des Suisses, 92000, Nanterre, France

L. Půst, L. Pešek and A. Radolfová
Institute of Thermomechanics, AS CR, v.v.i., Dolejškova 5, 182 00 Prague, Czech Republic

M. Zajíček and J. Dupal
Faculty of Applied Sciences, University of West Bohemia, Univerzitní 22, 306 14 Plzeň, Czech Republic

A. Jonášová, O. Bublík and J. Vimmr
European Centre of Excellence NTIS – New Technologies for the Information Society, Faculty of Applied Sciences, University of West Bohemia, Univerzitní 8, 306 14 Pilsen, Czech Republic

R. Jain and P. Tandon
Mechanical Engineering Discipline, PDPM Indian Institute of Information Technology, Design and Manufacturing Jabalpur Jabalpur-482005, Madhya Pradesh, India

M. Vasantha Kumar
Altair Engineering, Bengaluru, India

J. Šimek
TECHLAB Ltd., Sokolovsk´a 207, 190 00 Praha, Czech Republic

P. Lindovský
První brněnská strojírna V. Bíteš, a. s., Vlkovská 279, 595 12 Velká Bíteš, Czech Republic

P. Polach, M. Hajžman and O. Červená
Section of Materials and Mechanical Engineering Research, Výzkumný a zkušební ústav Plzeň s. r. o., Tylova 1581/46, 301 00 Plzeň, Czech Republic

Z. Šika and P. Svatoš
Department of Mechanics, Biomechanics and Mechatronics, Faculty of Mechanical Engineering, Czech Technical University in Prague, Technická 4, 166 07 Praha, Czech Republic

Index

Printed in the USA
CPSIA information can be obtained
at www.ICGtesting.com
JSHW051408221024
72173JS00006B/1321